Applied Statistics Manual

Also available from ASQ Quality Press:

Statistics for Six Sigma Black Belts
Matthew A. Barsalou

The ASQ Pocket Guide to Statistics for Six Sigma Black Belts
Matthew A. Barsalou

Linear Regression Analysis with JMP and R
Rachel T. Silvestrini and Sarah E. Burke

Introduction to 8D Problem Solving
Ali Zarghami and Don Benbow

The Probability Workbook
Mary McShane-Vaughn

The Certified Six Sigma Black Belt Handbook, Third Edition
T.M. Kubiak and Donald W. Benbow

The Certified Six Sigma Yellow Belt Handbook
Govindarajan Ramu

The Certified Six Sigma Green Belt Handbook, Second Edition
Roderick A. Munro, Govindarajan Ramu, and Daniel J. Zrymiak

Practical Engineering, Process, and Reliability Statistics
Mark Allen Durivage

The Quality Toolbox, Second Edition
Nancy R. Tague

Root Cause Analysis: Simplified Tools and Techniques, Second Edition
Bjørn Andersen and Tom Fagerhaug

The ASQ Quality Improvement Pocket Guide: Basic History, Concepts, Tools, and Relationships
Grace L. Duffy, editor

To request a complimentary catalog of ASQ Quality Press publications, call 800-248-1946, or visit our website at http://www.asq.org/quality-press.

Applied Statistics Manual

A Guide to Improving and Sustaining Quality with Minitab

Matthew A. Barsalou and Joel Smith

ASQ Quality Press
Milwaukee, Wisconsin

American Society for Quality, Quality Press, Milwaukee 53203

Printed in the United States of America.
23 22 21 20 19 18 5 4 3 2 1

Library of Congress Cataloging-in-Publication Data
Names: Barsalou, Matthew A., 1975– author. | Smith, Joel, 1980– author.
Title: Applied statistics manual : a guide to improving and sustaining quality with Minitab / Matthew A. Barsalou, Joel Smith.
Description: Milwaukee : ASQ Quality Press, [2018] | Includes bibliographical references and index.
Identifiers: LCCN 2018036196 | ISBN 9780873899758 (hardcover : alk. paper)
Subjects: LCSH: Statistics. | Social sciences—Statistical methods.
Classification: LCC HA29 .B367 2018 | DDC 005.5/5—dc23 LC record available at https://lccn.loc.gov/2018036196

Director, Quality Press and Programs: Ray Zielke
Managing Editor: Paul Daniel O'Mara
Sr. Creative Services Specialist: Randy L. Benson

ASQ Mission: The American Society for Quality advances individual, organizational, and community excellence worldwide through learning, quality improvement, and knowledge exchange.

Printed on acid-free paper.

ASQ

Quality Press
600 N. Plankinton Ave.
Milwaukee, WI 53203-2914
E-mail: authors@asq.org

The Global Voice of Quality™

This book is dedicated to my wife Silvana, who on many evenings shared my attention with a computer screen and never once complained.

Joel Smith

This book is dedicated to my son, Leander G. Barsalou.

Matthew A. Barsalou

Table of Contents

List of Figures and Tables

Preface

The book *Applied Statistics Manual: A Guide to Improving and Sustaining Quality with Minitab* was written to provide guidance for those who need to apply statistical methods for practical use. While the book provides detailed guidance on the use of Minitab for calculation, simply entering data into a software program is not sufficient to reliably gain knowledge from data. The software will provide an answer, but the answer may be wrong if the sample was not taken properly, the data was unsuitable for the statistical test that was performed, or the wrong test was selected. It is also possible that the answer will be correct, but misinterpreted. This book provides both guidance in applying the statistical methods described as well as instructions for performing calculations without a statistical software program such as Minitab.

One of the authors is a professional statistician who spent nearly 13 years working at Minitab and the other is an experienced and certified Lean-Six Sigma Black Belt. Together, they strive to present the knowledge of a statistician in a format that can be easily understood and applied by nonstatisticians facing real-world problems.

Their guidance is provided with the goal of making data analysis accessible and practical. Rather than focusing on theoretical concepts, the book delivers only the information that is critical to success for the practitioner. It is a thorough guide for those who have not yet been exposed to the value of statistics, as well as a reliable reference for those who have been introduced to statistics but are not yet confident in their abilities.

ACKNOWLEDGMENTS

Matthew Barsalou would like to thank Artur Wilk and Jean-Baptiste Selfa for providing helpful feedback and suggestions.

Introduction

The book is divided into chapters based on common statistical concepts and often times aligns with the menus in Minitab. Certain chapters require an understanding of information from previous chapters, so readers using the book as a foundation should review material sequentially.

When learning how to apply statistics, it is important to first establish a fundamental understanding of data and distributions. The first chapter explains critical statistical concepts such as the difference between samples and populations as well as sampling methods and types of data. Chapter 2 introduces concepts related to descriptive statistics, such as measures of central tendency, measures of variability, and measures of shape. Chapter 3 describes Exploratory Data Analysis, which is an important first step in analyzing data. Chapter 4 describes probability statistics and various continuous and discrete distributions.

Once the fundamentals are understood, a variety of tools are accessible to practitioners to help them make decisions based on data. Chapter 5 explains hypothesis testing—the basis for statistical decision-making—for means, proportions, and variances. This chapter describes how to perform the hypotheses tests using both the p-value method and the critical value method. The critical value method is taught more often for those performing tests by hand, while the p-value method is the method typically taught to those with access to Minitab or other statistical software. Chapter 6 describes how to determine sample sizes and confidence intervals for the previously described hypothesis tests.

Chapter 7 builds on the basic hypothesis tests learned in Chapter 5 but pertains to the distribution of data and includes the normal probability plot, transformations, and individual distribution identification. Chapter 8 then describes nonparametric tests and their uses.

To this point in the book, hypothesis tests have focused on attributes of single variables or the relationship between two variables. The next chapters expand on this to enable many variables to be modeled simultaneously. Chapter 9 teaches analysis of variance (ANOVA), including the General Linear Model (GLM), and Chapter 10 covers regression.

A critical concept to understand is that of measurement error, even when performing basic tests. Once the reader is familiar with modeling, various tools for assessing measurement error can be understood. Chapter 11 presents various types of measurement system analysis (MSA).

Assessing the quality of a product or process requires understanding how it behaves. The subsequent chapter describes different types of process capability studies used to quantify the quality of a process or product. Chapter 13 discusses statistical process control (SPC) and includes control charts for variable data, continuous data with weighted averages, and attribute data.

To establish likely causation between factors in a study and a response of interest, a controlled experiment must be performed. Chapter 14 concerns design of experiments (DOE) and covers factorial designs, response surface, and mixture designs.

Chapter 15 then covers reliability, including distribution identification, distribution overview, and distribution analysis. Other quality tools that Minitab can create, such as the cause and effect diagram, Pareto chart, and run chart, are covered in Chapter 16.

Appendix A presents statistical terminology written in layman's terms, and Appendix B presents a quick guide to common hypothesis tests. Appendix C includes various statistical tables. Appendix D contains control chart constants, and Appendix E presents a tour of Minitab.

Chapter 1
Sampling and Types of Data

SAMPLES AND POPULATIONS

A population consists of every potential unit of something, and a sample is a subset of the population (Moore 1991). The population is conceptual and might be thought of as an infinite quantity of units that we draw from, and statistics is the science of using a sample to estimate characteristics of that population. A sample from an item with a limited production run might consist of 8 randomly selected items (as shown in Figure 1.1) or all 441 parts produced.

Total production size can be incredibly large. There may be over one million units in existence for an item that has been in production for a period of years. It would be unrealistic to measure every one of these items, so a smaller sample would be used instead. A properly selected sample provides an opportunity to make inferences about the parent population.

Even measuring every single part produced is typically considered a sample of the theoretical population, which can only be estimated. It would be incorrect to assume that the characteristics of a population are known even after 100% sampling. To understand why this is erroneous, consider a perfectly fair six-sided die where each side has equal probability of landing up when rolled. The population of rolls of this die consists of an infinite number of potential rolls and has the characteristic of each number occurring with probability 1/6 on any given roll. We are unlikely to ever have our sample indicate precisely a 1/6 chance of each number occurring even if we record every single roll of the die. This means that even sampling 100% of available units is still providing a theoretical subset of the population.

As another example, think of a perfectly stable process where each part has a length pulled from a single distribution. The part comes from a single distribution because the process is perfectly stable. A stable process it is not undergoing changes over time and the true population mean will remain constant. Every one of the 500 parts ever made has been sampled and the mean has been calculated. The mean of all parts will change slightly with the inclusion of each part as subsequent parts are pulled and included in the calculation. However, the true population mean is not changing. This indicates that statistics based on 100% sampling do not imply population characteristics are known.

Characteristics that describe the distribution of a population numerically are called *parameters* and are typically unknown. Characteristics that describe a sample numerically are called *statistics* and may change if we take a new sample (Moore 1991). For example, suppose the population of a certain type of part produced in a

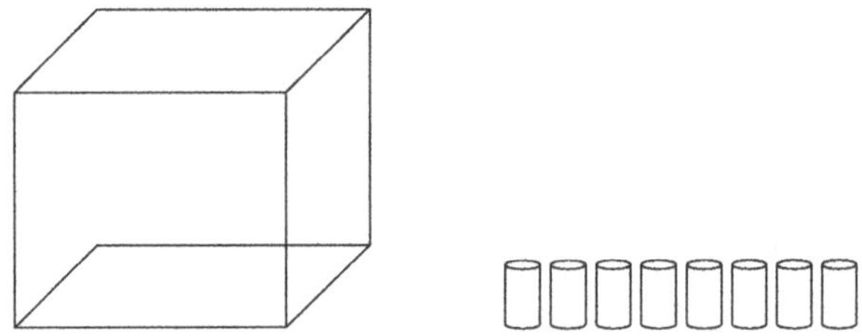

Figure 1.1 Two samples from a limited production run.

factory has a mean length of 34.504 millimeters (mm). This value is unknown and must be estimated using a sample. The quality manager wants to know this average length, so 10 parts are checked and found to have a sample mean of 34.623. The population parameter is 34.504, but the sample mean, which is the statistic used to describe the average length, is only 34.623. A second sample is taken and found to have a mean of 34.585, and a third sample has a mean of 34.524. The sample mean statistic changed with each new sample while the true population mean remains unchanged.

SAMPLING METHODS

A *sampling plan*, also known as *sampling design*, is the way in which samples are to be selected. Common sampling designs include simple random sampling, systematic random sampling, stratified random sampling, and quota sampling (Silver 1997). Sampling should be random, with each item having an equal chance of being selected. For example, suppose a quality engineer wanted to sample five parts out of a box of 100 parts to estimate the mean diameter of the parts. The easiest way to do this would be to open the top of the box and pull out the first five parts; however, these parts may not truly represent all the parts in the box. The production machine tooling may have been worn around the time the last parts were produced and placed in the box, resulting in diameters that do not represent those of all parts placed into the box. A good sampling plan can help to ensure such a mistake is not made.

A *simple random sampling* plan is used to ensure every object in a population has an equal chance of being selected. The simplest way to achieve this is by assigning each object a number and then using a random number generator to determine which objects will be selected. This may be difficult to implement in practice, such as when it would require emptying an entire box to assign a value to each individual part. It may also be difficult to implement simple random sampling when parts are being produced on a production line and individual part selection and removal is difficult. An easier-to-implement approach is a systematic random sampling design.

Systematic random sampling also uses the random selection of objects; however, the sample is taken only at a predetermined interval given as

$$\mathrm{k}=\left(\frac{N}{n}\right),$$

where k is the sampling interval, N is the total number of parts from which you want to sample, and n is the desired sample size. If 5 samples are required from a box of 105 parts, the sampling interval would be 105 divided by 5, or 21. This means every 21st part would be checked. This approach is easier to implement in a production operation where the production operator would be instructed to check every part at a predetermined interval, such as every fifth part produced. *Systematic random sampling* is also called *skip interval sampling*, and the sampling interval can be referred to as the skip interval. Although systematic random sampling is easier to implement than random sampling, there is a risk that the sample will not be truly representative of the population. For example, there may be a reason that the interval does not present the population, such as when there is a cyclic change in the process that causes every fifth part to be larger or smaller than the typical part.

Stratified random sampling separates the population into subpopulations based on known characteristics that differ between them. This form of sampling is often used for surveys, and Owen and Jones (1990) give an example of a survey of traffic on roads; such a survey may be biased toward vehicle owners, so the vehicle owners and non-vehicle owners would be separated into subpopulations based on their proportions of the entire population. If the entire population of interest consisted of 30,000 vehicle owners and 10,000 non-vehicle owners, then 25% of the entire population consists of non-vehicle owners and a sampling plan should result in 25% of the people surveyed being non-vehicle owners. Although often used in surveys, stratified random sampling can also be useful in industry, such as when a known subpopulation is the layer of parts in a box that is stacked eight layers deep.

Quota sampling is an easy-to-implement sampling method that requires the sampling of a predetermined number of items that fit a certain category. Quota sampling may include elements of stratified sampling, such as when a survey taker is instructed to interview 50 managers consisting of 30 males and 20 females, but units sampled within each category are not randomly selected. Although easy to implement, quota sampling is risky due to this lack of randomization. A production employee instructed to check a sample of 100 parts may simply check the most easily accessible 100 parts. Simply checking the most easily available parts is an example of convenience sampling and should be avoided whenever possible because the parts are not randomly selected and may not reflect the characteristics of the population.

TYPES OF DATA

Data can be classified as qualitative or quantitative. *Qualitative data* is of a descriptive nature, such as names and labels. *Quantitative data* is of a numerical nature and can be further divided into continuous and discrete data (Benbow and Kubiak 2009). *Continuous data*, also known as variable data, can be any value within a range of possible values. Examples of continuous data include measurements of length, weight, and temperature. *Discrete data*, also known as attribute data, can only take specific values such as a number of defects; there can be 7 defects, but not 7.5 defects. Types of data are shown in Figure 1.2.

An understanding of the types of data is essential for performing statistical tests, especially when assisted by a statistical software program. Using discrete data in a statistical operation that requires continuous data can yield a result; however, this result may be incorrect. Quantitative data typically contain more

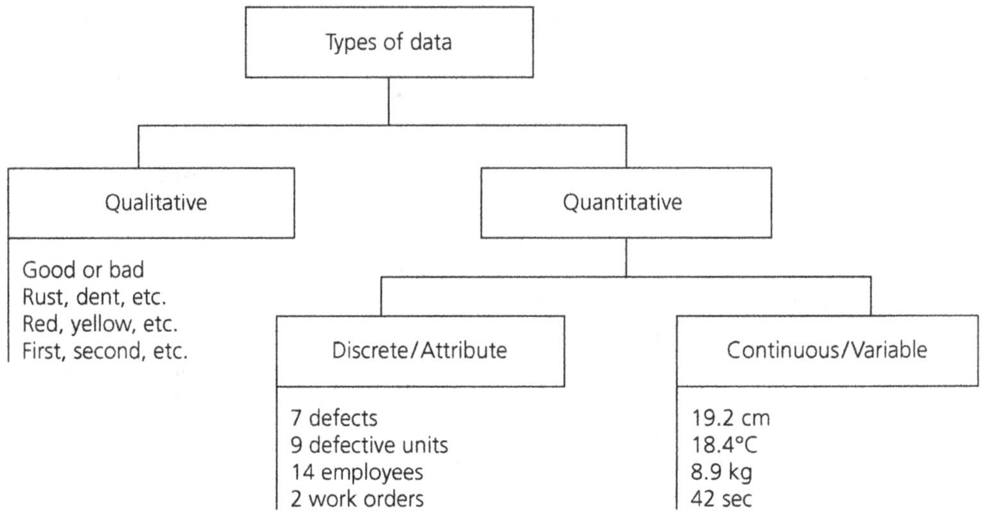

Figure 1.2 Types of data.

information than qualitative data and continuous data more than discrete data. Larger sample sizes are better than smaller sample sizes because more data result in a better characterization of the population.

A data collection plan should be created when collecting data for statistical analysis. The plan should describe what data is needed, the type of data, the units the data is measured in, how many will be measured, the measurement method, any boundary conditions such as a description of the types of items to collect or limitations, and where the data will be collected. Figure 1.3 shows a data collection plan for a large study involving multiple types of data, and Figure 1.4 shows a simpler plan for a smaller measurement study requiring only one type of measurement. The sampling plan provides guidance on what data are to be collected and also a form for recording the values. The large-scale study data collection plan may require multiple simpler data collection plans for recording the actual values.

It is critical to be aware of the pedigree of your data because any data given to you may not have been properly obtained or recorded. Incorrectly sampled or labeled data, for example, might not reflect the true population. Some things to consider when receiving data are:

- Where did it come from?
  - —Do the values represent real-world conditions or are they from controlled conditions?
- Is it properly labeled?
  - —Is the data labeled "day shift" really from the day shift?
- How was it collected?
  - —Was the sample random?
  - —Does the sample represent the process?
- How was it measured?
  - —Could there be rounding errors?
  - —Was the correct type of measuring device used and has the device been validated?

Data to Collect	Type of Data	Measurement Units	Number to Check	Measurement Method	Boundary Condition	Collected from/at
Part diameter	Continuous	mm	15	Micrometer	No failed parts on hand	Current stock
Contamination	Qualitative	Present or not present	1	Test per procedure 454a	n/a	Warehouse
Cast material	Qualitative	Material type	1 of each type	Per lab procedure	1 returned failed part and new part	Returned: quality dept. / New: warehouse
Number of failures in March	Discrete	Number of parts	One month	Check quality data for March 2018	Month of March only	Claims database

Figure 1.3 Data collection plan for a large-scale study.

Directions: Check every 10th part produced using calipers. Record the value here.	
Date and Shift	22 Jan. 2018
Caliper Number	327
Operator	George L.
Measurement Number	Length XY (22.4 ± 0.1 mm) per drawing number 684243b
1	
2	
3	
4	
5	
6	
7	
8	
9	
10	

Figure 1.4 Data collection plan for a simple measurement study.

QUESTIONS AND ANSWERS

Q: How many samples do I need?

A: This depends entirely on what you want to learn from the data. For most common hypothesis tests, there is a sample size calculator available in Minitab that can determine the correct size. For other tools such as capability analysis, guidelines are provided in the respective chapter.

Q: What is the difference between a sample and a population?

A: A population is a theoretical concept that can be thought of as an infinite quantity of an item of interest with some distribution of any given measurement. The items that are actually produced and measured are considered a sample, and these are used to make inferences about the population.

Q: What is better—qualitative or quantitative data? Why?

A: Quantitative data tend to be better, providing more information. For example, consider a foot race where data on the runners' finishing times can be provided in either of two ways. In the first, we simply know whether each runner finished in the first half of runners—a qualitative measure. In the second, we know the actual finishing times—a quantitative measure. The quantitative measure allows us to understand the relative speed of every runner whereas the qualitative is simply a rough grouping.

Q: I need the standard deviation of the sample to determine the sample size for a two sample t test; but what do I do if I have never taken any measurements? How many samples do I need to determine the standard deviation so that I can calculate the required sample size?

A: First, determine the margin of error that is permissible for the estimate and then use sample size for estimation to determine how many samples will be necessary.

Q: What sampling method should I use when I am sampling a process that produces parts that are a steady rate?

A: If tools that benefit from separating short-term and long-term variation are desired, such as control charts and capability analysis, then a random sampling plan with rational subgrouping is typically the best choice. Otherwise random sampling often provides the highest likelihood of the sample closely matching the population.

Q: What sampling method should I use when I am sampling several weeks' worth of production stored in a warehouse?

A: If inference is to be made about the population during that time period, stratified random sampling might be used to ensure that the product is tested over the entire period of time. Sampling from a single box or pallet likely does not include all sources of variation present over the time of production.

Chapter 2
Descriptive Statistics

Descriptive statistics "supply a number of tools, such as tables, graphs, and averages, for organizing and summarizing information about a collection of actual observations," which is in contrast to inferential statistics that "supply a number of tools for generalizing beyond actual observations" (Witte 1993, 4–5). Descriptive statistics provide a summary of the data, such as the mean and standard deviation, in comparison to inferential statistics such as hypothesis testing, which uses the data from a sample to make inferences about a population.

Descriptive statistics include *measures of central tendency* such as the mean, median, and mode, as well as *measures of variability* such as standard deviation and variance as well as the minimum and maximum values and the range of the data. *Measures of shape* are also a part of descriptive statistics, and two measures of shape are skewness and kurtosis.

DISTRIBUTIONS

The *distribution of a population* is the frequency in which values are expected to occur, and the *distribution of a sample* is the frequency in which values were observed. For continuous variables, the distribution is typically illustrated using a curve in which higher points on the curve indicate more likely values and lower points indicate less likely values (see Figure 2.1). Distributions are often described by their peak, which indicates the most likely values, and their tails, which indicate how unlikely values become the progressively farther they are from the peak. Specific distributions and their typical uses and their characteristics will be described in Chapter 4.

MEASURES OF CENTRAL TENDENCY

Measures of central tendency are also known as *measures of position* and include the mean, the median, and the mode (ReVelle 2004). The mean and the median may be in the same position if the distribution is symmetric; however, the two will not be the same value if the distribution is skewed.

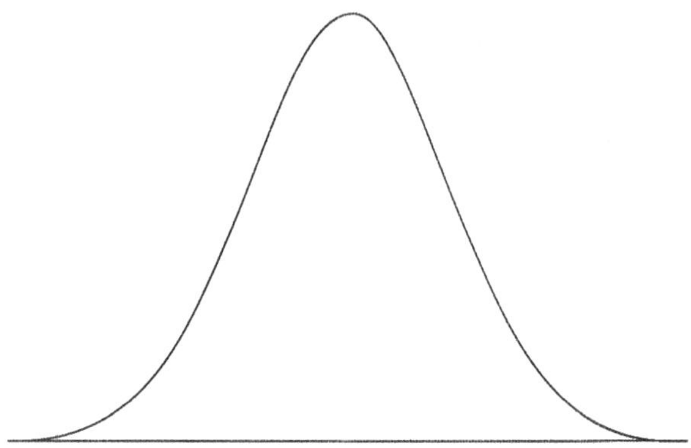

Figure 2.1 A distribution.

Mean

The mean, or average, of a data set is the sum of the individual items in the data set divided by the number of items in the data set. The formula for the sample mean is:

$$\bar{x} = \frac{\Sigma X_i}{n}$$

Suppose a student took five exams and got the following test scores: 72, 85, 88, 72, and 93. The mean would be:

$$\frac{72+85+88+72+93}{5} = 82.0$$

The mean of a data set can be misleading, such as when most values are clustered close together and a few values are far removed from the others. For example, the average income for 10 computer programmers might not accurately reflect the income of a typical programmer if nine of them are just beginning their careers and one is the owner of a leading software company.

Median

The median is the value in the middle of a data set when the data set is arranged from lowest to highest and represents the point that divides the data in half. The median is the mean of the two values in the middle of the data set if there are an even number of values in the data set. The median for 72, 72, 85, 88, and 93 is 85. The median may be a more useful measure of central tendency than the mean when the data set is heavily skewed.

Mode

The mode is the value in a data set that occurs the most frequently. The mode for 72, 72, 85, 88, and 93 is 72. It is possible for a data set to have multiple modes. For example, there are three modes in the data set 22, 14, 21, 17, 13, 19, 12, 12, 22, and 19. The numbers 12, 19, and 22 each occur twice in this example.

MEASURES OF VARIABILITY

Measures of variability provide an indication as to the dispersion of a data set (Witte 1993). The mean of a data set alone might not provide useful information if the spread of the data is not considered. For example, a process with a mean in the center of the specification limits can produce out-of-specification parts if the spread of the data is very high. Conversely, a process with an off-centered mean will not produce defective parts if the spread is sufficiently low.

The empirical rule shown in Figure 2.2 applies to approximately normal, symmetrical distributions and holds that 68.27% of all values are within one standard deviation above or below the mean. Plus or minus two standard deviations from the mean contains 95.45% of all values, and 99.73% of all values are within three standard deviations of the mean.

Variance and Standard Deviation

The notation used for a population differs from that of a sample. A population standard deviation is identified as the Greek letter σ (pronounced sigma) and the population mean is the Greek μ (pronounced mu). A sample standard deviation is identified as s, and a sample mean is $\bar{x}$ (pronounced X-bar).

The population variance is estimated by the summation of the squared differences between the population mean and the individual values, divided by the number of items in the data set. The formula for variance is:

$$\sigma^2 = \frac{\Sigma(x_i - \mu)^2}{n}$$

The standard deviation of a population is equal to the square root of the variance and the formula is:

$$\sigma = \sqrt{\sigma^2}$$

The easiest way to calculate variance of a population by hand is to create a table (such as the one shown in Table 2.1) for a population with values of 11, 9, 11, and 10.

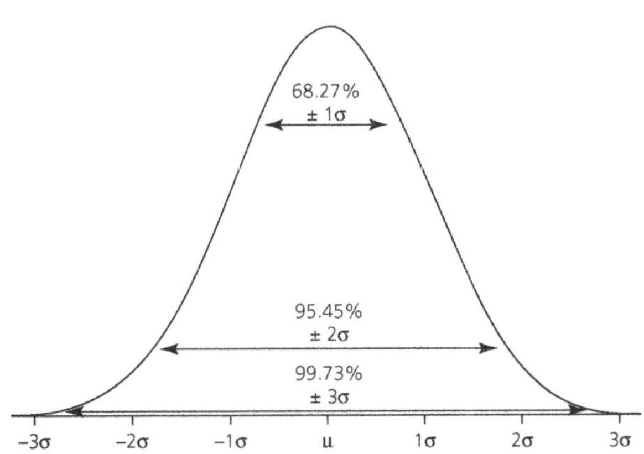

Figure 2.2 The empirical rule.

Table 2.1 Calculating the variance of a population.

x_i	$\bar{\mu}$	$x_i - \bar{\mu}$	$(x_i - \bar{\mu})^2$
11	10.25	0.75	0.5625
9	10.25	−1.25	1.5625
11	10.25	0.75	0.5625
10	10.25	−0.25	0.0625
		Summation of $(x_i - \bar{\mu})^2$	2.75

Table 2.2 Calculating the sample variance.

x_i	$\bar{x}$	$x_i - \bar{x}$	$(x_i - \bar{x})^2$
27	26.25	0.75	0.5625
25	26.25	−1.25	1.5625
29	26.25	2.75	7.5625
24	26.25	−2.25	5.0625
		Summation of : $(x_i - \bar{x})^2$	14.75

The resulting summation of $(x_i - \bar{\mu})^2$ is 2.75 and this value is divided by $n=4$, which is the number of data points. The resulting variance would be $2.75/4=0.6875$. The standard deviation is the square root of the variance squared; therefore, the standard deviation is:

$$\sigma = \sqrt{0.6875} = 0.829$$

A sample standard deviation is the square root of the sample variance. The formula for the sample standard deviation and sample variance is:

$$s = \sqrt{s^2} \text{ and the variance is } s^2 = \frac{\Sigma(x_i - \bar{x})^2}{n-1}$$

Like calculating the variance of a population, a table should be created when calculating the sample variance by hand. Table 2.2 shows an example using the values 27, 25, 29, and 24.

The sample variance is the summation of $(x_i - \bar{x})^2$, which is 14.75, and this number is divided by the sample size minus one; therefore, $14.75/(4-1)=4.92$. The sample standard deviation is then equal to

$$s = \sqrt{4.92} = 2.22.$$

In practice, it is much more common to calculate the sample variance and sample standard deviation because the population mean is typically an unknown value. The sample variance and sample standard deviation are then used as estimates of the corresponding population parameters.

Minimum and Maximum Values and Range

The minimum of a data set is the lowest number in the data set and the maximum is the highest number in the data set. The range is the difference between the two and can be found by subtracting the minimum from the maximum. The minimum of a data set consisting of 7, 9, 14, and 22 is 7 and the maximum 22; the range is $22-7=15$.

MEASURES OF SHAPE

Skewness and kurtosis are common measures of the shape of a distribution (Silver 1997). A value for *skewness* indicates how symmetrical a distribution is, and *kurtosis* tells us how flat or peaked a distribution is.

Skewness

Skewness is a measure of the symmetry of a distribution, and nonsymmetrical distributions are referred to as skewed (Clarke and Cooke 1992). A distribution is left-skewed (see Figure 2.3) when the longer tail extends to the left toward lower values and right-skewed (see Figure 2.4) when the longer tail extends to the right.

The formula for skewness is Pearson's coefficient of skewness, which is

$$\text{Sk} = \frac{3(\bar{x} - Md)}{S},$$

where Md is the median. An Sk of 0 indicates the distribution is symmetrical.

A negative value for skewness indicates the distribution is left-skewed and a positive value for skewness indicates the distribution is right-skewed. Values close to zero indicate a distribution that is symmetric.

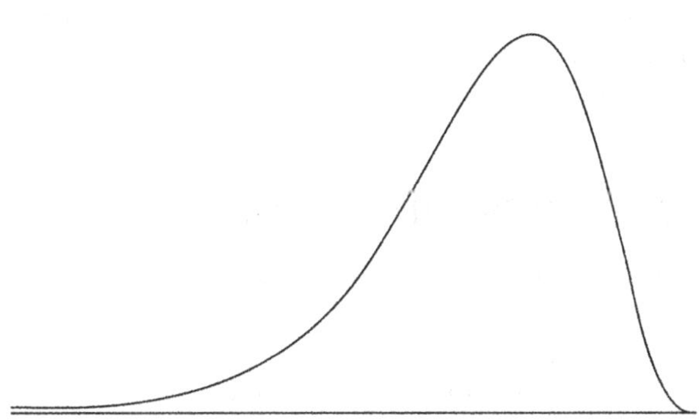

Figure 2.3 Left-skewed distribution.

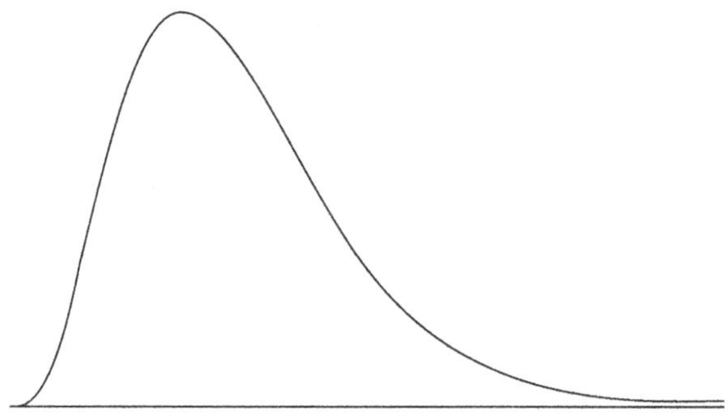

Figure 2.4 Right-skewed distribution.

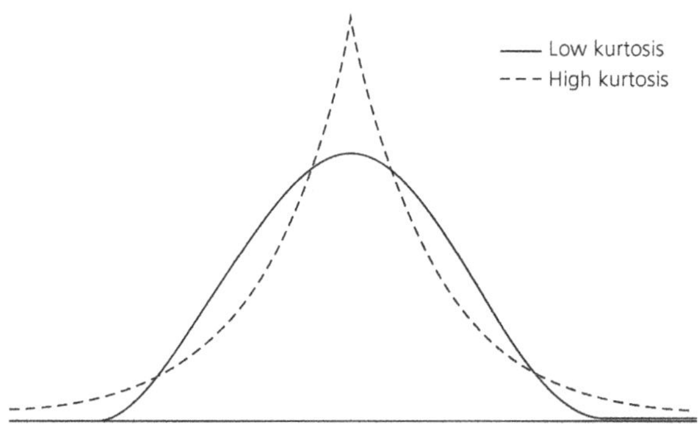

Figure 2.5 High and low kurtosis.

Kurtosis

Kurtosis is a measure of how "flat" or "peaked" a distribution is. The normal distribution has a kurtosis of 3 and is called *mesokurtic*. Kurtosis is equal to

$$\frac{n(n+1)}{(n-1)(n-2)(n-3)}\Sigma[(x_i-\bar{x})/s]^4-\frac{3(n-1)^2}{(n-2)(n-3)}.$$

A kurtosis greater than 3 indicates the distribution is more sharply peaked with heavier tails and is called *leptokurtic*. A distribution with a kurtosis less than 3 has a flatter peak and thinner tails and is called *platykurtic* (Silver 1997). Figure 2.5 shows both a high and low degree kurtosis.

DESCRIPTIVE STATISTICS WITH MINITAB

Various descriptive statistics can be quickly viewed by using the Minitab descriptive statistics function. Simply go to **Stat > Basic Statistics > Display Descriptive Statistics** (see Figure 2.6).

The Statistics subdialog can be used to select which statistics Minitab should display, such as the mode, skewness, and kurtosis (see Figure 2.7). Click **OK** in each dialog once completed.

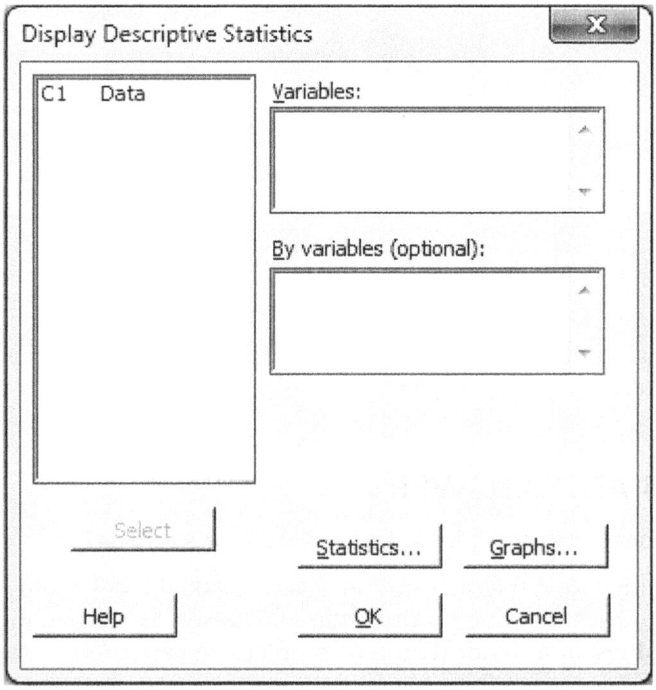

Figure 2.6 Minitab dialog box for Display Descriptive Statistics.

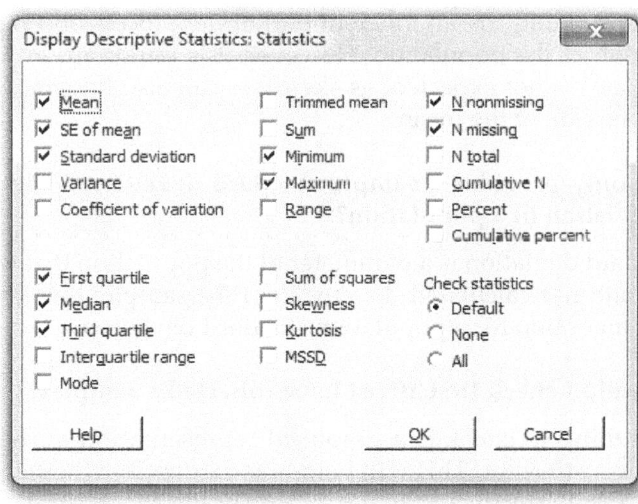

Figure 2.7 Statistics subdialog box.

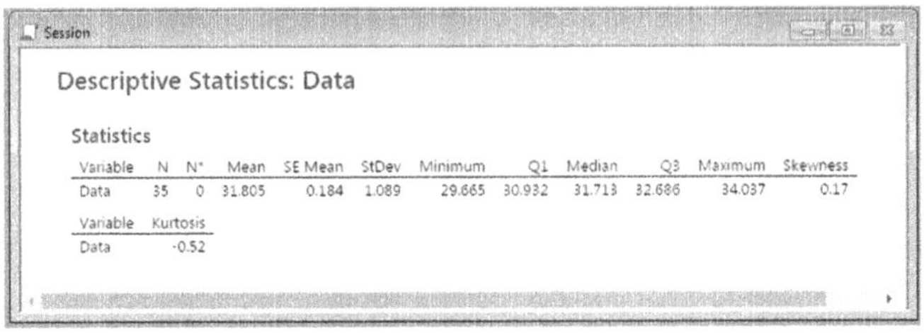

Session

Descriptive Statistics: Data

Statistics

Variable	N	N*	Mean	SE Mean	StDev	Minimum	Q1	Median	Q3	Maximum	Skewness
Data	35	0	31.805	0.184	1.089	29.665	30.932	31.713	32.686	34.037	0.17

Variable	Kurtosis
Data	-0.52

Figure 2.8 Descriptive statistics session window output.

The selected descriptive statistics then appear in the session window as shown in Figure 2.8.

QUESTIONS AND ANSWERS

Q: What does it mean if I have a high value for kurtosis?

A: This occurs when the distribution is less "peaked" and could indicate multiple conditions. One would be product that was sorted to remove more extreme high and low values, or another would be a process where those values are physically prevented from happening.

Q: Are heavily skewed data bad?

A: Often, skewed data are an inherent part of the process and represent the natural distribution of the population. However, it is important to understand when skewness is and is not expected, as skewness can also happen when outliers are present on one side of the mean.

Q: Minitab only provides a sample standard deviation. Where can I find the standard deviation of a population?

A: The standard deviation is a parameter of the population that is estimated using the sample and not calculated directly from the sample; therefore, Minitab does not provide an option for a population standard deviation.

Q: What should I check first after I have collected a sample?

A: The first thing to check is a graphical representation of the data, such as a histogram or scatterplot. This will help you to identify whether any points have been recorded incorrectly and to get a sense of the shape, spread, and location of the data.

Chapter 3
Exploratory Data Analysis

Exploratory Data Analysis (EDA) is "numerical detective work" (Tukey 1977, 1). De Mast and Trip explain "the purpose of EDA is the identification of dependent (Y) and independent (X) variables that may prove to be of interest for understanding or solving the problem under study" (2007, 301). Data are graphically analyzed to identify "salient features" that can be used as a basis for a hypothesis that can be evaluated using other methods (de Mast and Trip 2007, 304).

Firm conclusions should not be drawn from EDA; rather, the graphical exploration of data can provide insights that are worth pursuing with other methods. Common EDA tools include the *boxplot*, *scatterplot*, *stem-and-leaf plot*, and *multi-vari chart*. Graphing the data is useful at the start of analysis to quickly determine if there are outliers in the data; for example, a person 17 meters tall would stand out in a boxplot and give a quick indication that there may be an error in the data.

Prior to Minitab 17, these tools were located in an EDA menu, but these tools have now been incorporated into other menus in the software. However, if desired, the EDA menu can be reinstated using **Tools > Customize**.

BOXPLOT

A boxplot can be used to quickly compare data sets. The parts of a boxplot are shown in Figure 3.1. The box is formed between the lower quartile and upper quartile of the data. The lower quartile is the point in which 25% of the data falls below and 75% above, and the upper quartile is the point in which 75% falls below and 25% above (Tukey 1977). The median, which is the 50th percentile, is displayed using a line across the middle of the box. The box will generally contain 50% of all values; however, this might not be the case in the event of repeated values or certain sample sizes. Extending from each end of the box are "whiskers." The whiskers extend to the farthest point in the data set that is not more than 1.5 times the Interquartile Range (IQR) away from the box, with IQR being the height of the box. Outliers are values that fall more than 1.5 times the IQR from the box and they are denoted by asterisks.

An outlier may be due to measurement error, in which case it should be removed from the data set. Or, it could be due to a special event, in which case a decision must be made. The outlier can be removed if this is an event that is not expected to occur again, but it should be left in if comparable values could be expected to occur again in the future. When removing outliers, be sure to make a note that an outlier has been removed so that the value can be identified again later, if necessary, and

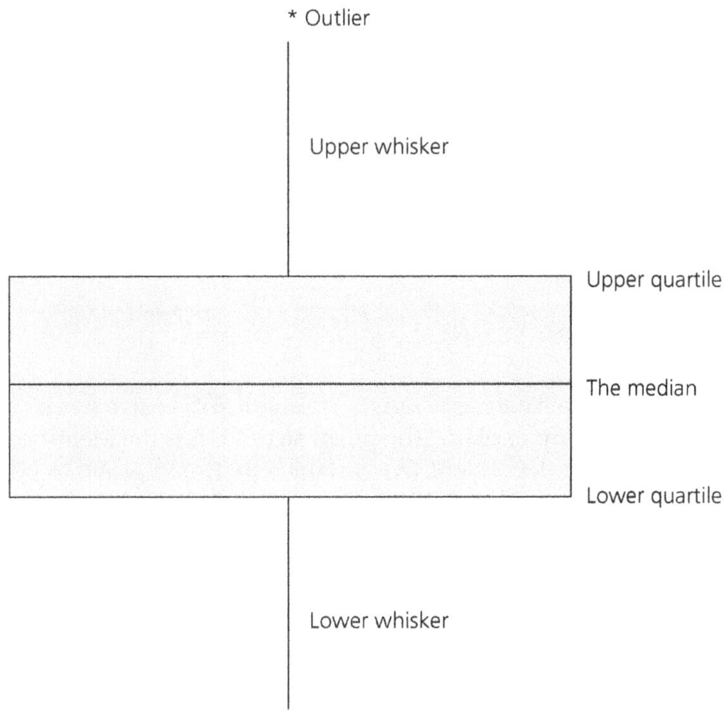

Figure 3.1 Anatomy of a boxplot.

to avoid the appearance of manipulating the data. An outlier is not necessarily an indication of a problem and may be an inherent part of the process that produced the data. Additionally, there is some probability of outliers occurring from most data sets, and as more data are collected, more outliers should be expected due simply to random chance.

In Minitab: Go to **Graph > Boxplot** and select **One Y > Sample** if there is only one data set or **Multiple Y's > Sample** if there is more than one data set (see Figure 3.2). Select the column or columns containing the data (see Figure 3.3) and click **OK**.

The results will appear as shown in Figure 3.4. Looking at these results, we can quickly see that the data in data set 2 stands out as having a wider distribution than the other two data sets. This may be a point to consider investigating further, such as by performing a test of equal variances.

SCATTERPLOT

Scatterplots can be used to quickly identify a potential correlation between continuous variables such as temperature of heat treatment and the resulting hardness of a material. A test of regression can tell if there is a statistically significant relationship present; however, a simple scatterplot can provide an indication that such a test is needed as well as give the practitioner an idea of the strength and nature of the relationship. As shown in Figure 3.5, correlations can be strong positive, weak positive, weak negative, or strong negative, or there may be no correlation present.

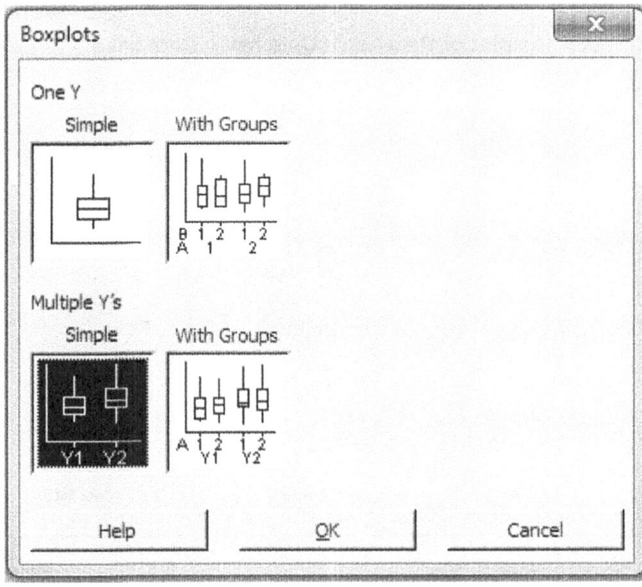

Figure 3.2 Boxplot menu.

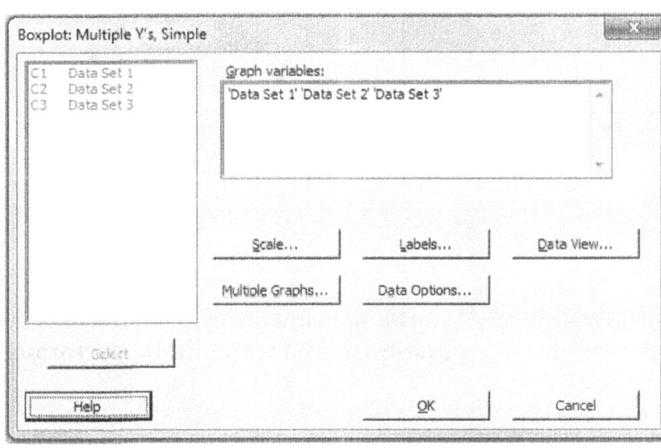

Figure 3.3 Minitab dialog box for boxplots.

The existence of a correlation does not necessarily mean either of the two variables are influencing the other but rather that there is a tendency for one variable to demonstrate higher or lower values as the other increases or decreases. Box, Hunter, and Hunter (2005) have shown a strong correlation between the population of a city and the local stork population; as it is unlikely that the higher human population is actually causing the higher stork population, the relationship may be due to a third, unstudied factor or just chance.

In Minitab: Go to **Graph > Scatterplot** and select the desired type of scatterplot from the gallery (Figure 3.6). Then select the columns containing the data for the Y and X variables and click **OK** (Figure 3.7).

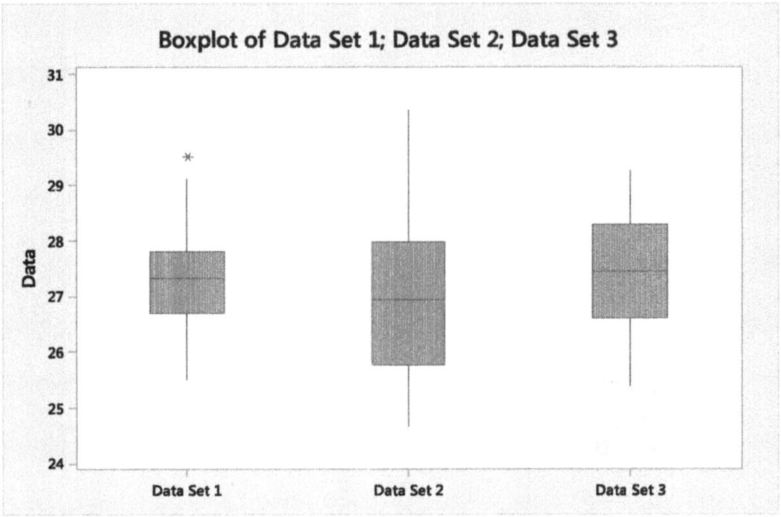

Figure 3.4 Boxplots.

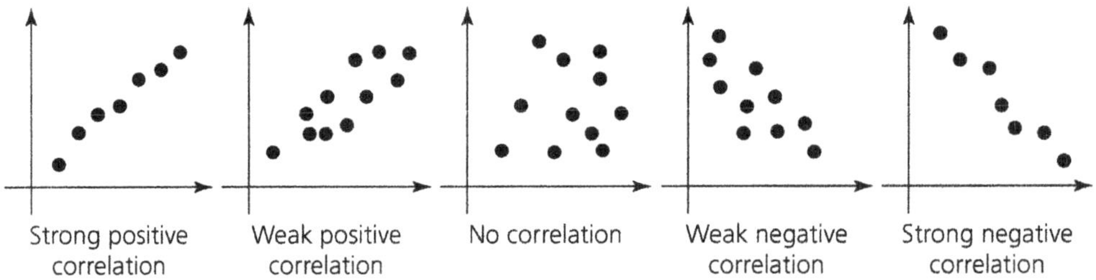

Figure 3.5 Examples of scatterplots.

The resulting scatterplot can be seen in Figure 3.8. A positive correlation appears to be present; however, a regression analysis should be performed to be sure.

STEM-AND-LEAF PLOT

A stem-and-leaf plot is used to graphically depict the distribution and occurrence of groups of values (Tukey 1977). The "leaf" portion displays the last digit in each number in the data set and the "stem" displays the first digit(s). All leaves corresponding to the same stem are listed beside it in order. For example, the numbers 44, 47, and 48 would all be placed on a single stem—4—with the leaves 478. A stem-and-leaf plot can be used to quickly display the distribution of data while also depicting the original values. The stem-and-leaf plot depicted in Figure 3.9 shows two occurrences with a first value of 7; these occurrences correspond to the numbers 74 and 76.

In Minitab: Choose **Graph > Stem-and-Leaf**. Enter the column containing the data and click **OK** (Figure 3.10). Selecting **Trim outliers** will result in outliers being removed from the stem-and-leaf plot and displayed separately. The results will then appear in the Minitab session window as shown in Figure 3.11.

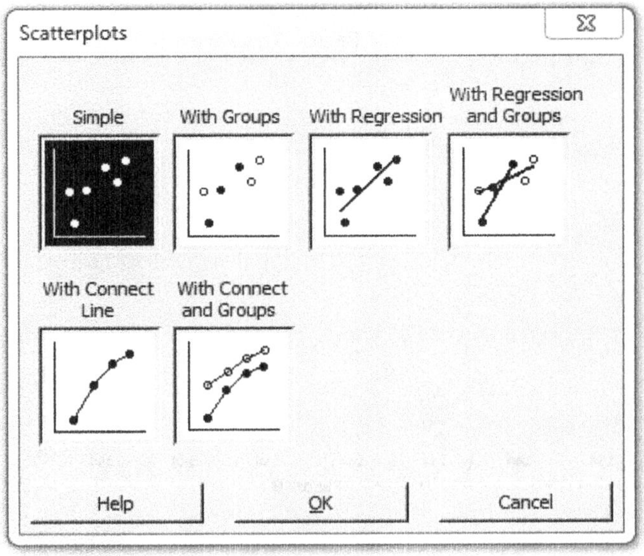

Figure 3.6 Scatterplot gallery.

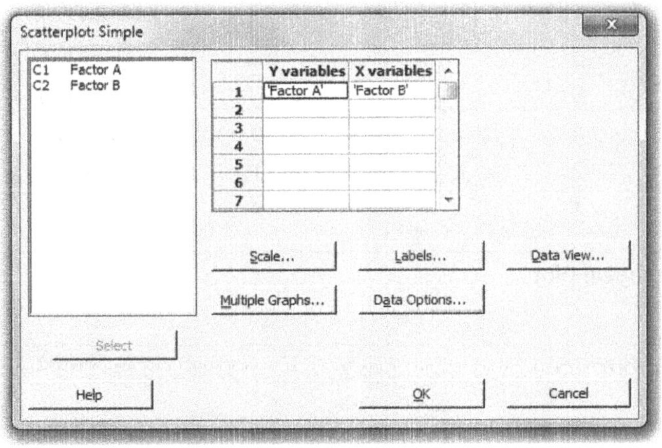

Figure 3.7 Minitab dialog box for scatterplots.

The value in parentheses in the left column of Figure 3.11 indicates the number of values in that row and that the row contains the median of the data set. The other values in the left column of the Minitab output correspond to the cumulative number of values in the row and all rows further from the median on that side. For example, there are 14 total values in the first four rows and 7 total values in the last four rows.

MULTI-VARI CHART

Multi-vari charts use a single graph to depict the potential relationships between a continuous variable and multiple categorical factors. These factors could be from

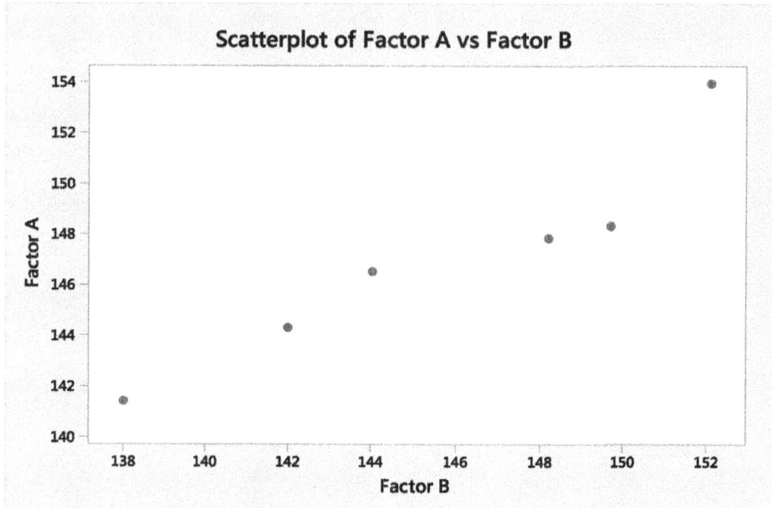

Figure 3.8 Scatterplot.

Number of occurrences	Stem	Leaf
2	7	46
3	8	257
5	9	13479
4	10	2367
2	11	58
1	12	2

Figure 3.9 Stem-and-leaf plot.

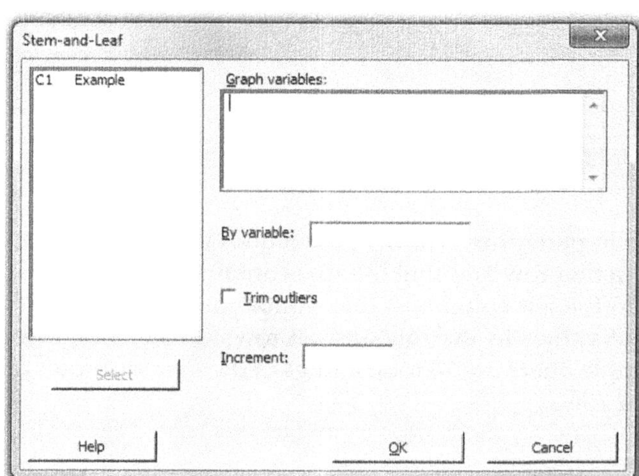

Figure 3.10 Minitab dialog box for stem-and-leaf plots.

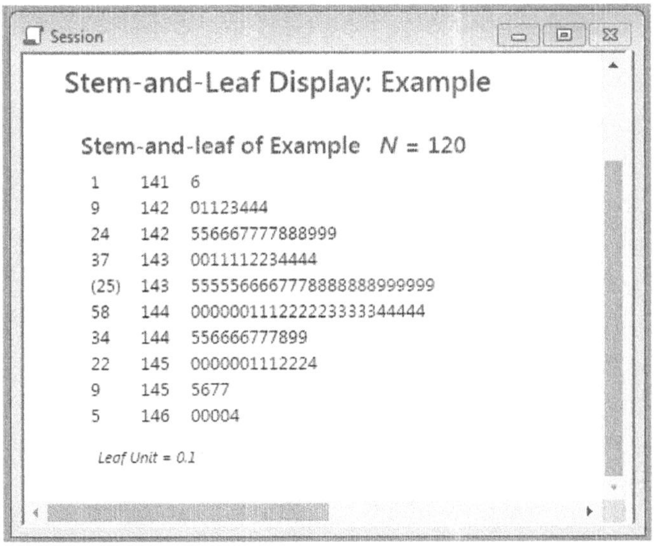

Figure 3.11 Minitab session window with a stem-and-leaf plot.

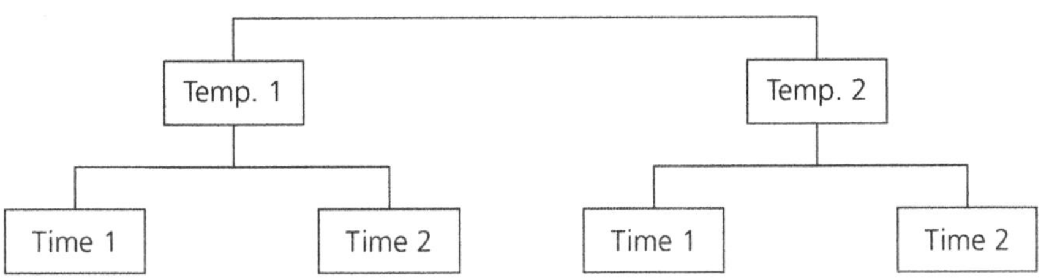

Figure 3.12 Multi-vari chart test plan.

different locations or times (Barsalou 2015a). A test plan such as the one in Figure 3.12 should be created before performing an experiment to collect data for a multi-vari chart. For example, a machine may have a high and low temperature setting and each temperature could have a time setting.

In Minitab: Suppose a quality engineer wanted to analyze how a response varied across three machines with three cells each (see Figure 3.13).

Go to **Stat > Quality Tools > Multi-Vari Chart** and enter the response and factors. Then click **OK** (see Figure 3.14). The multi-vari chart in Figure 3.15 shows more machine-to-machine variation than variation between nests in each machine. The dotted line connects the means of the second factor; in this case, that would be the means of each nest.

Worksheet 1 ***

↓	C1	C2	C3	C4	C5	C6	C7
	Machine	Nest	ok	nok	Data	Response	
1	1	1	3	5	0.625	14.0	
2	1	2	2	6	0.750	16.8	
3	1	3	5	3	0.375	8.4	
4	1	4	3	5	0.625	14.0	
5	1	5	5	3	0.375	8.4	
6	1	6	7	1	0.125	2.8	
7	1	7	6	2	0.250	5.6	
8	1	8	4	4	0.500	11.2	
9	1	9	7	1	0.125	2.8	
10	2	1	1	7	0.875	19.6	
11	2	2	7	1	0.125	2.8	
12	2	3	6	2	0.250	5.6	
13	2	4	7	1	0.125	2.8	
14	2	5	6	2	0.250	5.6	
15	2	6	6	2	0.250	5.6	
16	2	7	5	3	0.375	8.4	
17	2	8	6	2	0.250	5.6	
18	2	9	4	4	0.500	11.2	
19	3	1	2	6	0.750	16.8	
20	3	2	2	6	0.750	16.8	
21	3	3	6	2	0.250	5.6	
22	3	4	5	3	0.375	8.4	
23	3	5	6	2	0.250	5.6	
24	3	6	3	5	0.625	14.0	
25	3	7	7	1	0.125	2.8	
26	3	8	8	0	0.000	0.0	
27	3	9	8	0	0.000	0.0	
28	4	1	4	4	0.500	11.2	
29	4	2	6	2	0.250	5.6	
30	4	3	5	3	0.375	8.4	
31	4	4	5	3	0.375	8.4	
32	4	5	8	0	0.000	0.0	
33	4	6	7	1	0.125	2.8	
34	4	7	5	3	0.375	8.4	
35	4	8	3	5	0.625	14.0	
36	4	9	7	1	0.125	2.8	
37							

Figure 3.13 Minitab dialog box for multi-vari charts.

QUESTIONS AND ANSWERS

Q: What is the advantage of a boxplot over a histogram?

A: While a histogram is a great tool for assessing the distribution of a data set, a boxplot provides a statistical identification of outliers and may be a faster tool for assessing where the data are expected to lie.

Q: What is a multi-vari chart used for?

A: Multi-vari charts show the potential relationship between a continuous variable and multiple categorical factors.

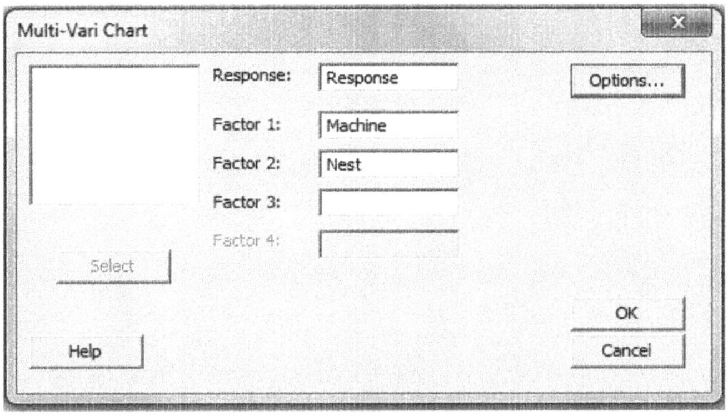

Figure 3.14 Minitab dialog box for multi-vari charts.

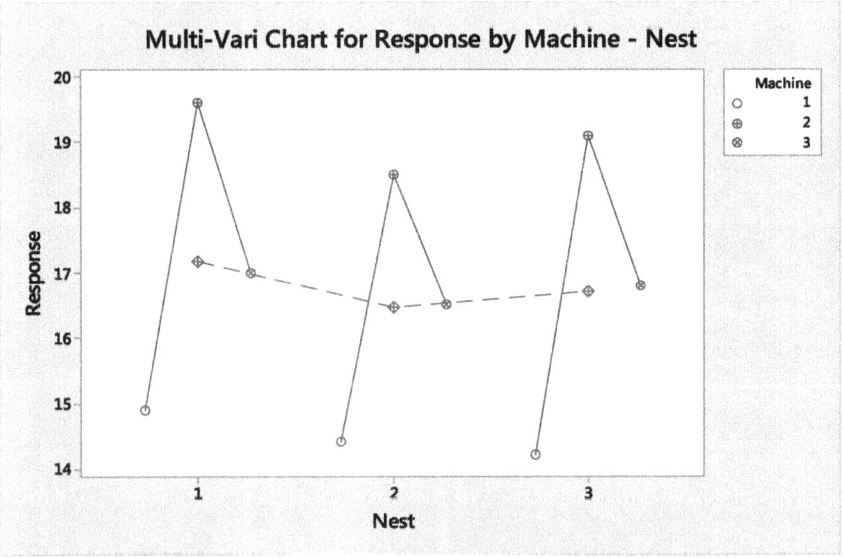

Figure 3.15 Multi-vari chart.

Q: When should I use a stem-and-leaf plot?

A: A stem-and-leaf plot is useful when you want to see the distribution of data while still keeping the specific data values. It also allows the median and mode of the data set to be easily identified on the plot.

Q: What should I do if there appears to be a correlation when I do a scatterplot?

A: Perform regression analysis to assess the data statistically.

Chapter 4
Probability

Probability is the likelihood that an event will occur. Witte defines probability as "the proportion or fraction of time that a particular event will occur" (1993, 188). Probabilities range from 0 to 1, with 0 corresponding to no chance of the event happening and 1 corresponding to a 100% chance of it happening. Observations are obtained by performing an experiment or making an observation and recording whether the event occurred or not; often an occurrence is referred to as a "success." In probability, the term "success" should not be confused with the common definition of the word that implies something good has happened. A success in probability may actually be something bad, such as a death or a defect, and the term simply means that the event of interest occurred. All possible outcomes of a trial or experiment are referred to as the sample space (Montgomery and Runger 1999). For example, a coin flipped two times has "heads then tails," "tails then heads," "heads then heads," or "tails then tails" as a sample space.

The probability of an event occurring is often denoted as P(event). For example, the probability of a flipped coin landing on heads would be denoted P(heads) = 0.5. Two or more events are considered mutually exclusive if they cannot both occur on the same trial. For the coin flip, the event of "lands on heads" and the event of "lands on tails" would be mutually exclusive because they cannot both occur on a single flip.

Table 4.1 shows probability symbols and meanings and Table 4.2 shows probability rules and formulas.

The conditional probability $P(A \mid B)$ can be calculated using Bayes' theorem if the individual probabilities are known. P(A) is the probability of A, $P(B \mid A)$ is the probability of B occurring given that A has occurred, $P(\sim A)$ is the probability of A not occurring, and $P(B \mid \sim A)$ is the probability that B occurs given that A has not occurred. The formula is:

$$P(A \mid B) = \frac{P(A)P(B \mid A)}{P(B \mid A)P(A) + P(B \mid \sim A)P(\sim A)}$$

Probability calculations can also be performed using discrete or continuous probability distributions. Table 4.3 lists common discrete probability distributions and their uses, and Table 4.4 lists common continuous probability distributions and their uses.

Table 4.1 Probability symbols and meanings.

Symbol	Meaning
$\cup$	Union: "or"
$\cap$	Intersection: "and"
~	Compliment of an event: "event has not occurred"
P(X|Y)	Conditional probability: "probability that X will occur given that Y has occurred"

Table 4.2 Probability rules and formulas.

Rule	Formula
Addition rule for union of two events that are not mutually exclusive	$P(A \cup B) = P(A) + P(B) - P(A \cap B)$
Addition rule for union of two mutually exclusive events	$P(A \cup B) = P(A) + P(B)$
Multiplication rule for the intersection of two dependent events	$P(A \cap B) = P(A) \times P(B \mid A)$
Multiplication rule for the intersection of two independent events	$P(A \cap B) = P(A) \times P(B)$

Table 4.3 Discrete probability distributions.

Distribution	Use
Binomial probability distribution	The number of events that occur in a given number of trials where each trial has only two possible outcomes such as event occurred or did not occur.
Poisson probability distribution	The count of events that occurred over a time interval or other continuum with no limit to the number possible.
Geometric probability distribution	The number of trials necessary before a success occurs.
Hypergeometric probability distribution	The probability of a certain number of events occurring in a number of trials. For trials with only two possible outcomes and a number of possible trials and successes.

DISCRETE PROBABILITY DISTRIBUTIONS

Discrete probability distributions can only take on certain values and give the probability of each value occurring. An example of a discrete probability distribution is the probability of pulling 0, 1, 2, 3, or 4 white pieces if four pieces are chosen randomly from a chess set. There are four common discrete probability distributions. They are the binomial probability distribution, the Poisson probability

Table 4.4 Continuous probability distributions.

Distribution	Use
Normal distribution	When data is clustered about the mean in a bell-shaped distribution.
Lognormal distribution	For when the logarithm of the data is normally distributed; typically used for risk assessment.
Student's *t* distribution	For normally teasting the mean of normally distributed data when the population standard deviation is unknown.
Chi-square distribution	For the distribution of sample variance.
F distribution	For the distribution of the ratio of two estimates of variances.
Exponential distribution	Time between random occurrences.
Weibull distribution	Used for time related data.

distribution, the geometric probability distribution, and the hypergeometric probability distribution.

Binomial Distribution

The binomial distribution is used for determining the probability of x events occurring in n trials with only two possible outcomes, such as success or failure, yes or no, and there or not there. The binomial distribution is used when the probability of the event occurring is the same for each trial. Calculating the probability of getting a specific number of tails when flipping a coin a given number of times is an example of the binomial distribution; the results of a coin toss can only be heads or tails (Silver 1997). The letter p represents the probability of success and the letter q represents the probability of failure, with $q=(1-p)$. The formula for the binomial distribution is $P(x) = C_x^n\,(p)^x\,(q)^{n-x}$, where n is sample size, x is the number of successes being considered, and C_x^n is the number of ways you can choose x items out of n and is equal to

$$C_x^n = \frac{n!}{x!(n-x)!}.$$

For n trials, Figure 4.1 shows the probability of each possible number of successes for $p=0.1$ and $p=0.4$.

Example

A process that produces light bulbs is known to have 10% of the bulbs defective. What is the probability that 3 light bulbs out of a sample of 10 will be defective? In this example, a success is a defect and a failure is defined as a part without a defect. The probability a part is OK is 100% − 10% = 90% = 0.90, and the probability of a defect is 10% = 0.10. Calculate $P(x=3) = C_x^n\,(p)^x\,(q)^{n-x} = \frac{10!}{3!7!}(0.1)^3(0.9)^{10-3} = 0.057$. There is a 5.7% chance of observing 3 defects in a sample of 10.

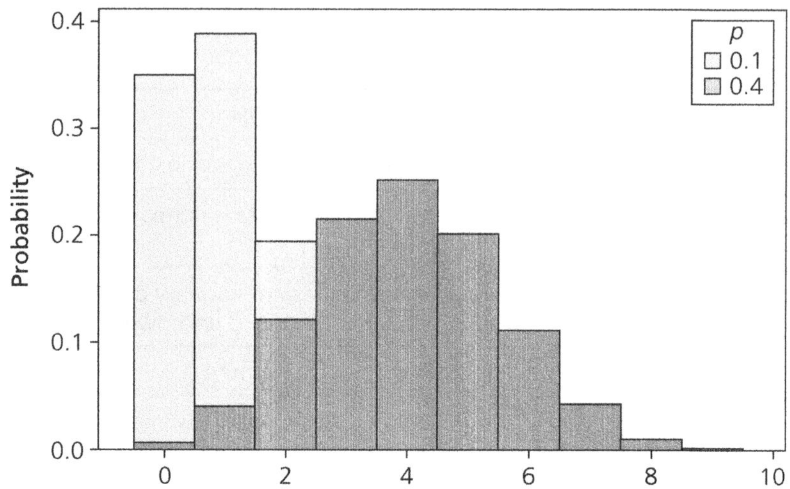

Figure 4.1 Binomial probability distribution for $n=10$ with $p=0.1$ and $p=0.4$.

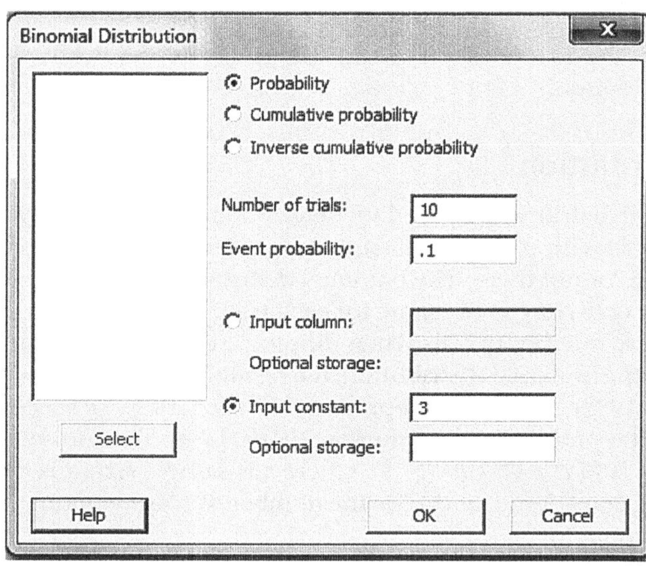

Figure 4.2 Minitab dialog box for a binomial distribution probability.

In Minitab: Go to **Calc > Probability Distributions > Binomial**. Select **Probability** and enter the number of trials and event probability (see Figure 4.2). Then select **Input constant** and enter the number of successes you want a probability for and click **OK** (see Figure 4.3).

Note: If you have multiple numbers of success of interest, you can enter them in a column and use Input column instead. You can then enter a column in Optional storage to store the individual probabilities when performing probability calculations with Minitab.

There is a 0.057 probability that 3 out of 10 light bulbs will be defective when n is 10, the probability of a defective light bulb is 0.1, and 3 light bulbs are checked (see Figure 4.3).

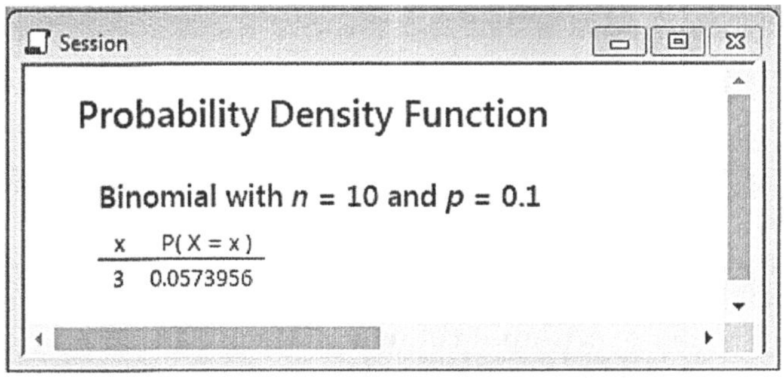

Figure 4.3 Minitab session window for a binomial distribution probability.

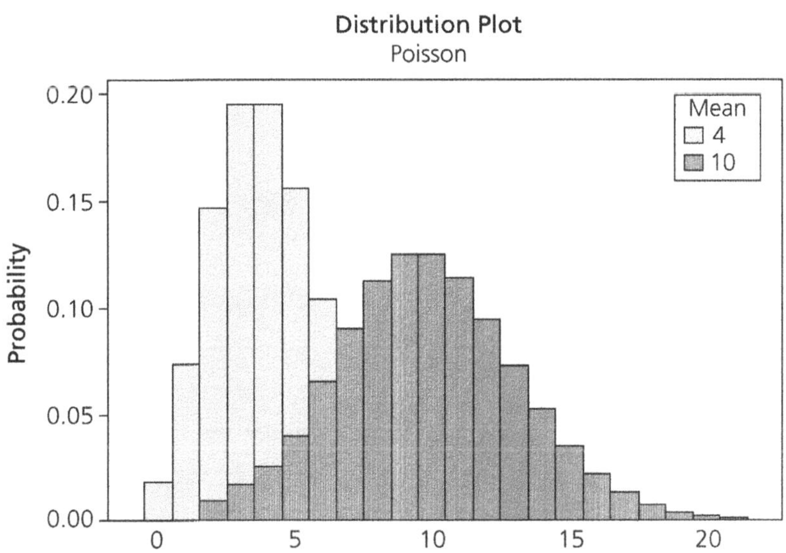

Figure 4.4 Poisson probability distribution for means of 4.0 and 10.0.

Poisson Distribution

The Poisson distribution is used for a number of events occurring over an interval or continuum where there is no specific number of opportunities (Montgomery, Runger, and Hubele 2001), such as the number of surface blemishes on a machined part. The Poisson distribution can be used in place of the binomial distribution for large sample sizes ($n > 20$) if the probability is low ($p \leq .05$). The Poisson distribution is equal to

$$P(x) = \frac{\mu^x e^{-\mu}}{x!},$$

where $p(x)$ is the probability of an event occurring, μ is the mean number of occurrences in a particular interval, and e is the constant 2.718282. The mean and variance of the Poisson distribution are equal. Two Poisson distributions can be seen in Figure 4.4; one has a mean of 4 and the other has a mean of 10.

Example

An average of 3 light bulbs are found defective every shift. What is the probability of 4 failures on any given shift? We set μ equal to 3 and X equal to 4 and then calculate $P(4) = \frac{3^4 e^{-3}}{4!} = 0.168$. There is a 16.8% chance of finding 4 failures on any given shift.

In Minitab: Go to **Calc > Probability Distributions > Poisson** and select **Probability**. Enter the mean, click on **Input constant** and enter the number of events of interest (optionally select **Input column** and input a column containing multiple numbers of events), and click **OK** (see Figure 4.5).

The probability of 4 failures in a Poisson process with a mean of 3 is 0.168031 (see Figure 4.6).

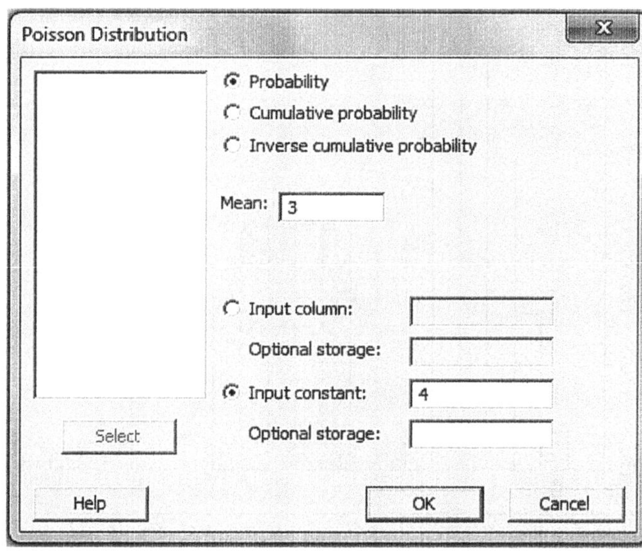

Figure 4.5 Minitab dialog box for a binomial Poisson probability.

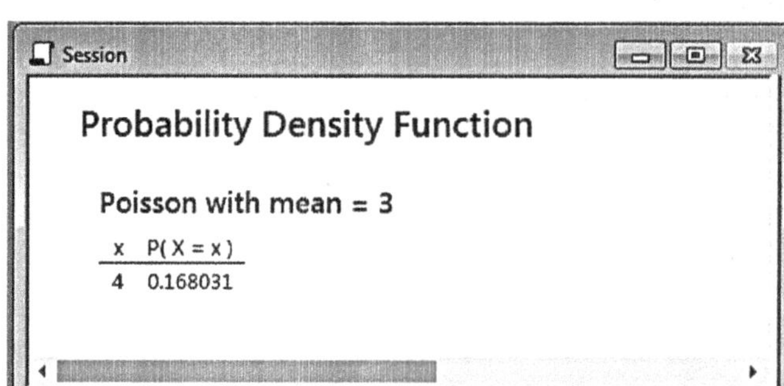

Figure 4.6 Minitab session window for a Poisson distribution probability.

Geometric Distribution

The geometric distribution is used to find the number of trials necessary before the first occurrence of an event (Clarke and Cooke 1992), such as the probability that 200 light bulbs checked will be found not defective before finding a defective one. The probability of the first success at the kth trial using the geometric distribution is $P(k, p) = p(1-p)^{k-1}$ where k is the number of trials until the first success, $k-1$ is the number of failures prior to the success, and p is the probability of success. The probability that more than n trails are needed before the first success using the geometric distribution is $P(k>n) = (1-p)^n$ where p is the probability of a success and n is the number of trials. Geometric distributions with a p of 0.1 and 0.4 can be seen in Figure 4.7.

Example

What is the probability the first defective light bulb will be the 50th checked if the probability of a defective light bulb is 0.06? Set $k=50$, $p=0.06$, and $q=0.94$. Then calculate $P(50, 0.06) = 0.06 \times 0.94^{50-1} = 0.0028935$. The probability that the 50th light bulb checked will be defective and the first 49 will be okay is 0.0028935.

In Minitab: Go to **Calc > Probability Distributions > Geometric Distribution** and select **Probability**. Then enter the event probability, select **Input constant**, enter the number of trials until the first success (alternatively enter multiple values of interest in a column and use **Input column**), and click **OK** (see Figure 4.8).

The probability that the 50th light bulb checked will be defective and the first 49 will be okay is 0.0028935, as shown in Figure 4.9.

Hypergeometric Distribution

The hypergeometric distribution is used in place of the binomial distribution when both the number of possible trials and the number of possible successes are finite. Unlike the binomial distribution, the results of trials following the hypergeometric distribution are dependent because the probability of a result is dependent on the previous results (McClave and Sincich 2009). The probability of success is the same

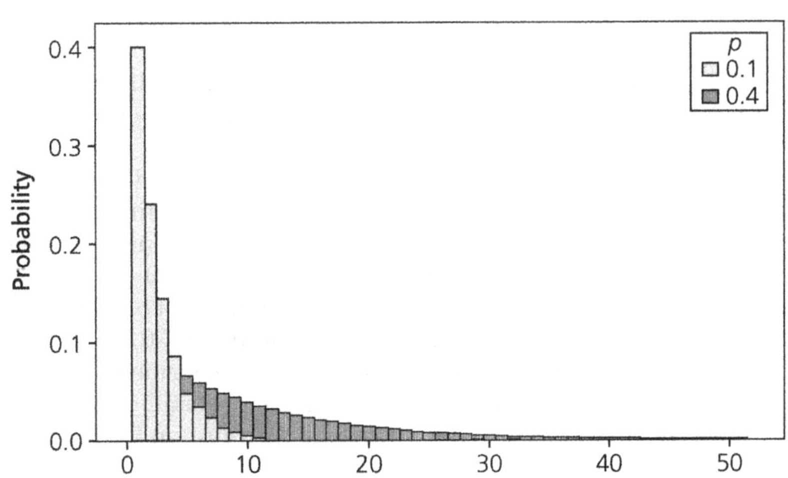

Figure 4.7 Geometric probability distribution for a p of 0.1 and 0.04.

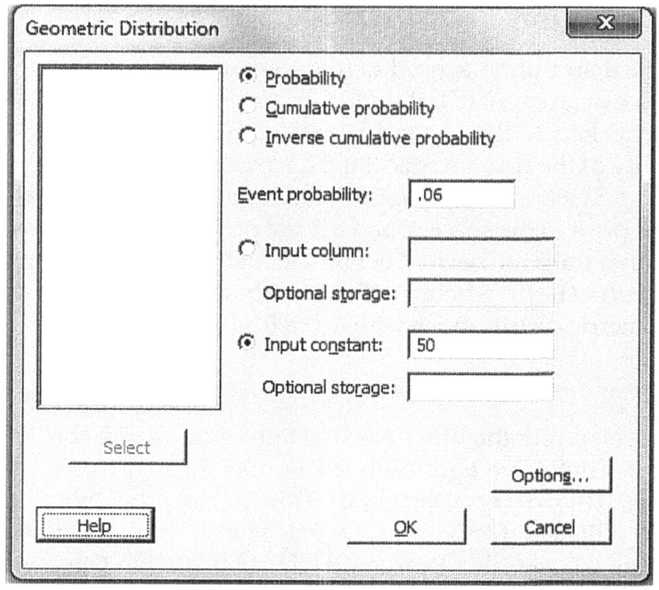

Figure 4.8 Minitab dialog box for a geometric distribution probability.

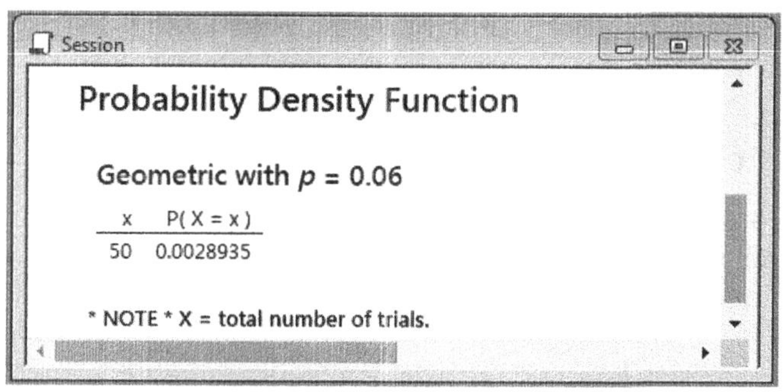

Figure 4.9 Minitab session window for a geometric distribution probability.

for each trial in a binomial experiment, whereas the probability in the second trial of a hypergeometric experiment changes because the number of remaining trials and successes has changed.

For example, suppose a box contained 100 light bulbs and 20 of them are defective. The probability of getting a defective light bulb on the first try is $20/100 = 0.2$. The probability of getting a defective light bulb on the second try depends on the results of the first try. If the first bulb was defective the chance of getting a defective light bulb on the second trial is $19/99 = 0.1919$. If the first bulb was not defective the chance of getting a defective bulb on the second trial is $20/99 = .2020$. The binomial distribution should be used if sampling with replacement is performed (i.e., the light bulb is always returned to the box) as the probability of success remains constant from trial to trial. The hypergeometric distribution formula is

$$\mathrm{P}(x) = \frac{C_x^k \, C_{n-x}^{N-k}}{C_n^N},$$

where k is the number of success, N is the population size, n is the number of trials, and x is the value between 0 and n that we selected and is $\leq k$. An example of hypergeometric distributions for 100 possible trials with 25 possible successes and with n equal to 5 and 10 can be seen in Figure 4.10.

Example

We select 10 light bulbs out of a box containing 250 and we know there are 30 defective light bulbs in the box. What is the probability that exactly 3 of the light bulbs we selected will be defective? Identify k, N, x, and n, which are given as 30, 250, 3, and 10, respectively. Then calculate

$$P(10) = \frac{C_x^k C_{n-x}^{N-k}}{C_n^N} = \frac{C_3^{30} C_{10-3}^{250-30}}{C_{10}^{250}} = 0.0833163.$$

The probability that exactly 3 out of 10 light bulbs are defective is 0.0833163.

In Minitab: Go to **Calc > Probability Distributions > Hypergeometric Distribution** and select **Probability**. Then enter the population size, count of events in the population, and sample size. Then select **Input constant** and enter 3 as the number of events and click **OK** (see Figure 4.11).

The probability that exactly 3 out of 10 light bulbs are defective is 0.0833163 (see Figure 4.12).

CONTINUOUS PROBABILITY DISTRIBUTIONS

Continuous distributions are used when a variable can take on any value in a specified range, such as the length of a component, time to fill a work order, or the weight of a part. Common distributions that describe the probability of values occurring from measurements of these types of characteristics include normal, lognormal,

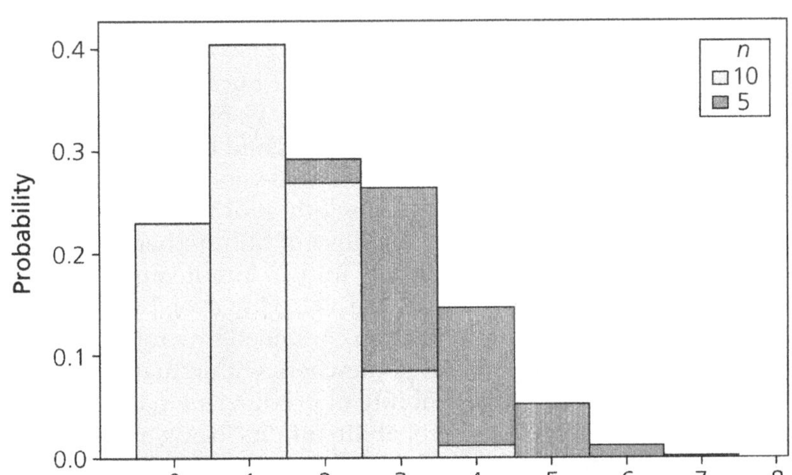

Figure 4.10 Hypergeometric probability distribution for 100 possible trials with 25 possible events and sample sizes of 10 and 5.

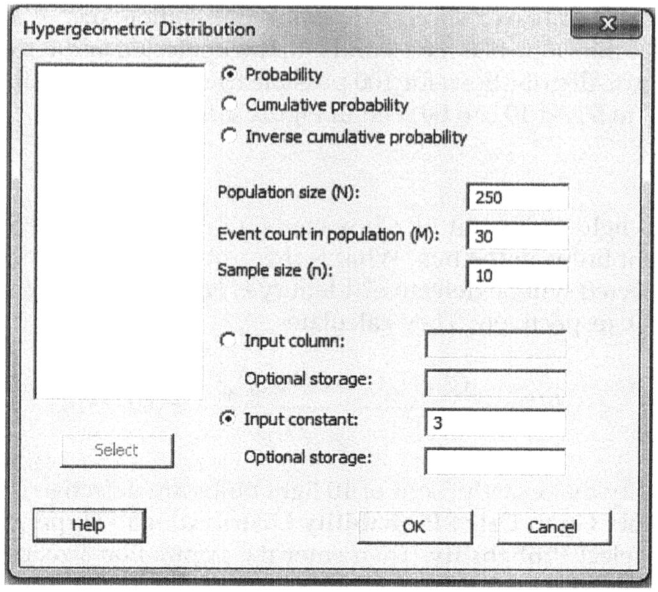

Figure 4.11 Minitab dialog box for a hypergeometric distribution probability.

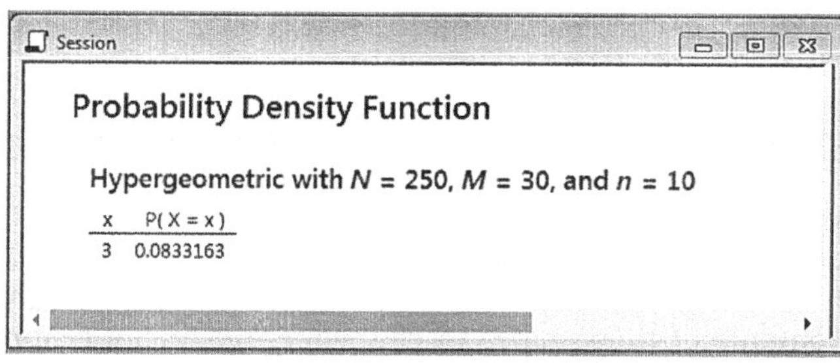

Figure 4.12 Minitab session window for a hypergeometric distribution probability.

Weibull, and exponential. Other continuous distributions are used for hypothesis testing rather than to describe natural processes. Common examples are the Student's *t* distribution, the Chi-square distribution, and the *F* distribution.

Because continuous probability distributions can take on an infinite number of values within a range of possible values, any individual value has a probability of 0. This is because the total probability of all numbers combined is 1 and this number must be divided in some way into an infinite number of possibilities, leaving the probability of any one of those as 0. However, by calculating the area under the probability distribution function contained by a range of values of interest, the probability of finding a value somewhere within that range may be determined. For example, while the probability of producing a part weighing precisely 97 grams is mathematically 0, the probability of producing a part that is 97 grams or more can be determined.

A continuous probability distribution is described using a mathematical formula that gives the probability density of the distribution for any value. This formula is known as the *probability density function*. While there is little value in ever determining the probability density for a single value, the function is critically important for two reasons. First, the function allows a plot of the distribution to be made that demonstrates which values are most likely to occur and which values are unlikely to occur. For example, the normal distribution described in the next section has a probability density function that looks like a bell when graphed and is commonly referred to as the "bell curve." Second, the function can be integrated over a range of values of interest to determine the probability of a random value from that distribution falling within that range. The function that describes the probability of all values less than or equal to a value is known as the *cumulative distribution function*.

Values of interest are typically computed using statistical software or referenced in tables for common values because many distributions have complex probability distribution functions that cannot be directly or easily integrated.

Normal Distribution

The normal distribution is a commonly used distribution for data that is centered around the mean and forms a symmetric "bell curve" as values move away from the mean. Most stable processes that create symmetric data, such as the diameter of CNC machined cylinders, will follow the normal distribution. The normal probability distribution is also called the *Gaussian distribution* (Silver 1997) and the probability density function is

$$f(x)\frac{1}{\sigma\sqrt{2\pi}}e^{-\frac{(x-\mu)^2}{2\sigma^2}},$$

with $e \approx 2.7182828$ and $\pi \approx 3.1416$. The shape and location of the distribution varies based on the mean and standard deviation. The formula $Z = X - \mu / \sigma$ can be used to convert the normal distribution into the simpler standard normal distribution, which has a mean of 0 and a standard deviation of 1. The standard normal distribution puts every value in terms of the number of standard deviations it is from the mean, a number called the Z-score of that value. Regardless of the original mean and standard deviation, standardizing means that any given Z-score always has the same probability on either side of it. These probabilities can be found in the tables in Appendix C. Examples of the normal distribution are shown in Figure 4.13.

Example

A light bulb burns an average of 4800 hours with a standard deviation of 150. What is the probability of a bulb burning at least 4600 hours? First calculate

$$Z = \frac{X-\mu}{\sigma} = \frac{4600-4800}{150} = -1.3333,$$

and then look up the Z-score for −1.333 in Table C.1 in Appendix C (as shown in Table 4.5). This table gives the cumulative probability of all values less than or equal to the Z-score of interest.

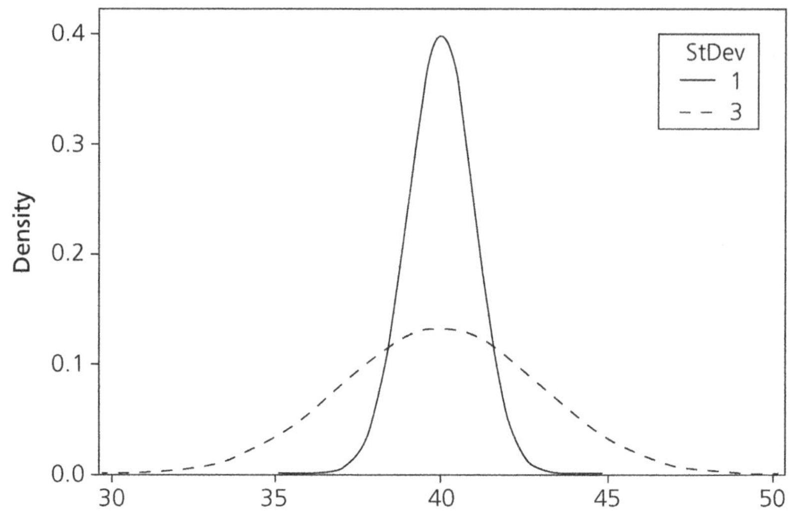

Figure 4.13 Normal distributions for a mean of 40 with standard deviations of 1.0 and 3.0.

Table 4.5 Z-score table for Z =−1.3333.

Z	0	0.01	0.02	0.03	0.04
0	0.5000	0.4960	0.4920	0.4880	0.4840
−0.1	0.4602	0.4562	0.4522	0.4483	0.4443
−0.2	0.4207	0.4168	0.4129	0.4090	0.4052
−0.3	0.3821	0.3783	0.3745	0.3707	0.3669
−0.4	0.3446	0.3409	0.3372	0.3336	0.3300
−0.5	0.3085	0.3050	0.3015	0.2981	0.2946
−0.6	0.2743	0.2709	0.2676	0.2643	0.2611
−0.7	0.2420	0.2389	0.2358	0.2327	0.2296
−0.8	0.2119	0.2090	0.2061	0.2033	0.2005
−0.9	0.1841	0.1814	0.1788	0.1762	0.1736
−1	0.1587	0.1562	0.1539	0.1515	0.1492
−1.1	0.1357	0.1335	0.1314	0.1292	0.1271
−1.2	0.1151	0.1131	0.1112	0.1093	0.1075
−1.3	0.0968	0.0951	0.0934	**0.0918**	0.0901
−1.4	0.0808	0.0793	0.0778	0.0764	0.0749

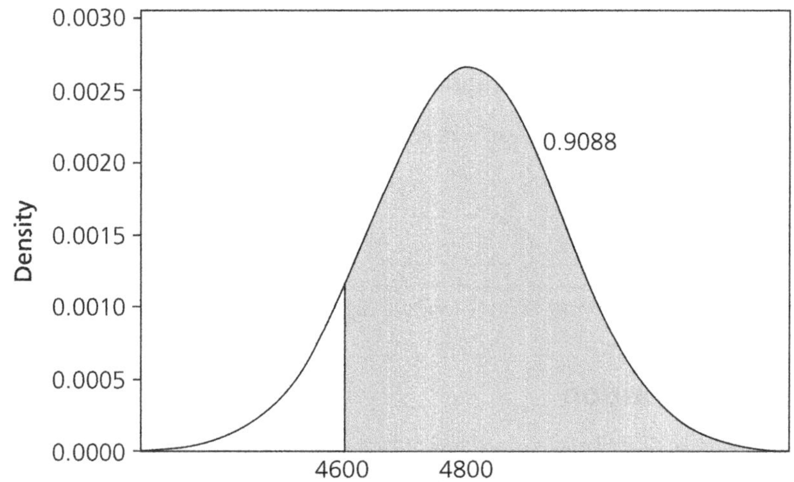

Figure 4.14 Normal distribution with mean 4800, standard deviation 150.

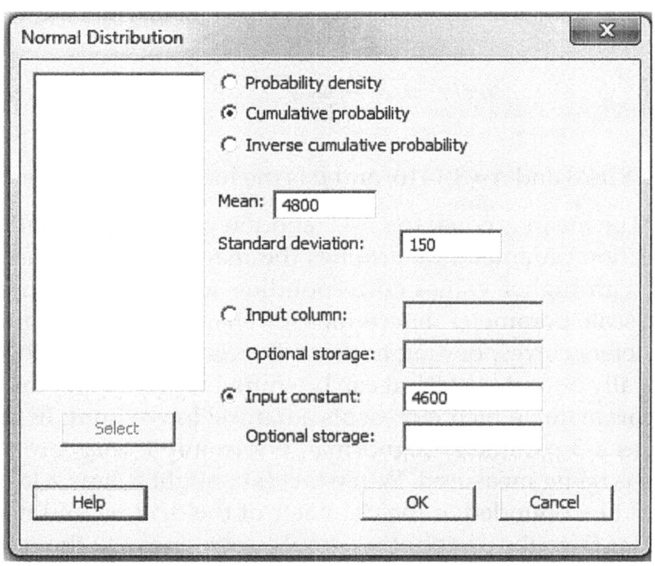

Figure 4.15 Minitab dialog box for a normal distribution probability.

The Z-score of –1.333 corresponds to a value of 0.0918 in the Z-score table. This tells us that there is a 9.18% probability of a bulb burning less than 4600 hours. The probability of it burning more than 4600 hours is 1 – 0.0918 = 0.9082 = 90.82% (see Figure 4.14).

In Minitab: Go to **Calc > Probability Distributions > Normal** and enter the mean and standard deviation. Next, check **Input constant** and enter the value of interest and click **OK** (see Figure 4.15). As with other probability distributions, you can optionally enter a column containing multiple values of interest and select a column where you want to store the results.

The results are shown in Figure 4.16.

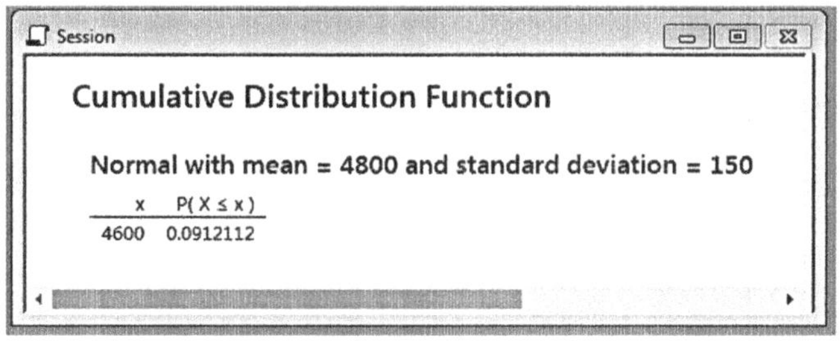

Figure 4.16 Minitab session window for a normal distribution probability.

Lognormal Distribution

The lognormal distribution is used when the natural log of the data is normally distributed. It often occurs when measuring time between repairs and the size of natural phenomena such as the length of scratches or the diameter of surface pits. The distribution is skewed to the right and useful when the data span orders of magnitude (Lawson and Erjavec 2001). The probability density function is

$$f(x) = \frac{1}{x\sigma\sqrt{2\pi}} e^{-\frac{1}{2}\left(\frac{\ln z - \mu}{\sigma}\right)^2} \quad x > 0,$$

with $e \approx 2.7182828$ and $\pi \approx 3.1416$, and μ is the location parameter and σ is the scale parameter. The mean is equal to $e^{\mu - \frac{1}{2}\sigma^2}$ and the variance is equal to $\left(e^{\sigma^2} - 1\right)e^{2\mu+\sigma^2}$.

The location parameter determines the magnitude of values the distribution will cover, with higher values corresponding to covering a larger magnitude of values. The scale parameter determines the shape of the distribution, with higher scale parameters corresponding to more skewed shapes. The lognormal distribution is generally bounded at 0 but can be shifted to any other bound by including a threshold parameter, which represents a natural lower limit. In this case it is often referred to as a 3-parameter lognormal. For example, suppose the diameter of a drilled hole is being measured. While the data might follow a lognormal distribution, it would be bounded at the diameter of the drill bit and not 0 and a threshold parameter fit to the distribution would correspond to the minimum diameter caused by the bit. Figure 4.17 depicts several lognormal distributions.

Example

A quality inspector measures the length of discolorations in carpet fibers. Discolorations longer than 1.5 centimeters (cm) are often visible to the customer and considered defects. A sample of discolorations show that the lengths follow the lognormal distribution with location 0.1 and scale 0.4. What proportion of discolorations will exceed 1.5 cm in length?

In Minitab, go to **Calc > Probability Distributions > Lognormal** and complete the dialog as shown in Figure 4.18. Click on **OK**.

The session window output is shown in Figure 4.19.

The output shows that the probability of a discoloration less than or equal to 1.5 cm is 0.775. Therefore, the probability of a discoloration of more than 1.5 cm is $1 - 0.775 = 0.225$.

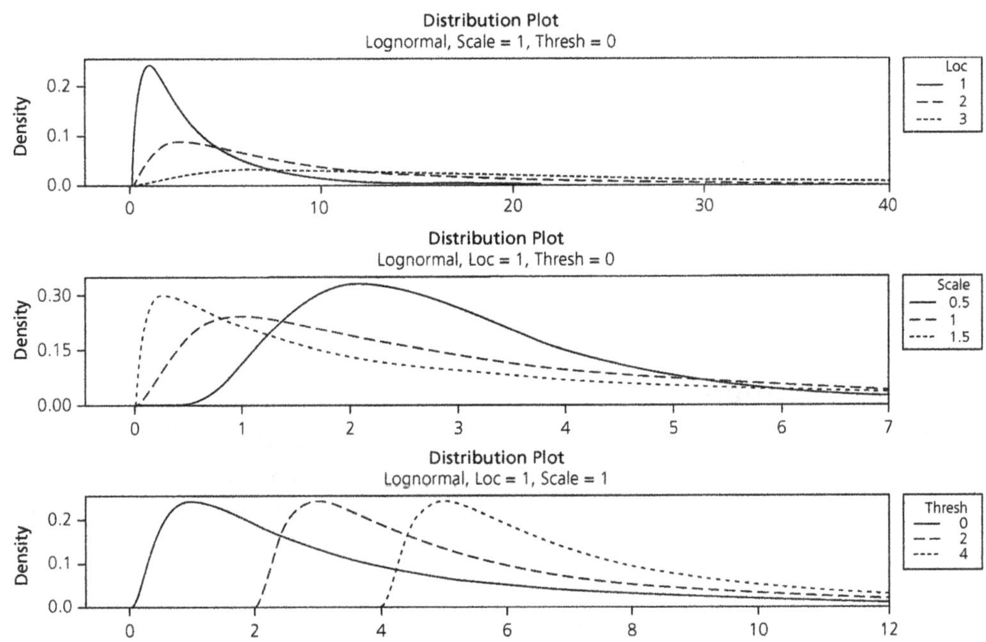

Figure 4.17 Lognormal distributions.

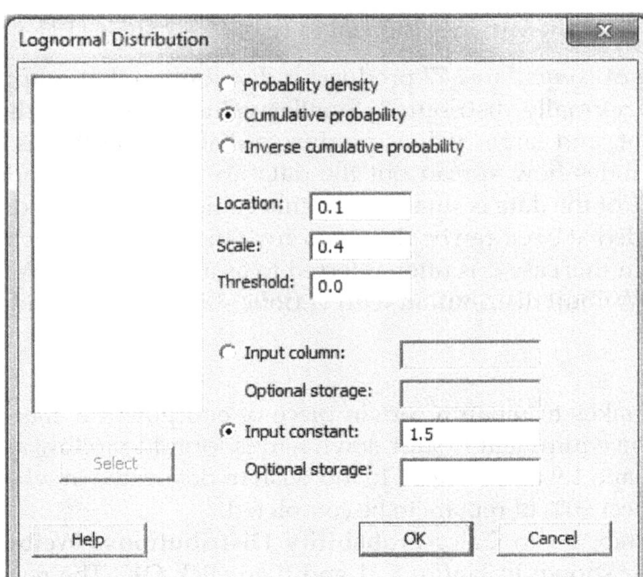

Figure 4.18 Dialog box for lognormal distribution.

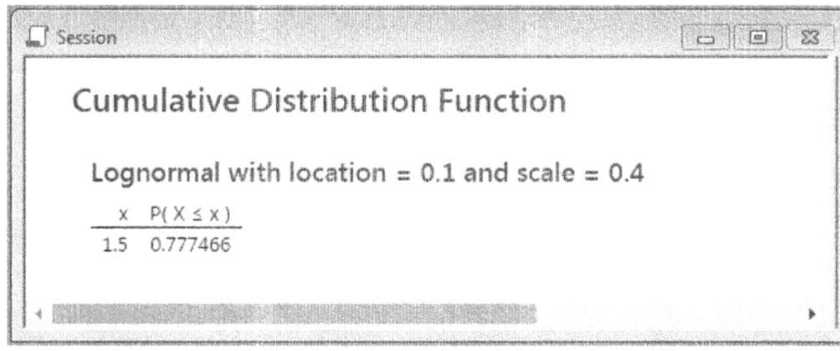

Figure 4.19 Session window output for lognormal.

Weibull Distribution

The Weibull distribution commonly occurs when measuring variables such as the time or stress applied until a part fails. It is often used in reliability engineering and can approximate other distributions when its parameters are adjusted (O'Connor and Kleyner 2012). The probability distribution function is

$$f(x) = \frac{\beta}{\alpha}\left(\frac{x}{\alpha}\right)^{\beta-1} e^{-\left(x/\alpha\right)^{\beta}} x > 0,$$

with $e \approx 2.7182828$, scale parameter α, and shape parameter β. The mean is equal to $\alpha\Gamma(1+1/\beta)$ and the variance is equal to $\alpha^2\left[\Gamma(1+2/\beta)+\left(\Gamma(1+1/\beta)\right)^2\right]$.

A value of β equal to 2.77 produces a distribution that is approximately symmetric and normally distributed. Smaller values produce a distribution that is skewed right, and larger values produce a distribution that is skewed left. The scale determines how spread out the data are and corresponds to the point in which 63.2% of the data is smaller than this value. The Weibull distribution is generally bounded at 0 but can be shifted to any other bound by including a threshold parameter. In this case it is often referred to as a 3-parameter Weibull. Figure 4.20 depicts the Weibull distribution with various shapes, scales, and thresholds.

Example

The time it takes to repair a certain piece of equipment is measured in minutes each time the equipment is shut down and is found to follow a Weibull distribution with shape 1.9 and scale 24.1. You wish to determine in what amount of time you can expect 90% of repairs to be completed.

In Minitab, go to **Calc > Probability Distributions > Weibull** and complete the dialog as shown in Figure 4.21 and then click **OK**. The results are shown in Figure 4.22.

Because we want to know what value corresponds to a cumulative probability, the Inverse cumulative probability option is chosen. The session window output indicates that based on the distribution, we can expect 90% of repairs to be completed within 37.4 minutes.

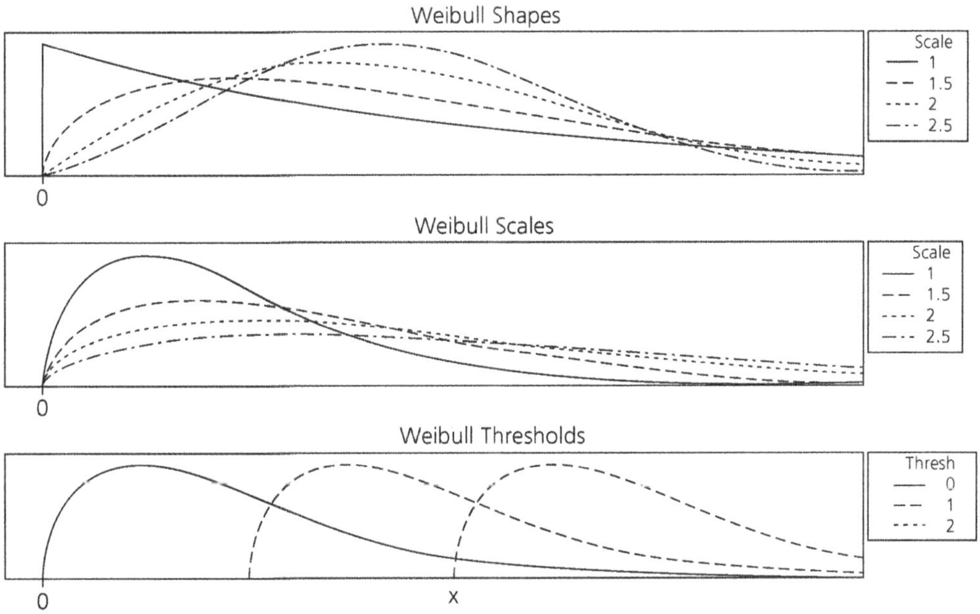

Figure 4.20 Weibull distribution with various shapes, scales, and thresholds.

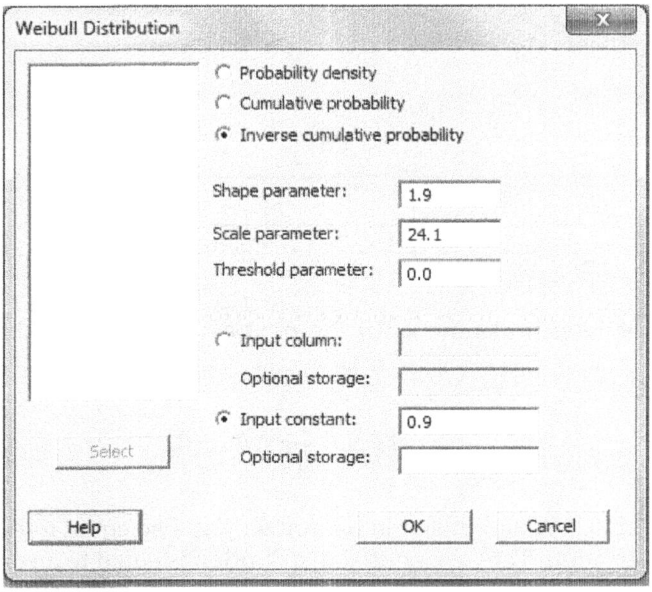

Figure 4.21 Minitab dialog box for a Weibull distribution probability.

Exponential Distribution

The exponential distribution is used for time between events in a Poisson process (Montgomery, Runger, and Hubele 2001), such as the time between workplace accidents at a light bulb factory. A Poisson process is one in which events occur completely randomly and the occurrence of an event has no

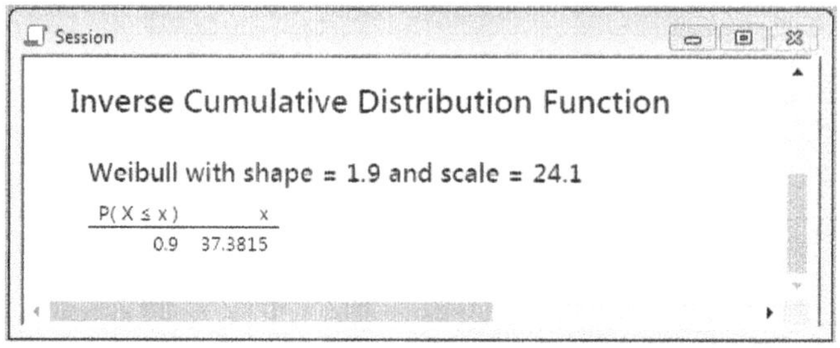

Figure 4.22 Minitab session window for a Weibull distribution probability.

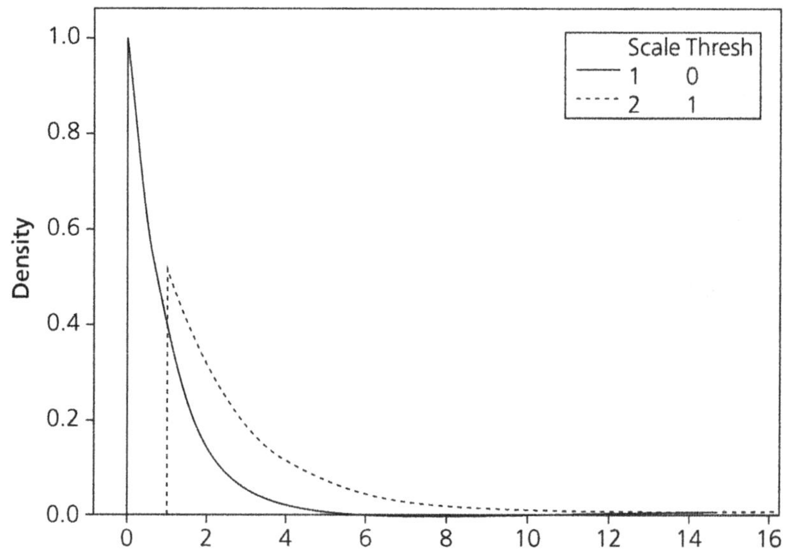

Figure 4.23 Exponential distributions with scale 1.0, threshold 0.0 and scale 2.0, threshold 1.0.

effect on the length of time until the next event. The probability distribution function is

$$f(x) = \frac{1}{\beta} e^{-x/\beta} \; x > 0,$$

where $e \approx 2.7182828$ and the scale parameter β is also equal to both the mean and standard deviation. The exponential distribution is equal to a Weibull distribution with a shape parameter of 1. It is generally bounded at 0 but can be shifted to any other bound by including a threshold parameter. In this case it is often referred to as a 2-parameter exponential (see Figure 4.23).

Example

The time and date are recorded each time an injury occurs in a factory. The time between injuries is found to follow an exponential distribution with a scale of 43 days. You wish to know the probability of an injury occurring within 7 days of another.

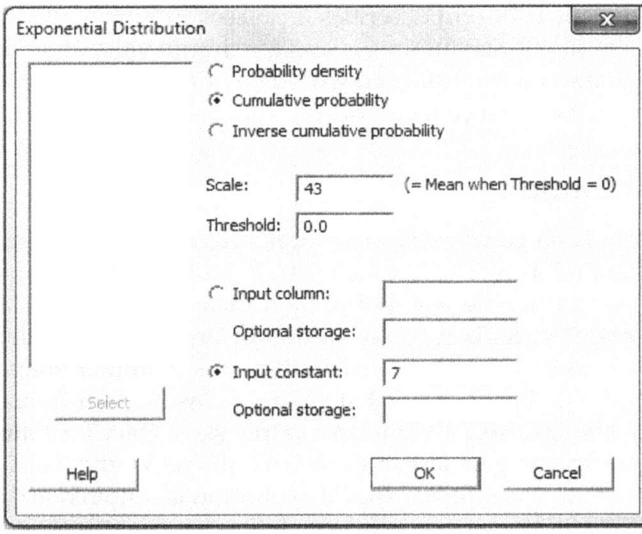

Figure 4.24 Minitab dialog box for an exponential distribution probability.

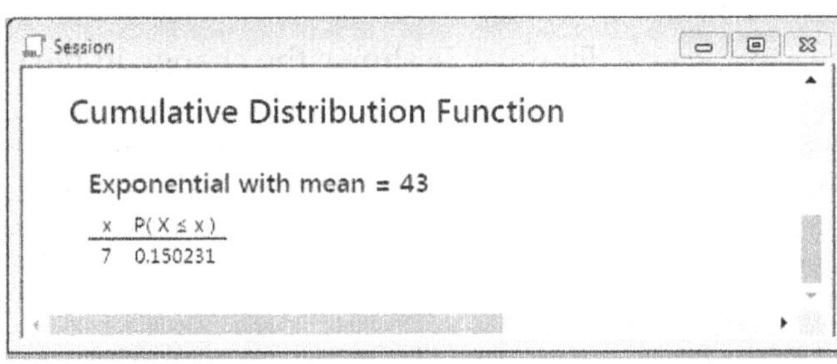

Figure 4.25 Minitab session window for an exponential distribution probability.

In Minitab, go to **Calc > Probability Distributions > Exponential** and complete the dialog as shown in Figure 4.24 and click **OK**. The results are shown in Figure 4.25.

The probability of an injury occurring within 7 days of the previous one is found to be 0.150231 or about 15%. Because the exponential distribution is so highly skewed, this proportion is much higher than many would expect given the average time of 43 days. Thus, although two injuries in one week would seem unexpectedly high, the statistics show it to be not all that unusual.

Other Natural Distributions

Although less common in practice, there are many other distributions that can occur in certain types of data. The smallest and largest extreme value distributions are used for the minimum or maximum values taken from a series of homogenous data sets. For example, the largest extreme value distribution might be used for the longest downtime a machine experiences each week and the smallest extreme value distribution might be used for the lowest weight recorded from lots of parts when multiple parts are sampled from each lot.

The uniform distribution describes a process in which all values are equally likely and there is a defined minimum and maximum value. For example, the time of day in which parts were produced would be uniformly distributed in a continuous process operating at a constant rate.

Student's *t* Distribution

When a sample is taken from a normally distributed population, the population mean μ is estimated using the sample mean $\bar{x}$. Although the sample mean is likely close to the true value, the true population mean is almost certainly a different value. The *t* distribution is the distribution of how likely possible values of the true population mean are. The *t* distribution has a similar shape as the normal distribution but with heavier tails that get lighter as the sample size increases. The *t* distribution is often used for smaller sample sizes (McClave and Sincich 2009), and as the sample size gets larger, especially above 30, the distribution is nearly equal to the normal distribution. The *t* distribution is centered at 0 and has a single parameter known as degrees of freedom, defined as

$$v = n - 1,$$

where n is the sample size. Figure 4.26 shows the change in the shape as degrees of freedom (df) go from 5 to 30.

The values in a *t*-score table such as the one in Table 4.6 are known as critical *t*-values. They correspond to the value in which a *t* distribution with the chosen degrees of freedom has the chosen probability of the true mean falling above it. These tables are found in Appendix C.

Chi-Square Distribution

The Chi-square distribution is used for estimating the true variance of a normally distributed population based on the sample variance (McClave and Dietrich 1991).

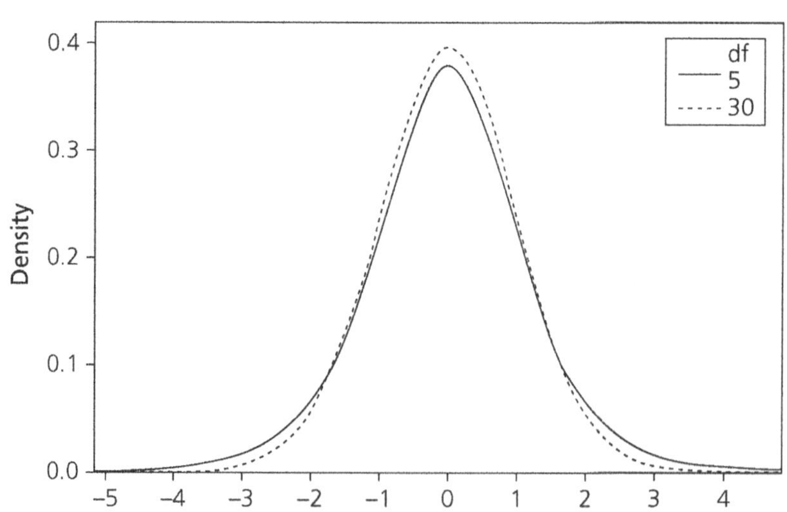

Figure 4.26 Student's *t* distribution (degrees of freedom = 5 and 30).

Table 4.6 *t*-score for 19 degrees of freedom and probability 0.05.

df	0.4	0.3	0.2	0.1	0.05	0.025
1	1.376	1.963	3.078	6.314	12.706	25.452
2	1.061	1.386	1.886	2.920	4.303	6.205
3	0.978	1.250	1.638	2.353	3.182	4.177
4	0.941	1.190	1.533	2.132	2.776	3.495
5	0.920	1.156	1.476	2.015	2.571	3.163
6	0.906	1.134	1.440	1.943	2.447	2.969
7	0.896	1.119	1.415	1.895	2.365	2.841
8	0.889	1.108	1.397	1.860	2.306	2.752
9	0.883	1.100	1.383	1.833	2.262	2.685
10	0.879	1.093	1.372	1.812	2.228	2.634
11	0.876	1.088	1.363	1.796	2.201	2.593
12	0.873	1.083	1.356	1.782	2.179	2.560
13	0.870	1.079	1.350	1.771	2.160	2.533
14	0.868	1.076	1.345	1.761	2.145	2.510
15	0.866	1.074	1.341	1.753	2.131	2.490
16	0.865	1.071	1.337	1.746	2.120	2.473
17	0.863	1.069	1.333	1.740	2.110	2.458
18	0.862	1.067	1.330	1.734	2.101	2.445
19	0.861	1.066	1.328	1.729	**2.093**	2.433
20	0.860	1.064	1.325	1.725	2.086	2.423

It is also used for calculating F-statistics and is the basis for many hypothesis tests. The Chi-square distribution has a single parameter known as degrees of freedom that is calculated as

$$\upsilon = n - 1,$$

where n is the sample size. For smaller values of v the distribution is skewed right and for larger values it is approximately symmetric. The mean of the distribution is υ.

The Chi-square distribution relates the sample variance to the population variance in that $\frac{(n-1)S^2}{\sigma^2}$ follows a Chi-square distribution with $n-1$ degrees of

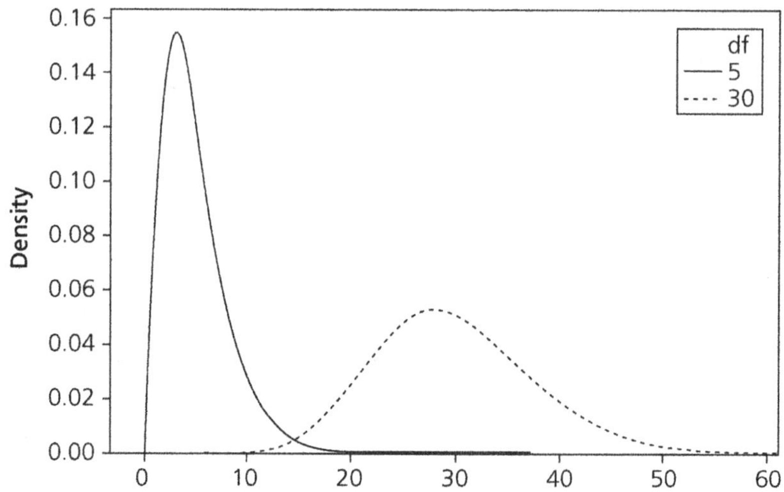

Figure 4.27 Chi-square distribution (degrees of freedom = 5 and 30).

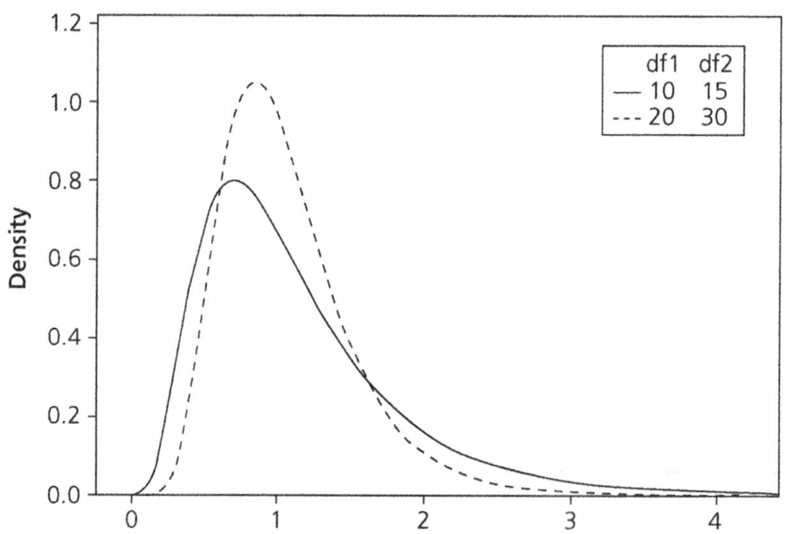

Figure 4.28 F distribution (v_1 10, v_2 15 and v_1 20, v_2 30).

freedom. The Chi-square distribution with 5 and 30 degrees of freedom is depicted in Figure 4.27.

F Distribution

The F distribution is used for comparing population variances using sample variances (McClave and Dietrich 1991). The F distribution has two parameters known as the *numerator degrees of freedom* and the *denominator degrees of freedom*, calculated as

$$\upsilon_1 = n_1 - 1,$$

$$\upsilon_2 = n_2 - 1,$$

where n_1 is the first sample size and n_2 is the second sample size. The F distribution is equal to the ratio of one Chi-square variable divided by its degrees of freedom and a second Chi-square variable divided by its degrees of freedom. Therefore, the ratio $\frac{S_1^2/\sigma_1^2}{S_2^2/\sigma_2^2}$ follows an F distribution. Examples of the F distribution are shown in Figure 4.28.

QUESTIONS AND ANSWERS

Q: What is the difference between the binomial distribution and the Poisson distribution?

A: The binomial distribution describes the numbers of occurrences of an event when a total number of opportunities is possible, whereas the Poisson distribution describes the number of occurrences when there is no finite number of opportunities possible.

Q: Which distribution should I use if my data consist for random occurrences over an interval of time?

A: The exponential distribution describes the time between randomly occurring events.

Q: What are the most common statistical distributions?

A: For continuous data, the most commonly occurring distributions are normal, lognormal, Weibull, and exponential.

Q: Which statistical distributions are typically used when there is a natural limit?

A: The most common would be the Weibull or 3-parameter Weibull distributions, which have a left bound. This typically occurs for time studies or measures of force, which cannot be negative.

Chapter 5
Hypothesis Testing for Means, Proportions, and Variances

Hypothesis testing is the method through which data is used as evidence of the likelihood of some condition. Hypothesis testing can be performed to decide whether a statement regarding a parameter should be rejected or not rejected (Montgomery, Runger, and Hubele 2001). For example, one might want to know if the mean weight of parts differs from some target value or whether the percentage of defective parts is greater than 2%. By collecting data, we can use a hypothesis test to determine whether we can conclude with high confidence that those conditions are true.

To perform a hypothesis test, the following steps are performed:

1. "Null" and "alternative" hypotheses are determined.
2. An alpha level is chosen.
3. An adequate amount of data is collected.
4. The appropriate test statistic is calculated and from it, a p-value. Alternatively, when performing a test by hand, critical values may be found in standard tables.
5. The p-value is compared to the alpha level or the test statistic is compared to the critical value and a decision is made.

PERFORMING A HYPOTHESIS TEST

"Null" and "Alternative" Hypotheses Are Determined

The alternative hypothesis, H_a, is the condition that would be of interest to be proven likely true, with the restriction that the alternative hypothesis can never be a statement of equality. For example, an alternative hypothesis cannot state that the mean breaking strength of rivets from one supplier is equal to that of another. The null hypothesis, H_0, is the complement to the alternative hypothesis and will be assumed to be true unless the data show it to be likely false in favor of the alternative hypothesis. All possible outcomes must be covered by either the null or the alternative hypothesis.

For example, suppose the alternative hypothesis is "The mean fill weight of bottles of juice is greater than the listed weight of 12 ounces." The null hypothesis must state that the mean fill weight is less than or equal to 12 ounces to cover all possible outcomes—that the mean weight is under, equal to, or over 12 ounces.

An Alpha Level Is Chosen

In the final step of performing a hypothesis test, a decision will be made to either reject the null hypothesis and conclude the alternative hypothesis is likely true, or to not reject the null hypothesis and conclude there is not enough evidence to accept the alternative hypothesis. If the decision is made to reject the null in favor of the alternative, there is some risk that the decision was incorrect and the null hypothesis is in fact true. This is commonly known as a *Type I error* or *alpha error*. The amount of risk that is taken when performing the test is known as the alpha level and is expressed as a proportion. An alpha level of 0.05 is most commonly used, which would correspond to incorrectly rejecting the null hypothesis 5% of the time and falsely accepting the alternative.

Although 0.05 is a reasonable alpha level for many applications, it is important to consider the context of the test being performed and what risk is appropriate. In agricultural experiments where there are numerous "noise" factors and performing a well-controlled experiment is difficult, an alpha level of 0.1 may be more reasonable. Conversely, in an application where accepting the alternative hypothesis has implications for human safety, a lower risk level may be desired, such as an alpha level of 0.01.

An Adequate Amount of Data Is Collected

The sample size calculation gives the best estimate of the amount of data that will be needed to achieve the desired power level. The chance of being able to reject the null hypothesis in favor of the alternative, if the alternative is true, may be too low if an insufficient amount of data is collected for the desired test. As a result, the test may come back as insignificant and an important conclusion would be missed. It is also important that the data be collected using sound sampling as described in Chapter 1.

Just as there is a risk that the null hypothesis will be rejected when it is true, there is risk that the null hypothesis will not be rejected when it is in fact false. This is commonly known as a *Type II error* or *beta error*. This risk is reduced as more data are collected to provide the evidence that the alternative hypothesis is true. The complement to a Type II error is known as *power*, and it is simply one minus the Type II error you are willing to accept. For example, if a 10% chance of not detecting that the alternative hypothesis is true is accepted, then the power to detect that it is true is 90%.

Sample size calculations are discussed in more detail in Chapter 6. The Power and Sample Size menu in Minitab calculates the sample size needed to obtain a desired power level for a variety of hypothesis tests. This is an important calculation to perform before collecting data to help ensure that enough data will be available to provide the desired power when the test is performed.

The Appropriate Test Statistic Is Calculated, and a p-Value Is Determined from the Test Statistic or a Critical Value Is Obtained

For any common scenario there exists a statistical test that includes calculating a test statistic. Which statistic is appropriate depends on the type of data, what is being compared, and other considerations that will be discussed in more detail in this chapter. The data collected are used to calculate the test statistic, which is a standard measure of how much the data disagree with the null hypothesis. The distribution of a given test statistic is the same regardless of the data that it was calculated from and is centered at the expected value if the null hypothesis were true.

While the exact calculations depend on the null and alternative hypotheses and specific test statistic, generally the p-value is the cumulative probability of all values equal to or more extreme than the test statistic. Because the distribution of the test statistic is standardized at what would be expected if the null hypothesis were true, the p-value provides the odds of seeing data at least as extreme as ours if the null hypothesis were true.

When performing a statistical test by hand, it may be simpler to determine a critical value than to obtain a p-value. The critical value provides a cutoff point for where a determination would be made to reject the null hypothesis. The critical value is the value in which if the null hypothesis is true, the cumulative probability of all values at least as extreme is equal to the alpha level.

The p-Value Is Compared to the Alpha Level or the Test Statistic Is Compared to a Critical Value and a Decision Is Made

If the p-value method is used, the p-value is compared to the alpha level selected previously in order to make a decision about whether to reject the null hypothesis. If the p-value is smaller than the alpha level, then the null hypothesis is rejected in favor of the alternative as the low p-value is an indication that the data were very unlikely to be produced if the null hypothesis were true. Recall that the alpha level chosen corresponds to the proportion of the time that the null hypothesis will be rejected when it is in fact true. For example, if an alpha level of 0.05 is chosen and the null hypothesis is true, 5% of the time data collected will produce a p-value less than .05 and the null hypothesis will be incorrectly rejected.

If the p-value is greater than the chosen alpha level, then the null hypothesis is not rejected. This should not be confused with accepting that the null hypothesis is true, but rather that the data do not provide sufficient evidence to determine that it is likely false. This is synonymous with the legal system in most countries, where defendants may be found *not guilty*—an indication that not enough evidence was produced to conclude guilt—rather than being found innocent (Kenney 1988).

If the critical value method is used, the test statistic is compared to the critical value, which is the value corresponding to the alpha level. If the test statistic is more extreme than the critical value, the null hypothesis is rejected as that signifies the p-value would be smaller than the alpha level. If the test statistic is less extreme than the critical value, the null hypothesis is not rejected as the p-value is larger than the alpha level.

Other Considerations

When the p-value for a test is smaller than alpha and the null hypothesis is rejected in favor of the alternative, it is often said that the result of the test is "statistically significant." This means that the data provided enough evidence to favor the alternative hypothesis of interest. However, this should not be confused with "practical significance," which indicates whether the data demonstrated a result that is meaningful in real terms. For example, an engineer may find a statistically significant result that a potential supplier offers engine belts 0.1 units stronger than the current supplier. However, the current supplier's belt strength is 7 units away from the specification. Although the hypothesis test found statistical significance, the strength improvement is not meaningful given the current performance and so it is likely not practically significant. Despite the test result, one would likely stay with the current supplier. If the potential supplier offered an improvement of 3 units, one might decide that the result is now practically significant as well and act accordingly.

It is important to be aware of what Simmons, Nelson, and Simonsohn call "researcher degrees of freedom" (2011, 1359). Often times in hypothesis testing there are decisions to be made, such as the appropriate test to use, and those decisions may not have a clear answer. This means the practitioner must make subjective decisions and is often biased toward the choice that will most likely provide the desired answer. It is important to objectively assess each choice, and it may be appropriate to present the results of more than one option.

Finally, most hypothesis tests are formed by starting with certain assumptions about the underlying process, such as the data will be normally distributed or different groups will have equal variances. Although many texts will teach these assumptions, it has been shown through research that in many cases violations of these assumptions do not meaningfully affect the accuracy of the test (Minitab "1-Sample t-Test"). It is important to learn the assumptions for each test as well as how sensitive the test is to violations of those assumptions. One assumption of virtually any test is that the data collected are independent and identically distributed (IID). This means that the process is stable while data are being collected and that the value of each data point is not affected by the previous; in other words, if a data point is taken that is somewhat low relative to the others, the next point taken isn't more likely to also be low. Chapter 13 discusses control charts and checking for stability of a process.

There are many hypothesis tests available. Table 5.1 shows tests of means for normally distributed data, and Table 5.2 shows tests of proportions. Table 5.3 depicts hypothesis tests that can be performed to assess variance.

TESTS FOR MEANS

1-Sample Z Test

A Z test is used to test the mean against some value of interest, μ_0, when the population standard deviation is known. Although it is a formal assumption that the underlying distribution is normal, the test performs well for other distributions and provides roughly the same Type I and Type II error rates. For highly skewed distributions, a sample size of at least 30 should be used. Before conducting the test the practitioner must decide which of three combinations of null and alternative hypothesis are of interest:

H_0: $\mu = \mu_0$ and H_a: $\mu \neq \mu_0$

H_0: $\mu \leq \mu_0$ and H_a: $\mu > \mu_0$

H_0: $\mu \geq \mu_0$ and H_a: $\mu < \mu_0$

Once data is collected, the test statistic is calculated as:

$$Z_{test} = \frac{\bar{x} - \mu_0}{\sigma/\sqrt{n}}$$

Table 5.1 Hypothesis tests for means.

Test name	Assumptions	Test statistic and two-tailed test
1-sample z test	For highly skewed data, sample size should be at least 30.	$Z = \dfrac{\overline{x} - u}{\sigma / \sqrt{n}}$ with sample size $= n$ H_0: $\mu = \mu_0$ H_a: $\mu \neq \mu_0$ Reject H_0 if $Z < Z_{\alpha/2}$ or $Z > Z_{1-\alpha/2}$
1-sample t test	For highly skewed data, sample size should be at least 30.	$t = \dfrac{\overline{x} - \mu}{s / \sqrt{n}}$ degrees of freedom $= n - 1$ H_0: $\mu = \mu_0$ H_a: $\mu \neq \mu_0$ Reject H_0 if $t < t_{\alpha/2,\, n-1}$ or $t > t_{1-\alpha/2,\, n-1}$
2-sample t test with unknown and unequal standard deviations (Note: Minitab uses the pooled standard deviation)	Populations should have similar shapes. If classical test is used, populations should have equal variances; otherwise, use Welch's test	$t = \dfrac{(\overline{x}_1 - \overline{x}_2) - \delta_0}{\sqrt{\dfrac{s_1^2}{n_1} + \dfrac{s_1^2}{n_2}}}$ degrees of freedom $= \dfrac{\left(\dfrac{s_1^2}{n_1} + \dfrac{s_2^2}{n_2}\right)^2}{\dfrac{\left(\dfrac{s_1^2}{n_1}\right)^2}{n_1 - 1} + \dfrac{\left(\dfrac{s_2^2}{n_2}\right)^2}{n_2 - 1}}$ H_0: $\mu_1 - \mu_2 = \delta_0$ where δ_0 is the hypothesized difference between μ_1 and μ_2 H_a: $\mu_1 - \mu_2 \neq \delta_0$ Reject H_0 if $t < t_{\alpha/2,\, n-1}$ or $t > t_{1-\alpha/2,\, n-1}$
2-sample t test with unknown and equal standard deviation		$t = \dfrac{(\overline{x}_1 - \overline{x}_2) - \delta_0}{\sqrt{\dfrac{(n_1 - 1)s_1^2 + (n_2 - 1)s_2^2}{n_1 + n_2 - 2}}\sqrt{\dfrac{1}{n_1} + \dfrac{1}{n_2}}}$ degrees of freedom $= n_1 + n_2 - 2$ H_0: is $\mu_1 - \mu_2 = \delta_0$ where δ_0 is the hypothesized difference between μ_1 and μ_2 H_a: $\mu_1 - \mu_2 \neq \delta_0$ H_a: $\mu_1 \neq \mu_2$ Reject H_0 if $t < t_{\alpha/2,\, n1+n2-2}$ or $t > t_{1-\alpha/2, n1+n2-2}$
paired t test	If differences are highly skewed, sample size should be at least 30.	$t = \dfrac{\overline{d}}{s_d / \sqrt{n}}$ where $\overline{d}$ is the mean of the differences between the two samples and s_d is the standard deviations of the differences and degrees of freedom $= n$ 1. H_0: $\mu_d = \mu_0$ H_a: $\mu_d \neq \mu_0$ Reject H_0 if $t < -t_{\alpha/2,\, n-1}$ or $t > t_{1-\alpha/2,\, n-1}$

Table 5.2 Hypothesis tests for proportions.

Test Name	Assumptions	Test statistic and two-tailed test
1 proportion test	To use the normal approximation, both occurrences and nonoccurrences should be at least 5.	$z = \frac{\hat{p} - p_0}{\sqrt{\frac{p_0(1-p_0)}{n}}}$ where p = is the sample proportion (and equal occurrences / opportunities) and p_0 = a hypothesized value for the proportion Two-tailed Null Hypothesis (H_0) $p = p_0$ Alternative Hypothesis (H_a): $p \neq p_0$ Reject H_0 if $t < t_{\alpha/2, n-1}$ or $t > t_{1-\alpha/2, n-1}$
2 proportion test	To use the normal approximation, both occurrences and nonoccurrences should be at least 5 for each proportion.	$Z = \frac{p_1 - p_2}{\sqrt{\hat{p}(1-\hat{p})\left(\frac{1}{n_1} + \frac{1}{n_1}\right)}}$ with $\hat{p} = \frac{n_1 p_1 + n_2 p_2}{n_1 + n_2}$ where p_1 is the proportion for sample one and p_2 is the proportion for sample two. Two-tailed Null Hypothesis (H_0) $p_1 - p_2 = 0$ Alternative hypothesis (H_a): $p_1 - p_2 > 0$ Reject H_0 if $Z < Z_{\alpha/2}$ or $Z > Z_{1-\alpha/2}$

Table 5.3 Hypothesis tests for variance.

Test Name	Assumptions	Test statistic and two-tailed test
Chi-square test of one sample variance	To use the Chi-square test population must be normally-distributed; otherwise, use Bonett's test.	Test statistic: $\chi^2 = (n-1)\frac{s^2}{\sigma_0^2}$ degrees of freedom = $n - 1$ H_0: $\sigma = \sigma_0$ H_a: $\sigma \neq \sigma_0$ Reject H_0 if $\chi^2 < \chi^2_{n-1, 1-\alpha/2}$ or $\chi^2 >_{n-1, 1-\alpha/2}$
F test of the variance of two samples	To use the F test populations must be normally-distributed; otherwise, use Levene's test for sample sizes less than 20 or highly skewed populations and Bonett's test for all other cases.	$F = \frac{s_1^2}{s_2^2}$ degrees of freedom = $n_1 - 1$, $n_2 - 1$ $H_0 : \sigma_1^2 = \sigma_2^2$ $H_a : \sigma_1^2 \neq \sigma_2^2$ Reject H_0 if $F < F_{\alpha/2, n1-1, n2-2}$ or $F > F_{1-\alpha/2, n1-1, n2-1}$

If $\mu = \mu_0$, then Z_{test} will follow a standard normal (Z) distribution. Therefore the probability of seeing the test statistic or one more extreme on a standard normal distribution represents the p-value.

The calculation of the p-value depends on which hypothesis was being tested. For each hypothesis combination above, the calculation is as follows:

For H_0: $\mu = \mu_0$ and H_a: $\mu \neq \mu_0$, $p = 2 * P(Z \leq -|Z_{test}|) = 2$ times the lower-tail probability when the test statistic is negative or 2 times the upper-tail probability when it is positive

For H_0: $\mu \leq \mu_0$ and H_a: $\mu > \mu_0$, $p = P(Z \geq Z_{test})$ = the upper-tail probability of the test statistic

For H_0: $\mu \geq \mu_0$ and H_a: $\mu < \mu_0$, $p = P(Z \leq Z_{test})$ = the lower-tail probability of the test statistic

That p-value is then compared to the desired alpha and if smaller, the null hypothesis is rejected in favor of the alternative hypothesis.

Alternatively, Z_{test} can be compared to the value of Z corresponding to rejecting the null hypothesis, known as $Z_{critical}$. The criteria for rejecting the null hypothesis for each combination of hypotheses is as follows:

For H_a: $\mu \neq \mu_0$, reject if $|Z_{test}| > Z_{1-\alpha/2}$ with the Z value corresponding to a lower-tail probability of $1 - \alpha/2$

For H_a: $\mu > \mu_0$, reject if $Z_{test} > Z_{1-\alpha}$ with the Z value corresponding to a lower-tail probability of $1 - \alpha$

For H_a: $\mu < \mu_0$, reject if $Z_{test} < Z_{\alpha}$ with the Z value corresponding to a lower-tail probability of α

Example

A sample of 38 parts is pulled and the process mean is hypothesized to be 18.5. The sample mean is 18.05 and the population standard deviation is 1.189. Is there sufficient evidence to conclude the mean differs from 18.5?

Because we desire to know if the mean is different from 18.5 in either direction, the null hypothesis is set as (H_0): $\mu = 18.5$ and the alternative is set as (H_a): $\mu \neq 18.5$. The alpha level is set as 0.05. The data have been collected and the sample size was determined to be adequately large. The test statistic is calculated as $Z_{test} = \dfrac{\bar{x} - u}{\sigma/\sqrt{n}} = \dfrac{18.05 - 18.5}{1.189/\sqrt{38}} = -2.33.$

The p-value can be determined by finding the lower-tail probability corresponding to a Z of −2.33 and multiplying by 2. The probability can be found in Table C.1 in Appendix C as 0.0099, as shown in Table 5.4.

The p-value is then calculated as:

$$p = 2 * P(Z \leq -|-2.33|) = 2 * .0099 = .0198$$

The p-value of .0198 is less than the alpha level of 0.05, so the null hypothesis is rejected.

$Z_{critical}$ is determined by finding $1 - {}^{\alpha}/_{2} = .975$ in Table C.2 in Appendix C, and then the corresponding value of Z as shown in Table 5.5.

Table 5.4 Z table for Z =–2.33.

Z	0	0.01	0.02	0.03	0.04
0	0.5000	0.4960	0.4920	0.4880	0.4840
–0.1	0.4602	0.4562	0.4522	0.4483	0.4443
–0.2	0.4207	0.4168	0.4129	0.4090	0.4052
–0.3	0.3821	0.3783	0.3745	0.3707	0.3669
–0.4	0.3446	0.3409	0.3372	0.3336	0.3300
–0.5	0.3085	0.3050	0.3015	0.2981	0.2946
–0.6	0.2743	0.2709	0.2676	0.2643	0.2611
–0.7	0.2420	0.2389	0.2358	0.2327	0.2296
–0.8	0.2119	0.2090	0.2061	0.2033	0.2005
–0.9	0.1841	0.1814	0.1788	0.1762	0.1736
–1	0.1587	0.1562	0.1539	0.1515	0.1492
–1.1	0.1357	0.1335	0.1314	0.1292	0.1271
–1.2	0.1151	0.1131	0.1112	0.1093	0.1075
–1.3	0.0968	0.0951	0.0934	0.0918	0.0901
–1.4	0.0808	0.0793	0.0778	0.0764	0.0749
1.5	0.0668	0.0655	0.0643	0.0630	0.0618
–1.5	0.0548	0.0537	0.0526	0.0516	0.0505
–1.7	0.0446	0.0436	0.0427	0.0418	0.0409
–1.8	0.0359	0.0351	0.0344	0.0336	0.0329
–1.9	0.0287	0.0281	0.0274	0.0268	0.0262
–2.0	0.0228	0.0222	0.0217	0.0212	0.0207
–2.1	0.0179	0.0174	0.0170	0.0166	0.0162
–2.2	0.0139	0.0136	0.0132	0.0129	0.0125
–2.3	0.01072	0.01044	0.01017	**0.00990**	0.00964
–2.4	0.00820	0.00798	0.00776	0.00755	0.00734

Table 5.5 Lower-tail probability of 0.975 for Z distribution.

Z	0	0.01	0.02	0.03	0.04	0.05	0.06	0.07
0	0.50000	0.50399	0.50798	0.51197	0.51595	0.51994	0.52392	0.52790
0.1	0.53983	0.54380	0.54776	0.55172	0.55567	0.55962	0.56356	0.56749
0.2	0.57926	0.58317	0.58706	0.59095	0.59483	0.59871	0.60257	0.60642
0.3	0.61791	0.62172	0.62552	0.62930	0.63307	0.63683	0.64058	0.64431
0.4	0.65542	0.65910	0.66276	0.66640	0.67003	0.67364	0.67724	0.68082
0.5	0.69146	0.69497	0.69847	0.70194	0.70540	0.70884	0.71226	0.71566
0.6	0.72575	0.72907	0.73237	0.73565	0.73891	0.74215	0.74537	0.74857
0.7	0.75804	0.76115	0.76424	0.76730	0.77035	0.77337	0.77637	0.77935
0.8	0.78814	0.79103	0.79389	0.79673	0.79955	0.80234	0.80511	0.80785
0.9	0.81594	0.81859	0.82121	0.82381	0.82639	0.82894	0.83147	0.83398
1	0.84134	0.84375	0.84614	0.84849	0.85083	0.85314	0.85543	0.85769
1.1	0.86433	0.86650	0.86864	0.87076	0.87286	0.87493	0.87698	0.87900
1.2	0.88493	0.88686	0.88877	0.89065	0.89251	0.89435	0.89617	0.89796
1.3	0.90320	0.90490	0.90658	0.90824	0.90988	0.91149	0.91309	0.91466
1.4	0.91924	0.92073	0.92220	0.92364	0.92507	0.92647	0.92785	0.92922
1.5	0.93319	0.93448	0.93574	0.93699	0.93822	0.93943	0.94062	0.94179
1.6	0.94520	0.94630	0.94738	0.94845	0.94950	0.95053	0.95154	0.95254
1.7	0.95543	0.95637	0.95728	0.95818	0.95907	0.95994	0.96080	0.96164
1.8	0.96407	0.96485	0.96562	0.96638	0.96712	0.96784	0.96856	0.96926
1.9	0.97128	0.97193	0.97257	0.97320	0.97381	0.97441	0.97500	0.97558
2	0.97725	0.97778	0.97831	0.97882	0.97932	0.97982	0.98030	0.98077

The absolute value of Z_{test} of 2.33 can be compared to the $Z_{critical}$ value of 1.96 to reach the conclusion that the null hypothesis is rejected in favor of the alternative hypothesis. We conclude that there is sufficient evidence to determine the true population mean is not equal to 18.5.

In Minitab: Go to **Stat > Basic Statistics > 1-Sample Z** and select **Summarized data** from the drop-down selections. The option to choose **One or more samples, each in a column** would be used if the raw data were entered into a column. Enter the sample size, sample mean, and known standard deviation. Then select **Perform hypothesis test** and enter the hypothesized mean (see Figure 5.1). The **Options** button could be used to select a confidence level, which is $(1-\alpha)*100\%$, and other alternative hypotheses.

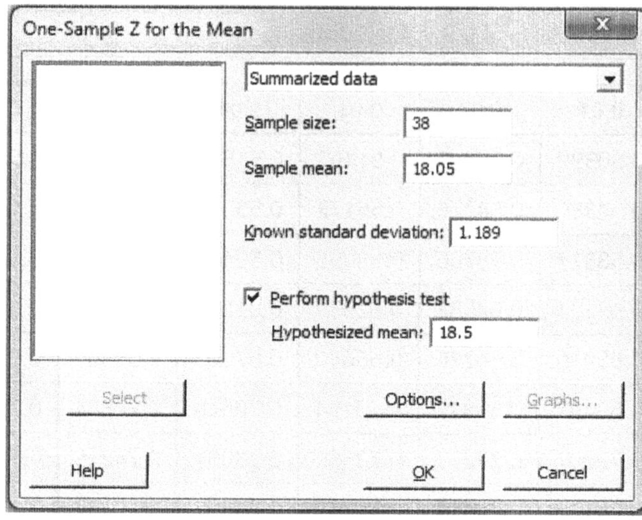

Figure 5.1 Dialog box for a 1-sample Z for the mean.

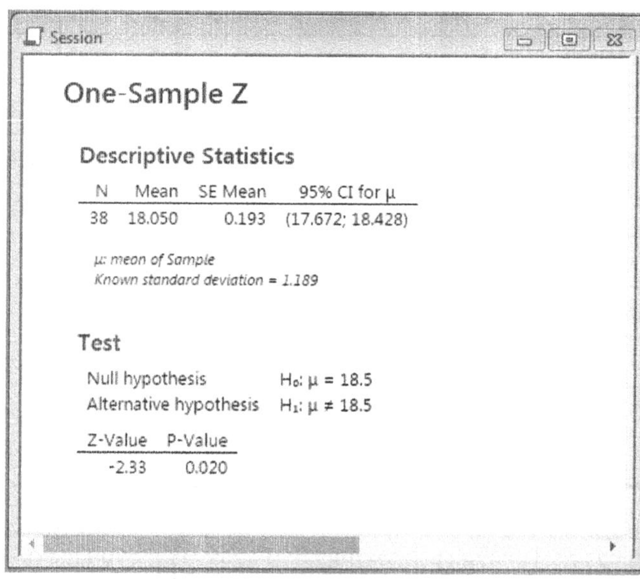

Figure 5.2 Session window for a 1-sample Z test.

The resulting p-value shown in Figure 5.2 is less than 0.05, so we reject the null hypothesis. The p-value differs slightly from that calculated by hand due to rounding. The Minitab session window is shown in Figure 5.2.

1-Sample t Test

The t test is used for testing the mean of a distribution and when the population standard deviation is unknown. As with the Z test, normality is a formal assumption, but the test performs well for other distributions (Minitab "1-Sample t-Test").

The t test performs well when data are symmetrical but not normally distributed (Efron 1969) as well as with slightly skewed data when the sample size is at least 20, and it also works well for highly skewed data when the sample size is at least 40 (Minitab "1-Sample t-Test").

The available hypotheses are the same as for the Z test. The test statistic is calculated as:

$$t_{test} = \frac{\overline{x} - \mu_0}{s/\sqrt{n}}$$

If $\mu=\mu_0$, then t_{test} will follow a t distribution with n − 1 degrees of freedom, where n is the sample size. The calculation of the p-value depends on which hypothesis was being tested. For each hypothesis combination, the calculation is as follows:

For H_0: $\mu=\mu_0$ and H_a: $\mu\neq\mu_0$, $p=2*P(t_{n-1}\geq|t_{test}|)$ = 2 times the lower-tail probability when the test statistic is negative or 2 times the upper-tail probability when it is positive

For H_0: $\mu\leq\mu_0$ and H_a: $\mu>\mu_0$, $p=P(t_{n-1}\geq t_{test})$ = the upper-tail probability of the test statistic

For H_0: $\mu\geq\mu_0$ and H_a: $\mu<\mu_0$, $p=P(t_{n-1}\leq t_{test})$ = the lower-tail probability of the test statistic

That p-value is then compared to the desired alpha and if smaller, the null hypothesis is rejected in favor of the alternative hypothesis.

Alternatively, t_{test} can be compared to the value of t corresponding to rejecting the null hypothesis, known as $t_{critical}$. The criteria for rejecting the null hypothesis for each combination of hypotheses are as follows:

For H_a: $\mu\neq\mu_0$, reject if $|t_{test}|>t_{n-1,1-\alpha/2}$ with the t value corresponding to a lower-tail probability of $1-\alpha/2$ with n − 1 degrees of freedom

For H_a: $\mu>\mu_0$, reject if $t_{test}>t_{n-1,1-\alpha}$ with the t value corresponding to a lower-tail probability of $1-\alpha$ with n − 1 degrees of freedom

For H_a: $\mu<\mu_0$, reject if $t_{test}<t_{\alpha}$ with the t value corresponding to a lower-tail probability of α with n − 1 degrees of freedom

Example

A sample of 16 parts is taken from a machining process. The sample mean is 38.61, and a manufacturing engineer wants to know if the process mean is not equal to the target of 39.4. The sample standard deviation is 1.385. Can we conclude that the process mean is not equal to 39.4?

Because we desire to know if the mean is different from 39.4 in either direction, the null hypothesis is set as (H_0): $\mu=39.4$ and the alternative is set as (H_a): $\mu\neq39.4$. The alpha level is set as 0.05. The data have been collected and the sample size was determined to be adequately large.

The test statistic is calculated as $t_{test}=\frac{38.61-39.4}{1.385/\sqrt{16}}=-2.28$. Because the test statistic was negative and the alternative hypothesis was (H_a): $\mu\neq39.4$, the p-value is

Table 5.6 *t*-score for 15 degrees of freedom and an alpha of 0.05.

df	0.4	0.3	0.2	0.1	0.05	0.025	0.01	0.005
1	–0.325	–0.727	–1.376	–3.078	–6.314	–12.706	–31.821	–63.657
2	–0.289	–0.617	–1.061	–1.886	–2.920	–4.303	–6.965	–9.925
3	–0.277	–0.584	–0.978	–1.638	–2.353	–3.182	–4.541	–5.841
4	–0.271	–0.569	–0.941	–1.533	–2.132	–2.776	–3.747	–4.604
5	–0.267	–0.559	–0.920	–1.476	–2.015	–2.571	–3.365	–4.032
6	–0.265	–0.553	–0.906	–1.440	–1.943	–2.447	–3.143	–3.707
7	0.263	–0.549	–0.896	–1.415	–1.895	–2.365	–2.998	–3.499
8	–0.262	–0.546	–0.889	–1.397	–1.860	–2.306	–2.896	–3.355
9	–0.261	–0.543	–0.883	–1.383	–1.833	–2.262	–2.821	–3.250
10	–0.260	–0.542	–0.879	–1.372	–1.812	–2.228	–2.764	–3.169
11	–0.260	–0.540	–0.876	–1.363	–1.796	–2.201	–2.718	–3.106
12	–0.259	–0.539	–0.873	–1.356	–1.782	–2.179	–2.681	–3.055
13	–0.259	–0.538	–0.870	–1.350	–1.771	–2.160	–2.650	–3.012
14	–0.258	–0.537	–0.868	–1.345	–1.761	–2.145	–2.624	–2.977
15	–0.258	–0.536	–0.866	–1.341	–1.753	–2.131	–2.602	–2.947
16	–0.258	–0.535	–0.865	–1.337	–1.746	–2.120	–2.583	–2.921

calculated by multiplying the lower-tail probability of the test statistic by 2. The lower-tail probability cannot be directly found in Table C.10 in Appendix C, but the range it lies within can be found.

First, the row corresponding to $n-1=15$ degrees of freedom is located in Table C.8 in Appendix C. Then, the closest values on either side of t_{test} in that row are found. Finally, the probabilities corresponding to those values are found at the top of the table are multiplied by 2. These final values represent the range that the p-value lies within. This method is illustrated in Table 5.6.

In the row of values corresponding to 15 degrees of freedom, the values the calculated t of –2.28 falls between are –2.131 and –2.602. The top row of the table indicates that these correspond to lower-tail probabilities of .01 and .025. These are multiplied by 2 to obtain a range of 0.02 and 0.05, so the p-value falls within this range.

The p-value is less than the alpha level of 0.05, so the null hypothesis is rejected and we conclude that there is sufficient evidence to determine the true population mean is not equal to 39.4.

Alternatively, the absolute value of $t=2.28$ could have been compared to the $t_{critical}$ of 2.131 to reach the same conclusion. The critical value is found in Table C.8

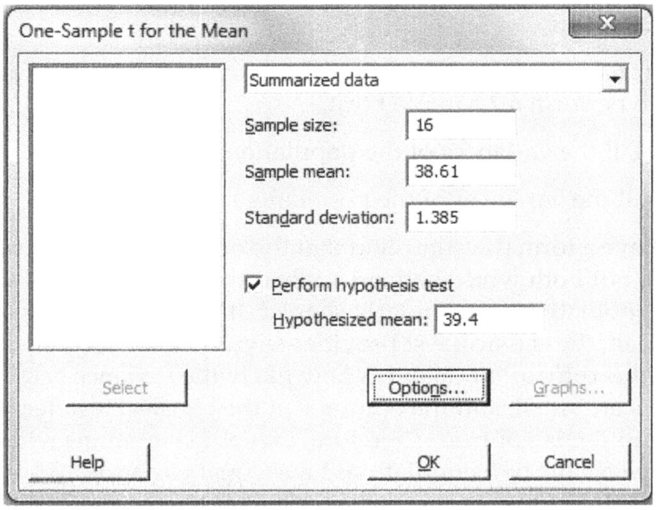

Figure 5.3 Dialog box for a 1-sample t for the mean.

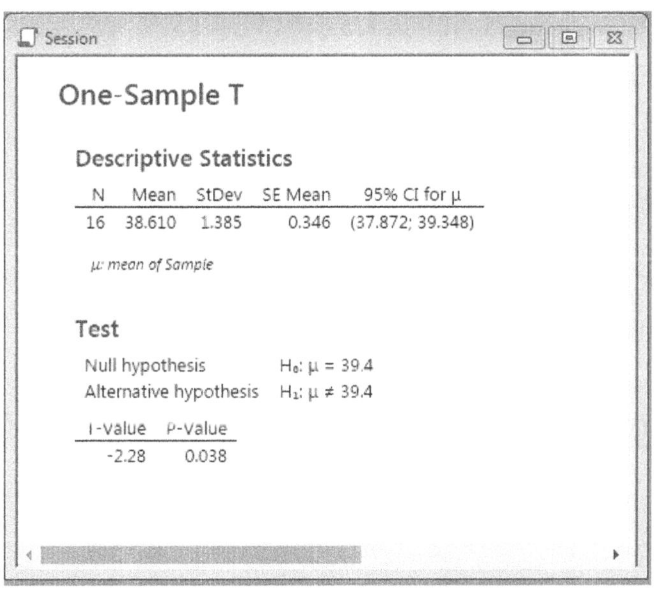

Figure 5.4 Session window for a 1-sample t test.

by finding the row corresponding to 15 degrees of freedom and reading the value in the column corresponding to $\alpha/2 = 0.025$.

In Minitab: Go to **Stat > Basic Statistics > 1-Sample t** and select **Summarized data**. Enter the sample size, sample mean, and sample standard deviation. Select **Perform hypothesis test** and enter the hypothesized mean (see Figure 5.3). Click **OK**.

The resulting p-value of 0.038 is less than 0.05, so we reject the null hypothesis (see Figure 5.4).

2-Sample t Test

A 2-sample t test is performed to compare the means of two populations. There are two ways to perform a 2-sample t test:

1. Classical, if the variances of the populations are equal
2. Welch's, if the variances of the populations are unequal

Both tests have a formal assumption that the underlying populations are normally distributed, but both work well even when this assumption is violated—as long as the two populations are similarly shaped. If the variances are equal and sample sizes are small, the classical test provides slightly more accurate results. However, in practice it is difficult to establish how likely the variances are to be equal when sample sizes are small, and the accuracy of the classical test declines considerably if the assumption is not true. Therefore, except in situations where there is strong reason to believe the two populations have equal variance, Welch's test is the preferable option (Minitab "2-Sample t-Test"). Both populations should have similar shapes.

There are three set of hypotheses that could be tested using a 2-sample t test:

H_0: $\mu_1 - \mu_2 = \Delta_0$ and H_a: $\mu_1 - \mu_2 \neq \Delta_0$

H_0: $\mu_1 - \mu_2 \leq \Delta_0$ and H_a: $\mu_1 - \mu_2 > \Delta_0$

H_0: $\mu_1 - \mu_2 \geq \Delta_0$ and H_a: $\mu_1 - \mu_2 < \Delta_0$

In these formulas, Δ_0 is a difference of interest between the two means, most commonly 0. The test statistics and degrees of freedom are:

$$t = \frac{(\bar{x}_1 - \bar{x}_2) - \Delta_0}{\sqrt{\frac{(n_1 - 1)s_1^2 + (n_2 - 1)s_2^2}{n_1 + n_2 - 2}}\sqrt{\frac{1}{n_1} + \frac{1}{n_2}}} \text{ and } df = n_1 + n_2 - 2 \text{ (Classical)}$$

$$t = \frac{(\bar{x}_1 - \bar{x}_2) - \Delta_0}{\sqrt{\frac{s_1^2}{n_1} + \frac{s_1^2}{n_2}}} \text{ and } df = \frac{\left(\frac{s_1^2}{n_1} + \frac{s_2^2}{n_2}\right)^2}{\frac{\left(\frac{s_1^2}{n_1}\right)^2}{n_1 - 1} + \frac{\left(\frac{s_2^2}{n_2}\right)^2}{n_2 - 1}} \text{ (Welch's)}$$

The calculation of the p-value depends on which hypothesis is being tested. For each hypothesis combination above, the calculation is as follows:

For H_a: $\mu_1 - \mu_2 \neq \Delta_0,$ $p = 2^* P(T_{df} \geq |t_{test}|) = 2$ times the lower-tail probability when the test statistic is negative or 2 times the upper-tail probability when it is positive

For H_a: $\mu_1 - \mu_2 > \Delta_0,$ $p = P(T_{df} \geq t_{test})$ = the upper-tail probability of the calculated statistic

For H_a: $\mu_1 - \mu_2 < \Delta_0,$ $p = P(T_{df} \leq t_{test})$ = the lower-tail probability of the calculated statistic

As with the 1-sample t test, the exact probability is often not found in the tables in Appendix C and the same method for finding the range that the p-value falls within is used. The critical values are the same as that of the 1-sample t test.

Example

A sample of 26 parts is pulled from machine A's production. The sample mean is 45.8 and the standard deviation of the sample is 0.4. Machine B produces the same part and a sample of 23 parts is pulled and found to have a sample mean of 45.3 with a sample standard deviation of 0.3. Can we conclude that the process means are different?

Because we desire to know if the means are different in either direction, the null hypothesis is set as H_0: $\mu_1 - \mu_2 = 0$ and the alternative is set as H_a: $\mu_1 - \mu_2 \neq 0$. The alpha level is set as 0.05. The data have been collected and determined to be adequately large. The Welch's test statistic is calculated as

$$t = \frac{(45.8 - 45.3) - 0}{\sqrt{\frac{0.4^2}{26} + \frac{0.3^2}{23}}} = 49.8,$$

and the degrees of freedom as

$$df = \frac{\left(\frac{.4^2}{26} + \frac{.3^2}{23}\right)^2}{\frac{\left(\frac{.4^2}{26}\right)^2}{26-1} + \frac{\left(\frac{.3^2}{23}\right)^2}{23-1}} = 45.$$

Because the chosen alternative hypothesis was H_a: $\mu_1 - \mu_2 \neq 0$ and the test statistic was positive, the p-value will be two times the upper-tailed probability of the test statistic with a t distribution with 45 degrees of freedom. In the tables in Appendix C, there is no row for 45 degrees of freedom. In cases where the exact degrees of freedom are not available in the table, the row corresponding to the next fewest degrees of freedom should be used. Looking across the row of t-values for 40 degrees of freedom in Table C.5, the calculated value of 4.98 is greater than the largest value of 3.551. That value corresponds to a p-value of .0005, which is multiplied by 2 to get 0.001. That represents the maximum of the range the p-value falls in.

The p-value is less than the alpha level of 0.05, so the null hypothesis is rejected and we conclude that there is sufficient evidence to determine that the true population means of parts produced by machine A and machine B are not equal.

Alternatively, the absolute value of $t = 4.98$ could have been compared to the $t_{critical}$ of 2.329 to reach the same conclusion (see Table 5.7).

In Minitab: Go to **Stat > Basic Statistics > 2-Sample t** and select **Summarized data**. Enter the sample sizes, sample means, and sample standard deviations (see Figure 5.5). The **Options** subdialog (Figure 5.6) allows you to select **Assume equal variances** to use the classical test instead of Welch's if desired. Click **OK**.

Table 5.7 *t* distribution with 40 degrees of freedom.

df	0.4	0.3	0.2	0.1	0.05	0.025	0.01
1	1.376	1.963	3.078	6.314	12.706	25.452	63.657
2	1.061	1.386	1.886	2.920	4.303	6.205	9.925
3	0.978	1.250	1.638	2.353	3.182	4.177	5.841
4	0.941	1.190	1.533	2.132	2.776	3.495	4.604
5	0.920	1.156	1.476	2.015	2.571	3.163	4.032
6	0.906	1.134	1.440	1.943	2.447	2.969	3.707
7	0.896	1.119	1.415	1.895	2.365	2.841	3.499
8	0.889	1.108	1.397	1.860	2.306	2.752	3.355
9	0.883	1.100	1.383	1.833	2.262	2.685	3.250
10	0.879	1.093	1.372	1.812	2.228	2.634	3.169
11	0.876	1.088	1.363	1.796	2.201	2.593	3.106
12	0.873	1.083	1.356	1.782	2.179	2.560	3.055
13	0.870	1.079	1.350	1.771	2.160	2.533	3.012
14	0.868	1.076	1.345	1.761	2.145	2.510	2.977
15	0.866	1.074	1.341	1.753	2.131	2.490	2.947
16	0.865	1.071	1.337	1.746	2.120	2.473	2.921
17	0.863	1.069	1.333	1.740	2.110	2.458	2.898
18	0.862	1.067	1.330	1.734	2.101	2.445	2.878
19	0.861	1.066	1.328	1.729	2.093	2.433	2.861
20	0.860	1.064	1.325	1.725	2.086	2.423	2.845
21	0.859	1.063	1.323	1.721	2.080	2.414	2.831
22	0.858	1.061	1.321	1.717	2.074	2.405	2.819
23	0.858	1.060	1.319	1.714	2.069	2.398	2.807
24	0.857	1.059	1.318	1.711	2.064	2.391	2.797
25	0.856	1.058	1.316	1.708	2.060	2.385	2.787
26	0.856	1.058	1.315	1.706	2.056	2.379	2.779
27	0.855	1.057	1.314	1.703	2.052	2.373	2.771
28	0.855	1.056	1.313	1.701	2.048	2.368	2.763
29	0.854	1.055	1.311	1.699	2.045	2.364	2.756
30	0.854	1.055	1.310	1.697	2.042	2.360	2.750
40	0.851	1.050	1.303	1.684	2.021	2.329	2.704
50	0.849	1.047	1.299	1.676	2.009	2.311	2.678

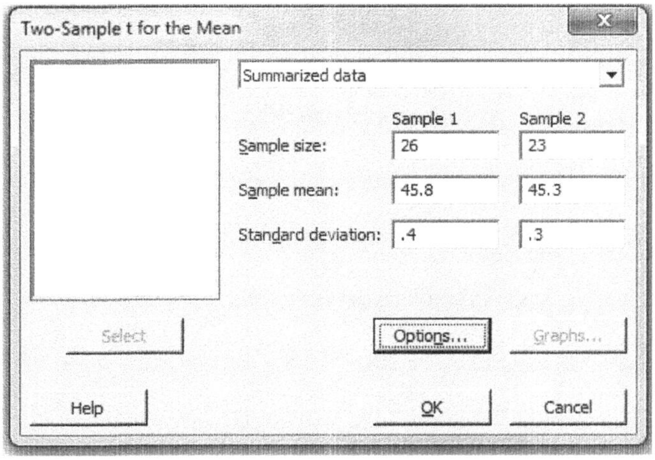

Figure 5.5 Dialog box for a 2-sample t test.

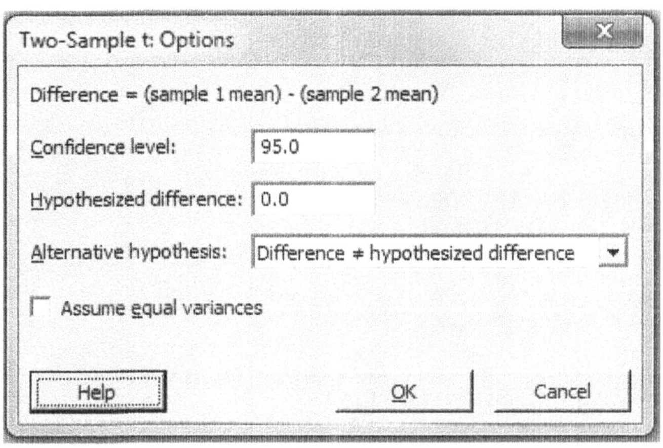

Figure 5.6 Options for a 2-sample t test.

The resulting p-value of 0.000 is less than 0.05, so we reject the null hypothesis (see Figure 5.7).

Paired t Test

A paired t test is used to compare the means of two groups when they represent data collected on the same items. Common examples would include data collected on the same set of parts before and after some treatment is applied. The sample sizes must be equal as the data points are paired. Because the two samples are not independent, they represent a single sample from which two pieces of data have been taken and a 2-sample t test cannot be performed (McClave and Sincich 2009). Instead, a paired t test is equivalent to taking the differences between the two columns and doing a 1-sample t test on the mean difference.

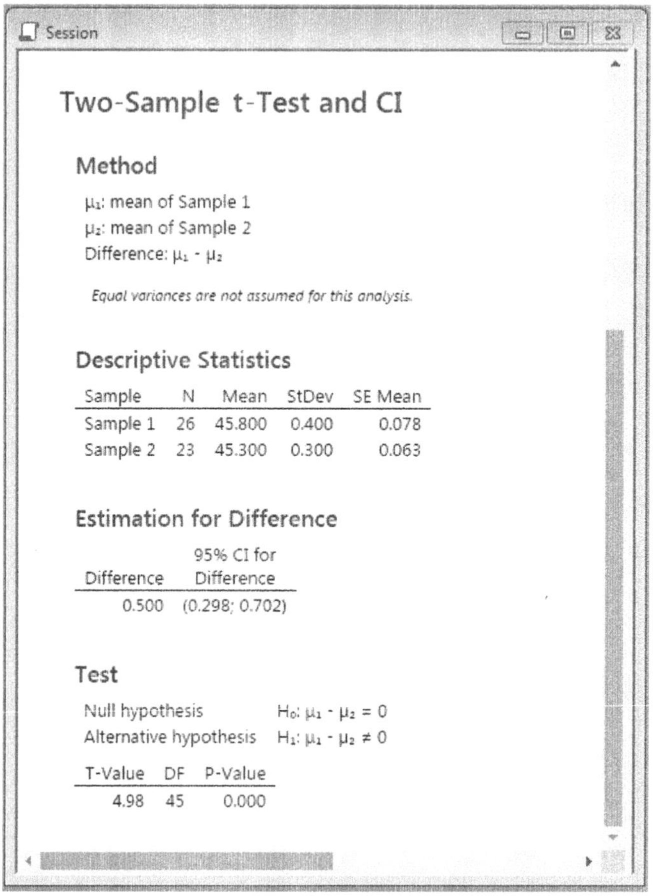

Session

Two-Sample t-Test and CI

Method

μ₁: mean of Sample 1
μ₂: mean of Sample 2
Difference: μ₁ - μ₂

Equal variances are not assumed for this analysis.

Descriptive Statistics

Sample	N	Mean	StDev	SE Mean
Sample 1	26	45.800	0.400	0.078
Sample 2	23	45.300	0.300	0.063

Estimation for Difference

Difference	95% CI for Difference
0.500	(0.298; 0.702)

Test

Null hypothesis H₀: μ₁ - μ₂ = 0
Alternative hypothesis H₁: μ₁ - μ₂ ≠ 0

T-Value	DF	P-Value
4.98	45	0.000

Figure 5.7 Session window for a 2-sample t test.

Three combinations of hypotheses are possible:

H_0: $\mu = \mu_0$ and H_a: $\mu \neq \mu_0$

H_0: $\mu \leq \mu_0$ and H_a: $\mu > \mu_0$

H_0: $\mu \geq \mu_0$ and H_a: $\mu < \mu_0$

Here, μ is the mean of the differences and μ_0 is a difference of interest, commonly 0. The calculations for the test statistic and p-value are the same as that of a 1-sample t test.

Example

Five machine operators are trained in a new method that is intended to increase their productivity. The number of parts they produce per hour are recorded to compare before and after values as seen in Table 5.8.

We want to know if there a difference in productivity using the new method as compared to the old one. Because it would be of interest to know if production has changed in either direction, the null hypothesis is set as $\mu_D = \mu_0$ and the alternative hypothesis as $\mu_D \neq \mu_0$. The alpha level is set as 0.05. The data have been

Table 5.8 Before and after values.

Operator	Before	After	Difference
1	215	213	2
2	210	215	-5
3	217	211	6
4	205	218	-13
5	216	224	-8

Table 5.9 *t* distribution with 4 degrees of freedom.

df	0.4	0.3	0.2	0.1	0.05	0.025	0.01
1	1.376	1.963	3.078	6.314	12.706	25.452	63.657
2	1.061	1.386	1.886	2.920	4.303	6.205	9.925
3	0.978	1.250	1.638	2.353	3.182	4.177	5.841
4	0.941	1.190	1.533	2.132	2.776	3.495	4.604
5	0.920	1.156	1.476	2.015	2.571	3.163	4.032

collected and although the sample size is small, the data represent all operators available.

The mean difference is calculated as –3.6 and the standard deviation of the differences is calculated as 7.64. The test statistic is calculated as $t = \frac{-3.6-0}{7.64/\sqrt{5}} = -1.05$ with $5-1=4$ degrees of freedom. Because the test statistic is negative and the chosen alternative hypothesis is $\mu_D \neq \mu_0$, the p-value is two times the lower-tail probability of the test statistic. In the lower-tail probabilities table in Appendix C, the row corresponding to 4 degrees of freedom shows values of –1.533 and -0.741 falling on either side of –1.05. These values correspond to probabilities of 0.1 and 0.25, which are multiplied by two to determine that the p-value falls between 0.2 and 0.5.

The p-value is larger than the alpha level of 0.05, so the null hypothesis is not rejected and we cannot conclude that there is sufficient evidence to determine the mean productivity has changed. Alternatively, the absolute value of $t=1.05$ could have been compared to the $t_{critical}$ of 3.495 to reach the same conclusion (see Table 5.9).

Note that because of the small sample size, there was low power to detect a difference unless it was substantial. As a result, it is important to conclude that we do not have evidence to reject the null hypothesis but also do not accept it as true based on the test results.

In Minitab: Enter the data into the Minitab worksheet as shown in Figure 5.8.

In Minitab: Go to **Stat > Basic Statistics > Paired t** and select **Each sample is in a column**; the drop-down provides the option to enter summarized data if it is

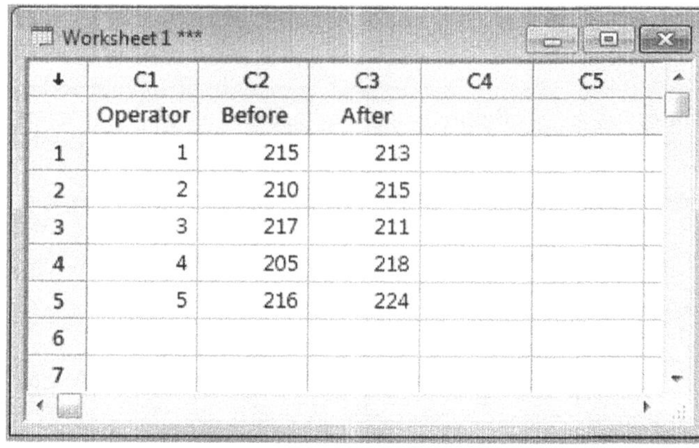

Figure 5.8 Minitab worksheet with before and after data.

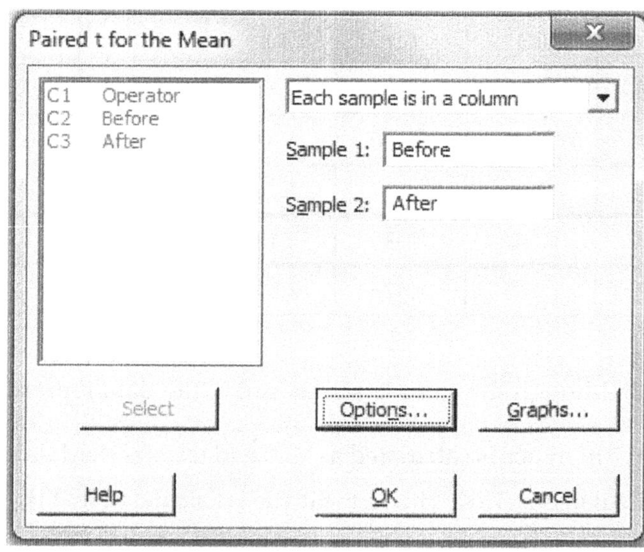

Figure 5.9 Dialog box for a paired t for the mean.

available. There is also an option under **Options** to enter a hypothesized difference (see Figure 5.9).

The resulting p-value of 0.351 is greater than 0.05, so we fail to reject the null hypothesis (see Figure 5.10).

TESTS FOR PROPORTIONS

A proportion (p) is the number of occurrences of an event divided by the number of opportunities for that event. For example, a production lot of 1000 rivets contains 980 good rivets and 20 defective ones. The p for defective rivets is 20/1000 = 0.02. Proportions are equivalent to percentages in that a proportion represents the number of events per opportunity, while percentages represent the number of events

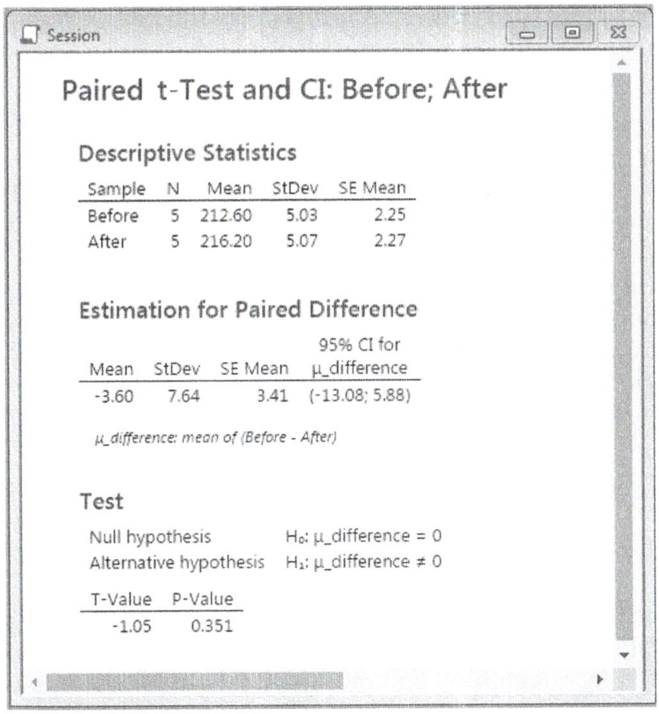

Session

Paired t-Test and CI: Before; After

Descriptive Statistics

Sample	N	Mean	StDev	SE Mean
Before	5	212.60	5.03	2.25
After	5	216.20	5.07	2.27

Estimation for Paired Difference

Mean	StDev	SE Mean	95% CI for μ_difference
-3.60	7.64	3.41	(-13.08; 5.88)

μ_difference: mean of (Before - After)

Test

Null hypothesis	H₀: μ_difference = 0
Alternative hypothesis	H₁: μ_difference ≠ 0

T-Value	P-Value
-1.05	0.351

Figure 5.10 Session window for a paired t test.

per hundred opportunities. A proportion can be any value from 0 through 1. Hypothesis tests can usually be performed on proportions using a normal approximation to the binomial distribution (Vining and Kowalski 2006).

1 Proportion Test

The 1 proportion test is used to test a proportion against a target proportion. The possible combinations of hypotheses are:

H_0: $P = P_0$ and H_a: $P \neq P_0$

H_0: $P \leq P_0$ and H_a: $P > P_0$

H_0: $P \geq P_0$ and H_a: $P < P_0$

There are two methods for calculating the test statistic. In all cases an exact method can be used, but the formula is complex and requires the use of a computer. Alternatively, if both the number of occurrences and number of nonoccurrences are at least 5, then a normal approximation can be used. Although this approximation is simpler to calculate, it is less powerful and accurate than the exact test. The formula for the normal approximation, which calculates a Z-statistic, is:

$$Z = \frac{\hat{p} - p_0}{\sqrt{\dfrac{p_0(1 - p_0)}{n}}}$$

In the formula, p_0 is the hypothesized proportion and n is the number of opportunities. The p-value and critical value are calculated the same as in the 1-sample Z test.

Example

A sample of 215 parts is taken and 38 of these parts are found to be defective, resulting in a sample proportion of 0.177. The process is known to have a historical defect rate of 15% (p_0=0.15). Can we conclude there has been a change in the defect rate?

Because it would be of interest to know if the defect rate has changed in either direction, the null hypothesis is set as P=0.15 and the alternative hypothesis as P≠0.15. The alpha level is set as 0.05. The data have been collected and the sample size is sufficiently large. The sample proportion is calculated as 0.177. Using the normal approximation, the test statistic is calculated as:

$$Z_{test} = \frac{0.177 - 0.15}{\sqrt{\frac{0.15(1-0.15)}{215}}} = 1.10$$

Because the alternative hypothesis was P≠0.15 and the test statistic was positive, the p-value is calculated by finding the upper-tail probability of the test statistic in Table C.1 in Appendix C and multiplying by two. The upper-tail probability corresponding to 1.10 is found to be 0.1357 as shown in Table 5.10.

Table 5.10 Upper-tail Z-score table for a Z of 1.10.

Z	0	0.01
0	0.5000	0.4960
0.1	0.4602	0.4562
0.2	0.4207	0.4168
0.3	0.3821	0.3783
0.4	0.3446	0.3409
0.5	0.3085	0.3050
0.6	0.2743	0.2709
0.7	0.2420	0.2389
0.8	0.2119	0.2090
0.9	0.1841	0.1814
1	0.1587	0.1562
1.1	0.1357	0.1335
1.2	0.1151	0.1131

Table 5.11 $Z_{critical}$ value.

Z	0	0.01	0.02	0.03	0.04	0.05	0.06	0.07
0	0.50000	0.50399	0.50798	0.51197	0.51595	0.51994	0.52392	0.52790
0.1	0.53983	0.54380	0.54776	0.55172	0.55567	0.55962	0.56356	0.56749
0.2	0.57926	0.58317	0.58706	0.59095	0.59483	0.59871	0.60257	0.60642
0.3	0.61791	0.62172	0.62552	0.62930	0.63307	0.63683	0.64058	0.64431
0.4	0.65542	0.65910	0.66276	0.66640	0.67003	0.67364	0.67724	0.68082
0.5	0.69146	0.69497	0.69847	0.70194	0.70540	0.70884	0.71226	0.71566
0.6	0.72575	0.72907	0.73237	0.73565	0.73891	0.74215	0.74537	0.74857
0.7	0.75804	0.76115	0.76424	0.76730	0.77035	0.77337	0.77637	0.77935
0.8	0.78814	0.79103	0.79389	0.79673	0.79955	0.80234	0.80511	0.80785
0.9	0.81594	0.81859	0.82121	0.82381	0.82639	0.82894	0.83147	0.83398
1	0.84134	0.84375	0.84614	0.84849	0.85083	0.85314	0.85543	0.85769
1.1	0.86433	0.86650	0.86864	0.87076	0.87286	0.87493	0.87698	0.87900
1.2	0.88493	0.88686	0.88877	0.89065	0.89251	0.89435	0.89617	0.89796
1.3	0.90320	0.90490	0.90658	0.90824	0.90988	0.91149	0.91309	0.91466
1.4	0.91924	0.92073	0.92220	0.92364	0.92507	0.92647	0.92785	0.92922
1.5	0.93319	0.93448	0.93574	0.93699	0.93822	0.93943	0.94062	0.94179
1.6	0.94520	0.94630	0.94738	0.94845	0.94950	0.95053	0.95154	0.95254
1.7	0.95543	0.95637	0.95728	0.95818	0.95907	0.95994	0.96080	0.96164
1.8	0.96407	0.96485	0.96562	0.96638	0.96712	0.96784	0.96856	0.96926
1.9	0.97128	0.97193	0.97257	0.97320	0.97381	0.97441	0.97500	0.97558
2	0.97725	0.97778	0.97831	0.97882	0.97932	0.97982	0.98030	0.98077

This is multiplied by two to get a p-value of 0.2714. The p-value is larger than the alpha level of 0.05, so the null hypothesis is not rejected and we cannot conclude that the proportion defective has likely changed. Alternatively, the absolute value of $Z = 1.10$ could have been compared to the $Z_{critical}$ of 1.96 to reach the same conclusion (see Table 5.11).

In Minitab: Go to **Stat > Basic Statistics > 1-Proportion** and select **Summarized data**. Enter the number of events and the number of trials and then select **Perform hypothesis test** and enter the hypothesized proportion and click **OK** (see Figure 5.11).

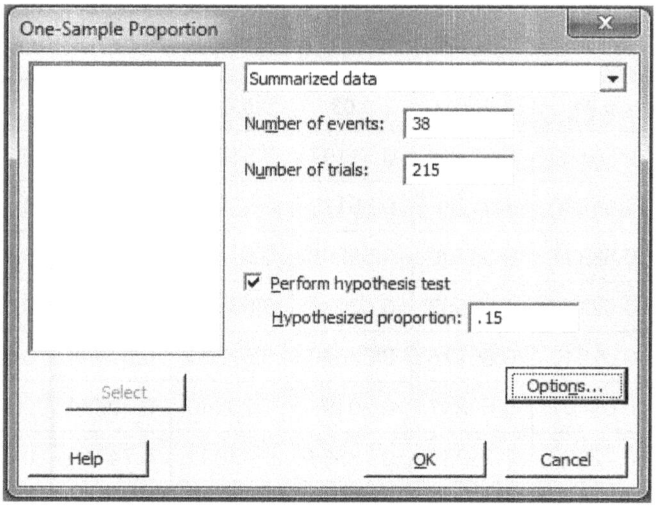

Figure 5.11 Dialog box for a 1 proportion Z test.

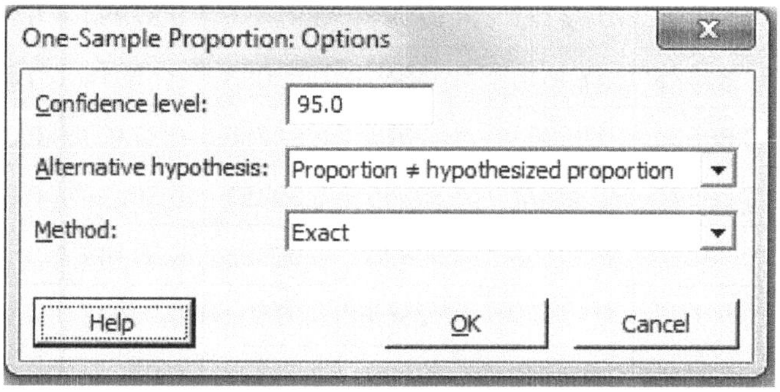

Figure 5.12 Options for a 1 proportion test.

Select **Options** and then choose **Exact** test from the method drop-down box (see Figure 5.12).

The resulting p-value of 0.272 is greater than 0.05, so we fail to reject the null hypothesis (see Figure 5.13). Because the exact test was used, the p-value does not exactly match the hand calculation using the normal approximation.

2 Proportion Test

A 2 proportion test is used to test two proportions against each other. The possible hypotheses are:

H_0: $P_1 - P_2 = \Delta_0$ and H_a: $P_1 - P_2 \neq \Delta_0$

H_0: $P_1 - P_2 \leq \Delta_0$ and H_a: $P_1 - P_2 > \Delta_0$

H_0: $P_1 - P_2 \geq \Delta_0$ and H_a: $P_1 - P_2 < \Delta_0$

Here Δ_0 is a difference of interest and it is often 0. When this is the case, there is an exact test that is the preferable test to use as it is accurate for all sample sizes

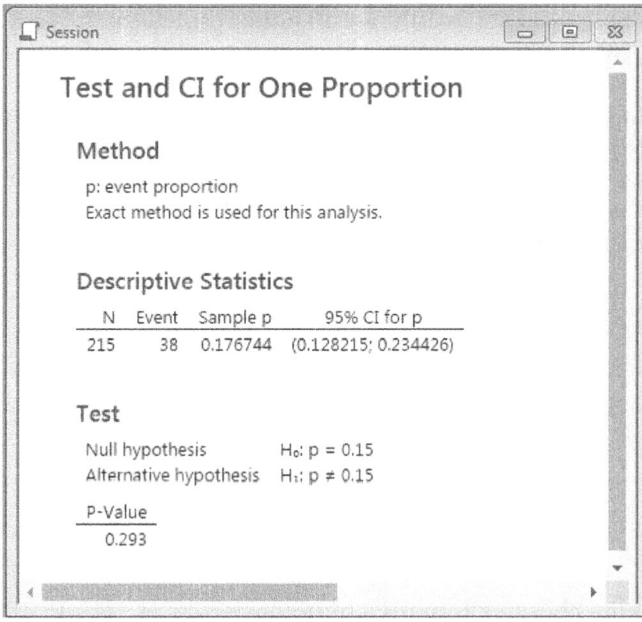

Figure 5.13 Session window for a 1 proportion test.

and event counts. As with the 1 proportion test, the calculations are complicated but can be done by Minitab. Otherwise, for any difference of interest and when the number of events and nonevents for both proportions are all at least 5, there is a normal approximation using a Z-statistic. One method of calculating Z uses separate estimates for each proportion to estimate the variance and a second uses a pooled estimate, only available when the difference of interest is 0. The pooled estimate combines the two proportions into a single overall proportion and is used because a hypothesized difference of 0 would indicate that both samples were pulled from populations with the same true proportion. The formulas are:

$$Z_{test} = \frac{P_1 - P_2 - \Delta_0}{\sqrt{\dfrac{P_1(1-P_1)}{n_1} + \dfrac{P_2(1-P_2)}{n_2}}} \text{ (Proportions used separately)}$$

$$Z_{test} = \frac{P_1 - P_2 - \Delta_0}{\sqrt{P_{overall}(1-P_{overall})\left(\dfrac{1}{n_1} + \dfrac{1}{n_2}\right)}} \text{ (Pooled estimate used)}$$

The formulas for p-values and $Z_{critical}$ are the same as used in a 1-sample Z test.

Example

A sample of 84 parts is taken from one supplier's production and a sample of 79 parts is taken from a second supplier's. The first sample has 17 defective parts and the second sample has 12 defective parts. Is there a statistically significant difference between the defect rates of the two suppliers? Because it would be of interest to know if the defect rate is different in either direction, the null hypothesis is set as H_0: $P_1 - P_2 = 0$ and the alternative hypothesis as H_a: $P_1 - P_2 \neq 0$. The alpha level is set as 0.05.

The data have been collected and the sample size is sufficiently large. Also, the counts of events (17 and 12) and nonevents (67 and 67) for the two samples are greater than or equal to 5, so the normal approximation can be used. The sample proportions are calculated as 0.202 and 0.152. Using the normal approximation and pooling the proportions as

$$P_{overall} = \frac{17+12}{84+79} = 0.178,$$

the test statistic is calculated as

$$Z_{test} = \frac{0.202 - 0.152}{\sqrt{0.178(1-0.178)\left(\frac{1}{84}+\frac{1}{79}\right)}} = 0.84.$$

Because the chosen alternative hypothesis was H_a: $P_1 - P_2 \neq 0$ and the test statistic is positive, the p-value is equal to two times the upper-tail probability of the Z distribution for Z_{test}. The value of 0.84 is found in Table C.4 in Appendix C, and the corresponding upper-tail probability is 0.2005. This value is multiplied by two to obtain the p-value of 0.401.

The p-value of 0.401 is larger than the alpha level of 0.05, so the null hypothesis is not rejected and we cannot conclude that the proportion defective is different between the two suppliers. Alternatively, the absolute value of $Z = 0.84$ could have been compared to the $Z_{critical}$ of 1.96 to reach the same conclusion (see Table 5.12).

In Minitab: Go to **Stat > Basic Statistics > 2-Proportions** and select **Summarized data.** Enter the number of events and the number of trials for each sample (see Figure 5.14). Then go to **Options** and select **Use the pooled standard deviation of the proportion** (as shown in Figure 5.15), and then click **OK.**

The resulting p-value of 0.397 is greater than 0.05, so we fail to reject the null hypothesis (see Figure 5.16). This value differs slightly from the hand calculation due to rounding.

TESTS FOR VARIANCE

Tests for variance are performed to compare either a variance to a target variance using the Chi-square test or to compare two variances using an F test (Vining and Kowalski 2006). Both tests use the ratio of the two variances for comparison purposes.

1 Variance Test

A 1 variance test is performed to test whether the population variance σ^2 likely differs from a target value, written as σ_0^2. The test uses the ratio of the true value

Table 5.12 Z-score table.

Z	0	0.01	0.02	0.03	0.04	0.05	0.06	0.07
0	0.50000	0.50399	0.50798	0.51197	0.51595	0.51994	0.52392	0.52790
0.1	0.53983	0.54380	0.54776	0.55172	0.55567	0.55962	0.56356	0.56749
0.2	0.57926	0.58317	0.58706	0.59095	0.59483	0.59871	0.60257	0.60642
0.3	0.61791	0.62172	0.62552	0.62930	0.63307	0.63683	0.64058	0.64431
0.4	0.65542	0.65910	0.66276	0.66640	0.67003	0.67364	0.67724	0.68082
0.5	0.69146	0.69497	0.69847	0.70194	0.70540	0.70884	0.71226	0.71566
0.6	0.72575	0.72907	0.73237	0.73565	0.73891	0.74215	0.74537	0.74857
0.7	0.75804	0.76115	0.76424	0.76730	0.77035	0.77337	0.77637	0.77935
0.8	0.78814	0.79103	0.79389	0.79673	0.79955	0.80234	0.80511	0.80785
0.9	0.81594	0.81859	0.82121	0.82381	0.82639	0.82894	0.83147	0.83398
1	0.84134	0.84375	0.84614	0.84849	0.85083	0.85314	0.85543	0.85769
1.1	0.86433	0.86650	0.86864	0.87076	0.87286	0.87493	0.87698	0.87900
1.2	0.88493	0.88686	0.88877	0.89065	0.89251	0.89435	0.89617	0.89796
1.3	0.90320	0.90490	0.90658	0.90824	0.90988	0.91149	0.91309	0.91466
1.4	0.91924	0.92073	0.92220	0.92364	0.92507	0.92647	0.92785	0.92922
1.5	0.93319	0.93448	0.93574	0.93699	0.93822	0.93943	0.94062	0.94179
1.6	0.94520	0.94630	0.94738	0.94845	0.94950	0.95053	0.95154	0.95254
1.7	0.95543	0.95637	0.95728	0.95818	0.95907	0.95994	0.96080	0.96164
1.8	0.96407	0.96485	0.96562	0.96638	0.96712	0.96784	0.96856	0.96926
1.9	0.97128	0.97193	0.97257	0.97320	0.97381	0.97441	0.97500	0.97558
2	0.97725	0.97778	0.97831	0.97882	0.97932	0.97982	0.98030	0.98077

to the target value, which is equal to 1 when the true value equals the target. The possible hypotheses are:

$$H_0: \frac{\sigma^2}{\sigma_0^2} = 1 \text{ and } H_a: \frac{\sigma^2}{\sigma_0^2} \neq 1$$

$$H_0: \frac{\sigma^2}{\sigma_0^2} \leq 1 \text{ and } H_a: \frac{\sigma^2}{\sigma_0^2} > 1$$

$$H_0: \frac{\sigma^2}{\sigma_0^2} \geq 1 \text{ and } H_a: \frac{\sigma^2}{\sigma_0^2} > 1$$

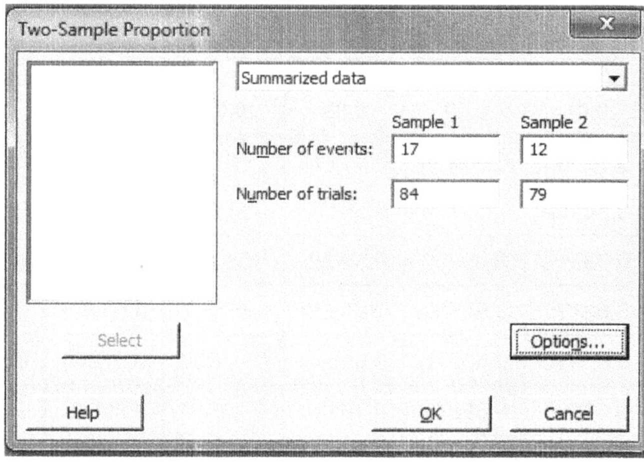

Figure 5.14 Dialog box for a 2 proportion test.

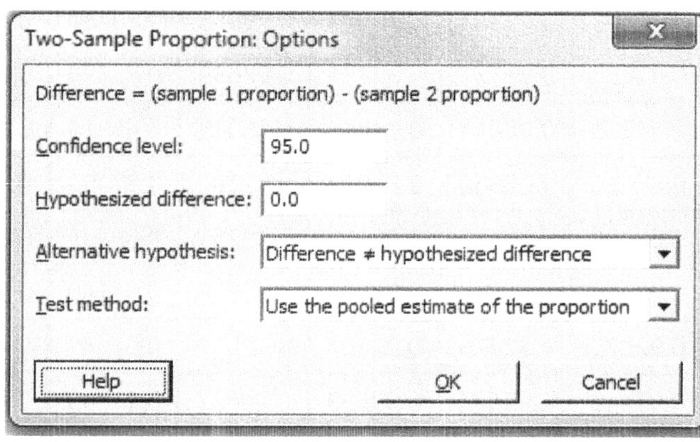

Figure 5.15 Options box for a 2 proportion test.

When the data are normally distributed, the test statistic is known as Chi-square and is calculated as

$$\chi^2_{n-1} = (n-1)\frac{s^2}{\sigma_0^2},$$

where s^2 is the sample variance calculated from the data.

The p-value is calculated using the Chi-square distribution with $n-1$ degrees of freedom for each possible alternative:

For $H_a: \frac{\sigma^2}{\sigma_0^2} \neq 1 p = 2 * \text{minimum}\ [P(\chi^2_{n-1} < \chi^2), P(\chi^2_{n-1} > \chi^2)] = 2$ times the smaller of the lower-tail and upper-tail probability of the test statistic

For $H_a: \frac{\sigma^2}{\sigma_0^2} < 1$, $p = P(\chi^2_{n-1} < \chi^2)$ = the lower-tail probability of the calculated value

For $H_a: \frac{\sigma^2}{\sigma_0^2} > 1$ $p = P(\chi^2_{n-1} > \chi^2)$ = the upper-tail probability of the calculated value

The Chi-square test does not perform well for data that do not follow the normal distribution. When data are non-normal, an alternative known as Bonett's test is preferable and the p-value is calculated using a series of equations that should be solved using statistical software such as Minitab due to the complexity of the calculation when done by hand.

Figure 5.17 shows a χ^2 distribution with 2 degrees of freedom, 5 degrees of freedom, and 10 degrees of freedom; the χ^2 distribution approaches the shape of the normal distribution as the number of degrees of freedom is increased.

Example

An engineer checks a sample of 12 parts and finds the sample deviation is 0.935. The population standard deviation is believed to be 0.9. Is the difference between the sample standard deviation the same as the population standard deviation

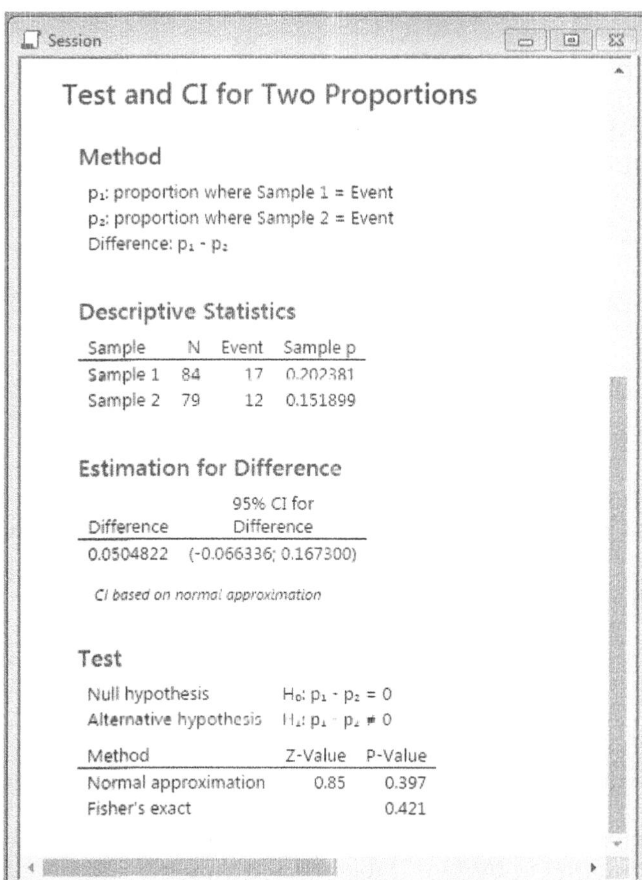

Figure 5.16 Session window for a 2 proportion test.

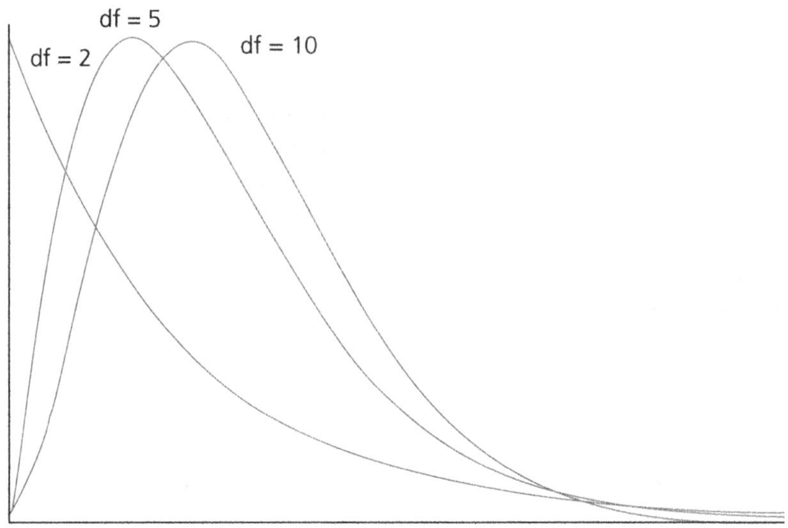

Figure 5.17 Chi-square distribution with 2, 5, and 10 degrees of freedom.

statistically different enough to decide that the true value is not 0.9? The data are normally distributed.

Because it would be of interest to know if the standard deviation is different in either direction, the null hypothesis is set as

$$H_0: \frac{\sigma^2}{\sigma_0^2} = 1,$$

and the alternative hypothesis is set as

$$H_a: \frac{\sigma^2}{\sigma_0^2} \neq 1.$$

The alpha level is set as 0.05.

The data have been collected and the sample size is adequate to detect moderately large differences with high power.

The Chi-square statistic is calculated as:

$$\chi^2_{12-1} = (12-1)\,\frac{0.935^2}{0.9^2} = 11.87$$

Because the chosen alternative hypothesis was $H_a: \frac{\sigma^2}{\sigma_0^2} \neq 1$, the p-value is found by determining the smaller of the upper- and lower-tail probabilities of the test statistic and multiplying by two. Table C.13 in Appendix C does not provide exact probabilities, so the range of probabilities that the test statistic falls within must be used. To find the range, the row corresponding to 11 degrees of freedom is found in Table C.13 in Appendix C. Reading across the row, we find the values that the calculated statistic of 11.87 would fall between 10.34 and 13.70. From the top of the

Table 5.13 Chi-square table.

df	0.995	0.990	0.975	0.900	0.750	0.500	0.250	0.100	0.050	0.025
1	0.00004	0.0002	0.001	0.02	0.10	0.45	1.32	2.71	3.84	5.02
2	0.01	0.02	0.05	0.21	0.58	1.39	2.77	4.61	5.99	7.38
3	0.07	0.11	0.22	0.58	1.21	2.37	4.11	6.25	7.81	9.35
4	0.21	0.30	0.48	1.06	1.92	3.36	5.39	7.78	9.49	11.14
5	0.41	0.55	0.83	1.61	2.67	4.35	6.63	9.24	11.07	12.83
6	0.68	0.87	1.24	2.20	3.45	5.35	7.84	10.64	12.59	14.45
7	0.99	1.24	1.69	2.83	4.25	6.35	9.04	12.02	14.07	16.01
8	1.34	1.65	2.18	3.49	5.07	7.34	10.22	13.36	15.51	17.53
9	1.73	2.09	2.70	4.17	5.90	8.34	11.39	14.68	16.92	19.02
10	2.16	2.56	3.25	4.87	6.74	9.34	12.55	15.99	18.31	20.48
11	2.60	3.05	3.82	5.58	7.58	10.34	13.70	17.28	19.68	21.92
12	3.07	3.57	4.40	6.30	8.44	11.34	14.85	18.55	21.03	23.34

columns these values fall in, we find they correspond to upper-tail probabilities of 0.5 and 0.25, respectively. Note that the lower-tail probability does not need to be found because it would equal one minus the upper-tail probability of 0.25 to 0.5 and therefore would be larger than the upper-tail probability. The range of 0.25 to 0.5 is multiplied by two and we determine that the p-value is somewhere between 0.5 and 1.

The p-value is larger than the alpha level of 0.05, so the null hypothesis is not rejected and we cannot conclude that the standard deviation differs from 0.9. Because the sample size only provided enough power to detect moderately large differences, taking a larger sample should be considered. Alternatively, the actual value could be compared to the critical value as shown in Table 5.13.

In Minitab: Go to **Stat > Basic Statistics > 1 Variance** and select **Sample standard deviation**. Enter the sample size and sample standard deviation. Then select **Perform hypothesis test** and **Hypothesized standard deviation**. Enter the hypothesized standard deviation in the value box and click **OK** (see Figure 5.18).

The resulting p-value of 0.747 is greater than 0.05, so we fail to reject the null hypothesis (see Figure 5.19). Minitab also gives results for the Bonett method when data are entered in columns. For data that are not normally distributed, Bonett's test provides more accurate Type I and Type II error rates and should be used (Minitab, "1-Sample Standard Deviation Test").

2 Variance Test

A 2 variance test is used to compare two variances or standard deviations to one another by taking the ratio of the two. If the two variances are equal, their ratio is one. The possible hypotheses are:

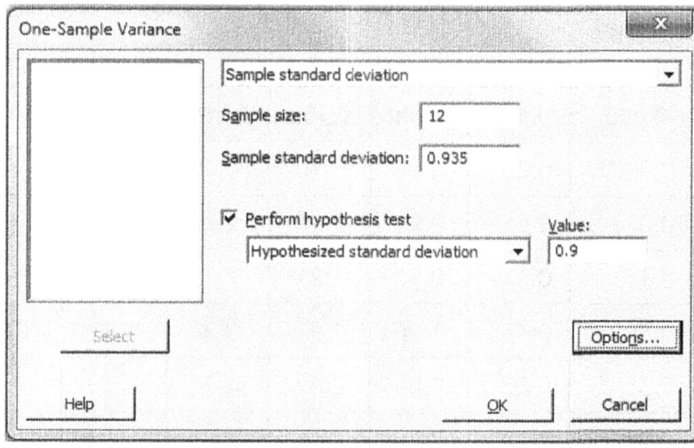

Figure 5.18 Dialog box for a Chi-square test of standard deviation.

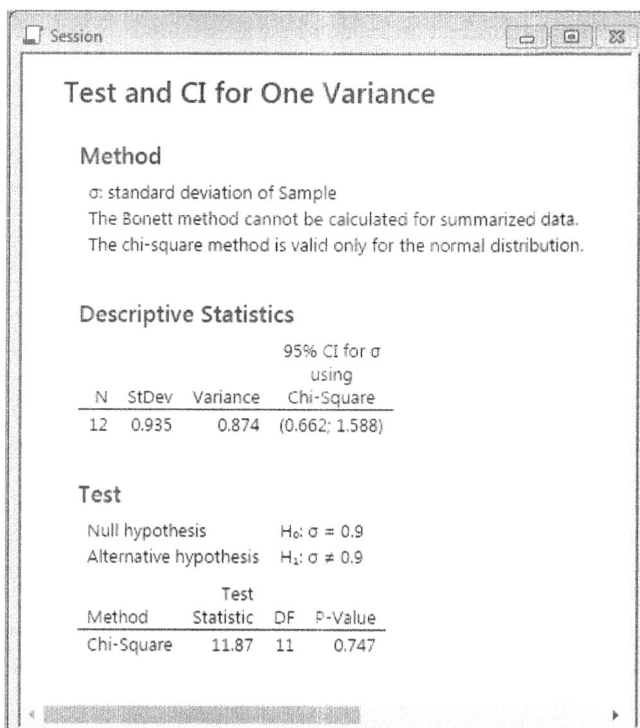

Session

Test and CI for One Variance

Method

σ: standard deviation of Sample
The Bonett method cannot be calculated for summarized data.
The chi-square method is valid only for the normal distribution.

Descriptive Statistics

N	StDev	Variance	95% CI for σ using Chi-Square
12	0.935	0.874	(0.662; 1.588)

Test

Null hypothesis	H_0: σ = 0.9
Alternative hypothesis	H_1: σ ≠ 0.9

Method	Test Statistic	DF	P-Value
Chi-Square	11.87	11	0.747

Figure 5.19 Session window for a Chi-square test of standard deviation.

$$H_0: \frac{\sigma_1^2}{\sigma_2^2} = 1 \text{ and } H_a: \frac{\sigma_1^2}{\sigma_2^2} \neq 1$$

$$H_0: \frac{\sigma_1^2}{\sigma_2^2} < 1 \text{ and } H_a: \frac{\sigma_1^2}{\sigma_2^2} \geq 1$$

$$H_0: \frac{\sigma_1^2}{\sigma_2^2} > 1 \text{ and } H_a: \frac{\sigma_1^2}{\sigma_2^2} \leq 1$$

When the data are from normally distributed populations the test statistic, F, is calculated as

$$F_{test} = \frac{s_1^2}{s_2^2},$$

where the F statistic follows an F distribution with numerator degrees of freedom of $n_1 - 1$ and denominator degrees of freedom of $n_2 - 1$ when the null hypothesis is true. The p-value for each alternative hypothesis is calculated as follows:

For $H_a: \frac{\sigma_1^2}{\sigma_2^2} \neq 1$, $p = 2 * minimum\left[P(F_{n_1-1,n_2-1} > F), P(F_{n_1-1,n_2-1} < F)\right] = 2$ times the smaller of the lower-tail and upper-tail probabilities of the calculated value. When $F < 1$, the lower-tail probability will be the smaller and when $F > 1$ the upper-tail probability will be smaller.

For $H_a: \frac{\sigma_1^2}{\sigma_2^2} \geq 1$, $p = P(F_{n_1-1,n_2-1} > F) =$ the upper-tail probability of the calculated value.

For $H_a: \frac{\sigma_1^2}{\sigma_2^2} \leq 1$, $p = P(F_{n_1-1,n_2-1} < F) =$ the lower-tail probability of the calculated value.

For data that are not normally distributed, the F test does not provide precise results and Bonett's test may be used as an alternative. The calculations for the p-value in the Bonett method are complex but can be performed by statistical software such as Minitab.

Example

An engineer pulls samples from two production lines and wants to know if the difference in variation in a particular metric between the production lines is statistically significant. The first production line's standard deviation is 1.08 and the sample size is 18. The second production line's standard deviation is 1.37 and the sample size 15.

Because it would be of interest to know if the standard deviation is different in either direction, the null hypothesis is set as $H_0: \frac{\sigma_1^2}{\sigma_2^2} = 1$ and the alternative hypothesis as $H_a: \frac{\sigma_1^2}{\sigma_2^2} \neq 1$. The alpha level is set as 0.05. The data have been collected and the sample size is adequate to detect moderate differences with high power.

The F statistic is calculated as $F_{test} = \frac{1.08^2}{1.37^2} = 0.621$. To calculate the p-value, the lower-tail probability of the test statistic must be found because the statistic

Table 5.14 F test table.

Denominator degrees of freedom	Numerator degrees of freedom													
	1	2	3	4	5	6	7	8	9	10	12	15	20	24
1	0.0015	0.026	0.0573	0.0818	0.0999	0.1135	0.1239	0.1321	0.1387	0.1442	0.1526	0.1613	0.1703	0.1749
2	0.0013	0.0256	0.0623	0.0939	0.1186	0.1377	0.1529	0.165	0.175	0.1833	0.1962	0.2099	0.2242	0.2315
3	0.0012	0.0255	0.0648	0.1002	0.1288	0.1515	0.1698	0.1846	0.1969	0.2072	0.2235	0.2408	0.2592	0.2687
4	0.0011	0.0255	0.0662	0.1041	0.1354	0.1606	0.1811	0.1979	0.212	0.2238	0.2426	0.2629	0.2845	0.2959
5	0.0011	0.0254	0.0672	0.1068	0.1399	0.167	0.1892	0.2076	0.223	0.2361	0.257	0.2796	0.304	0.317
6	0.0011	0.0254	0.0679	0.1087	0.1433	0.1718	0.1954	0.215	0.2315	0.2456	0.2682	0.2929	0.3197	0.3339
7	0.0011	0.0254	0.0684	0.1102	0.1459	0.1756	0.2002	0.2208	0.2383	0.2532	0.2773	0.3036	0.3325	0.348
8	0.001	0.0254	0.0688	0.1114	0.148	0.1786	0.2041	0.2256	0.2438	0.2594	0.2848	0.3126	0.3433	0.3598
9	0.001	0.0254	0.0691	0.1123	0.1497	0.181	0.2073	0.2295	0.2484	0.2646	0.291	0.3202	0.3525	0.37
10	0.001	0.0254	0.0694	0.1131	0.1511	0.1831	0.21	0.2328	0.2523	0.269	0.2964	0.3268	0.3605	0.3788
12	0.001	0.0254	0.0698	0.1143	0.1533	0.1864	0.2143	0.2381	0.2585	0.2762	0.3051	0.3375	0.3737	0.3935
15	0.001	0.0254	0.0702	0.1155	0.1556	0.1898	0.2189	0.2438	0.2653	0.284	0.3147	0.3494	0.3886	0.4103
20	0.001	0.0253	0.0706	0.1168	0.158	0.1935	0.2239	0.25	0.2727	0.2925	0.3254	0.3629	0.4058	0.4297

is less than one. That probability is then multiplied by two. Tables C.15–C.19 in Appendix C must be used by finding the tables containing the values on either side of 0.621 for the numerator degrees of freedom of 18 − 1 = 17 and denominator degrees of freedom of 15 − 1 = 14. These values will correspond to the range of values that the lower-tail probability of 0.621 falls within. Because the tables do not contain all possible degrees of freedom, each should be rounded down to the nearest available degrees of freedom in the table. For this example, that corresponds to a numerator degrees of freedom of 15 and a denominator degrees of freedom of 12.

Table C.18 corresponding to a lower-tail probability of 0.1 provides a value of 0.4958, and Table C.19 corresponding to a lower-tail probability of 0.2 provides a value of 0.6336. Multiplying these by two, we determine the p-value falls between 0.2 and 0.4. The p-value is larger than the alpha level of 0.05, so the null hypothesis is not rejected and we cannot conclude that the variances are different. Alternatively, the actual value could be compared to the critical value of 0.3375 from Table C.16 as shown in Table 5.14 to reach the same conclusion.

In Minitab: Go to **Stat > Basic Statistics > 2 Variances** and select **Sample standard deviations**. Then enter the sample sizes and standard deviations as shown in Figure 5.20. Click **OK**.

If data are used instead of summary statistics, the **Options** subdialog contains an option to base the test on the normal distribution and use the F test. Otherwise Bonett's test and Levene's test will be used. Levene's test provides consistent and accurate Type I error rates and performs better than the Bonett method with small or heavily skewed samples. However, it is less powerful generally (Minitab, "Tests for Standard Deviations (Two or More Samples)"). For samples of less than 20 observations each or with at least one highly skewed distribution, Levene's test should be used. Otherwise, Bonett's test is preferable.

The resulting p-value of 0.349 is greater than 0.05, so we fail to reject the null hypothesis (see Figure 5.21).

Minitab also provides a graphical illustration of a confidence interval (CI) for the ratio of standard deviations as shown in Figure 5.22. The plot on top provides a confidence interval of the ratio of the two variances and overlaps 1, which is an

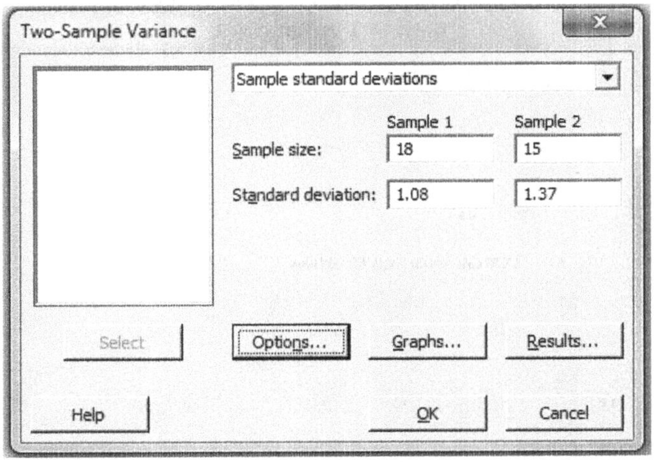

Figure 5.20 Dialog box for a 2 variance test.

Session

Test and CI for Two Variances

Method

σ_1: standard deviation of Sample 1
σ_2: standard deviation of Sample 2
Ratio: σ_1/σ_2
F method was used. This method is accurate for normal data only.

Descriptive Statistics

Sample	N	StDev	Variance	95% CI for σ
Sample 1	18	1.080	1.166	(0.810; 1.619)
Sample 2	15	1.370	1.877	(1.003; 2.161)

Ratio of Standard Deviations

Estimated Ratio	95% CI for Ratio using F
0.788321	(0.463; 1.308)

Test

Null hypothesis H_0: $\sigma_1 / \sigma_2 = 1$
Alternative hypothesis H_1: $\sigma_1 / \sigma_2 \neq 1$
Significance level $\alpha = 0.05$

Method	Test Statistic	DF1	DF2	P-Value
F	0.62	17	14	0.349

Figure 5.21 Session window output for a 2 variance test.

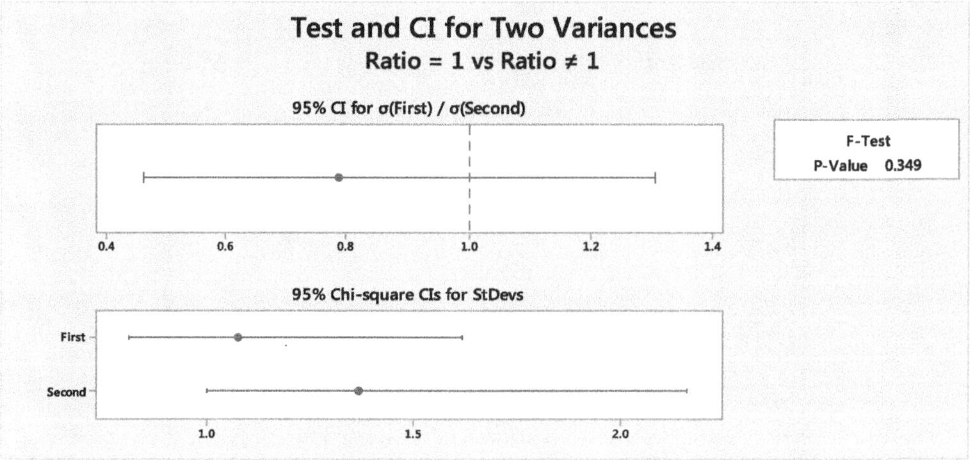

Figure 5.22 Test and confidence interval for two variances.

indication that there is no evidence the two variances differ. Similarly, the second plot provides confidence intervals on the individual variances and these also have substantial overlap.

Test for Equal Variances

The test for equal variances tests whether there is evidence that at least one of three or more groups have a different variance than the others. For example, you may want to determine whether delivery times are more or less consistent among multiple suppliers. There are many different statistical tests to check for equal variances, but Minitab provides two tests known as Levene's and multiple comparisons.

If the samples being compared have fewer than 20 observations each or appear to be heavily skewed or "heavy-tailed," then the p-value provided by Levene's test is more reliable and should be used. Otherwise, the multiple comparisons test is more powerful and that p-value should be used (Minitab, "Understanding Test for Equal Variances"). Neither test requires normality.

The null hypothesis for the test is H_0. All variances are equal, and the alternative hypothesis is H_a. At least one variance is different. A p-value less than the desired alpha level indicates that the null hypothesis should be rejected in favor of the alternative hypothesis.

Example

Metal pins are made using four different lathes. Data are collected by randomly selecting 10 pins over time made by each of the lathes, and you wish to determine if the variance in pin diameters is different for any of the lathes in an effort to reduce overall variation. The data are presented in Table 5.15.

Table 5.15 Pin diameters for four lathes.

Lathe A	Lathe B	Lathe C	Lathe D
53.9	53.8	53.9	54.2
54.1	53.5	54.3	54.2
54.4	54.7	53.9	53.9
54.2	54.9	53.7	53.8
53.9	53.5	53.9	54.1
53.6	55.4	54.1	54.2
54.2	54.5	54.0	54.3
54.3	52.9	53.9	54.2
54.0	53.5	54.0	53.7
53.9	53.6	53.9	54.2

In Minitab: Go to **Stat > ANOVA > Test for Equal Variances**, complete the dialog as shown in Figure 5.23, and click **OK**.

The session window output is shown in Figure 5.24 and the graphical output is shown in Figure 5.25.

Both the multiple comparisons test and Levene's test show very low p-values, indicating that at least one variance is different from the others. The graphical output clearly demonstrates that lathe B has much higher variance than the other three lathes, which show no differences among them. Lathe B should be investigated to determine why the pins are being made so inconsistently.

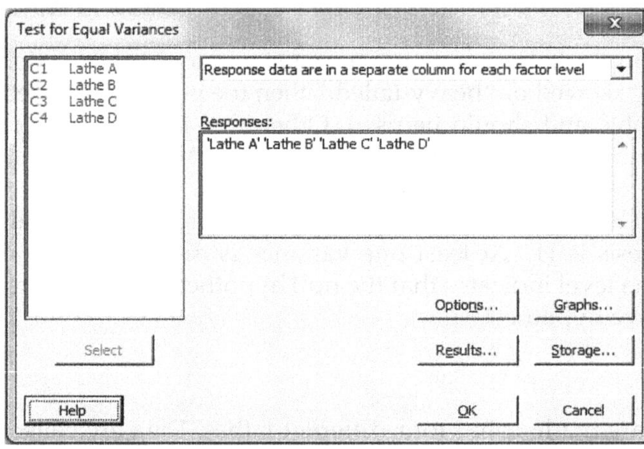

Figure 5.23 Minitab dialog box for test for equal variances.

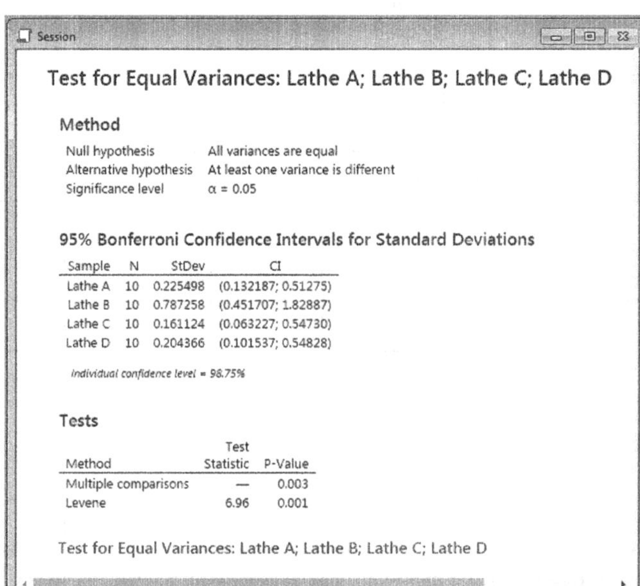

Session

Test for Equal Variances: Lathe A; Lathe B; Lathe C; Lathe D

Method

Null hypothesis	All variances are equal
Alternative hypothesis	At least one variance is different
Significance level	α = 0.05

95% Bonferroni Confidence Intervals for Standard Deviations

Sample	N	StDev	CI
Lathe A	10	0.225498	(0.132187; 0.51275)
Lathe B	10	0.787258	(0.451707; 1.82887)
Lathe C	10	0.161124	(0.063227; 0.54730)
Lathe D	10	0.204366	(0.101537; 0.54828)

Individual confidence level = 98.75%

Tests

Method	Test Statistic	P-Value
Multiple comparisons	—	0.003
Levene	6.96	0.001

Test for Equal Variances: Lathe A; Lathe B; Lathe C; Lathe D

Figure 5.24 Session window output for test for equal variances.

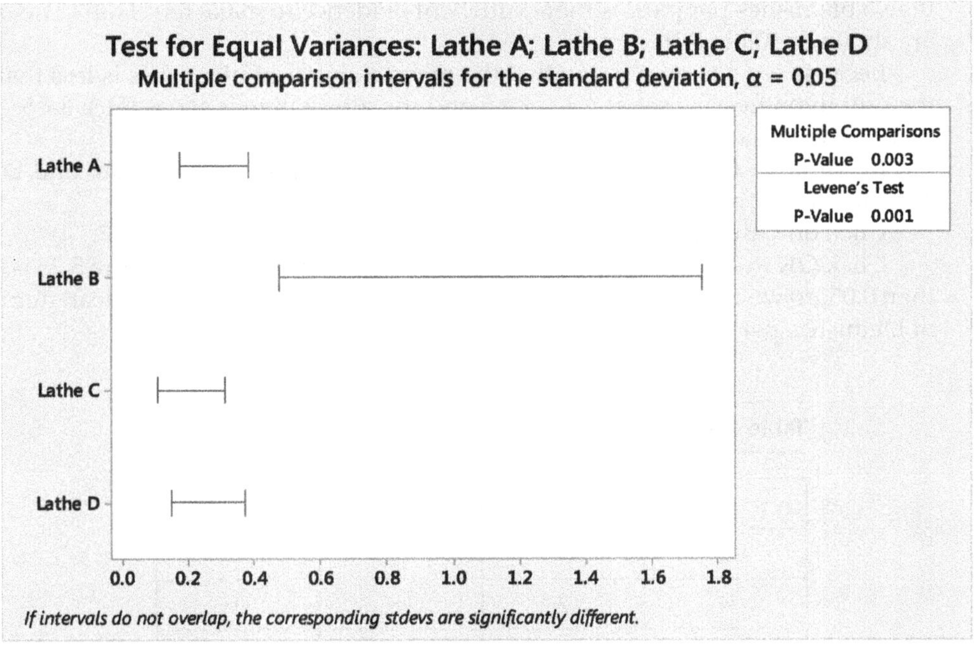

Figure 5.25 Graphical output for test for equal variances.

POISSON RATE TEST

Poisson rate tests are used to compare rates of occurrences and determine a range of values that likely includes the population rate of occurrences.

1-Sample Poisson Rate Test

A 1-sample Poisson rate test is used to test the mean of a Poisson-distributed variable against some value of interest, μ_0. Recall that the Poisson distribution is used for counts of items with no limit (Montgomery, Runger, and Hubele 2001), such as number of surface defects or server crashes per week. Before conducting the test the practitioner must decide which of three combinations of null and alternative hypothesis are of interest:

H_0: $\lambda = \lambda_0$ and H_a: $\lambda \neq \lambda_0$

H_0: $\lambda \leq \lambda_0$ and H_a: $\lambda > \lambda_0$

H_0: $\lambda \geq \lambda_0$ and H_a: $\lambda < \lambda_0$

The calculation of the p-value depends on which hypothesis was being tested. Because the calculation is difficult, software such as Minitab should be used. That p-value is then compared to the desired alpha and if smaller, the null hypothesis is rejected in favor of the alternative hypothesis.

Example

Twenty parts from a blow molding process are randomly sampled and inspected for surface blemishes. The customer would like to be assured that there are fewer than 5 blemishes per part. Is there sufficient evidence to make the claim? The data are shown in Table 5.16.

Because we desire to prove that the mean number of blemishes is less than 5, the null hypothesis is set as (H_0): $\lambda \geq 5$ and the alternative is set as (H_a): $\lambda < 5$. The alpha level chosen is 0.05.

In Minitab: Go to **Stat > Basic Statistics > 1-Sample Poisson Rate** and complete the dialog as shown in Figure 5.26.

Click on **Options** and complete the dialog as shown in Figure 5.27.

Click **OK** in each dialog box. The resulting p-value shown in Figure 5.28 is less than 0.05, so we reject the null hypothesis and can conclude that the mean number of blemishes per part is less than 5.

Table 5.16 Surface blemishes per part.

4	4	2	2
2	4	2	0
4	2	4	1
1	0	7	5
1	5	3	2

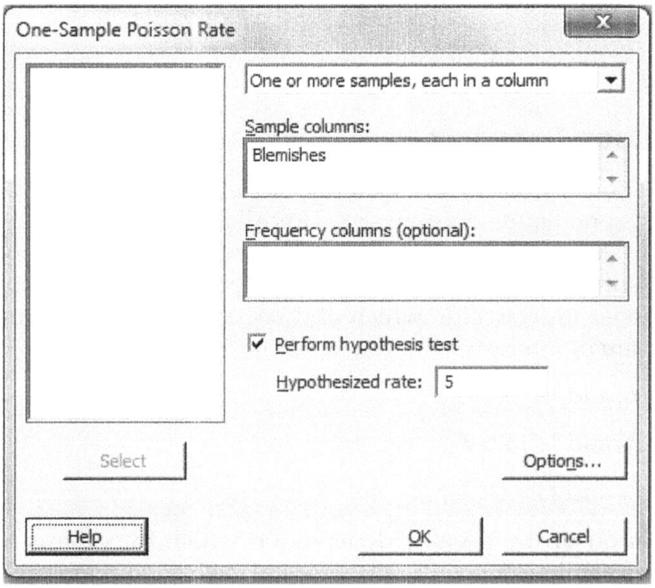

Figure 5.26 1-sample Poisson rate dialog box.

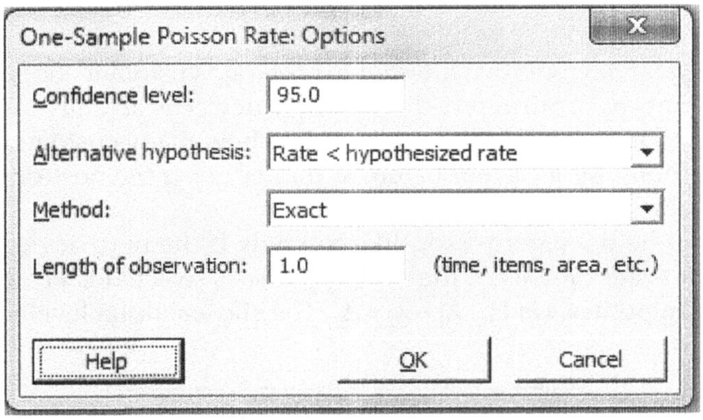

Figure 5.27 1-sample Poisson rate options dialog box.

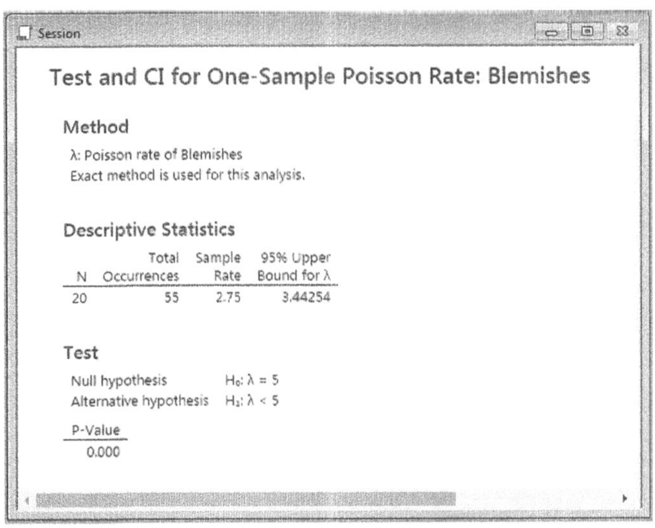

Figure 5.28 Session window output for 1-sample Poisson rate test.

2-Sample Poisson Rate Test

The 2-sample Poisson rate test is used to compare the means of two Poisson-distributed variables. For example, you could compare the count of orders received per hour before and after a new email campaign.

There are three sets of hypotheses that could be tested using a 2-sample t test:

H_0: $\lambda_1 - \lambda_2 = \Delta_0$ and H_a: $\lambda_1 - \lambda_2 \neq \Delta_0$

H_0: $\lambda_1 - \lambda_2 \leq \Delta_0$ and H_a: $\lambda_1 - \lambda_2 > \Delta_0$

H_0: $\lambda_1 - \lambda_2 \geq \Delta_0$ and H_a: $\lambda_1 - \lambda_2 < \Delta_0$

Here Δ_0 is a difference of interest between the two means, most commonly 0.

The p-value calculated is difficult by hand and should be done by software such as Minitab. Once determined, it is compared to the desired alpha level and if smaller, the null hypothesis is rejected and the alternative hypothesis is accepted.

Example

A new scanning device is being tested to speed up customer checkout at a store. The manager wants to confirm whether the new device is actually faster before investing in it for all registers. Data are collected on how many customers are checked out every 15 minutes by a cashier before and after using the new device. The data are shown in Table 5.17.

Because the manager would like to know if the new device makes checkout times either faster or slower, the null hypothesis selected is H_0: $\lambda_1 - \lambda_2 = \Delta_0$ and the alternative hypothesis is H_a: $\lambda_1 - \lambda_2 \neq \Delta_0$. The chosen alpha level is 0.05.

Table 5.17 Customers checked out before and after new device.

Customers before	Customers after
8	8
4	11
9	9
5	10
3	5
6	10
14	14
11	8
7	10
8	5

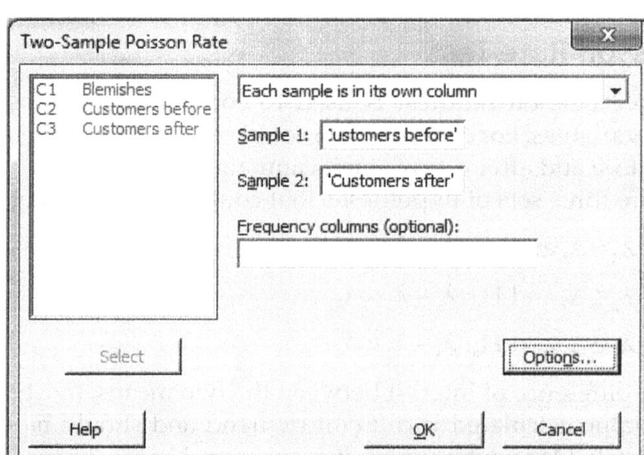

Figure 5.29 Minitab dialog box for 2-sample Poisson rate test.

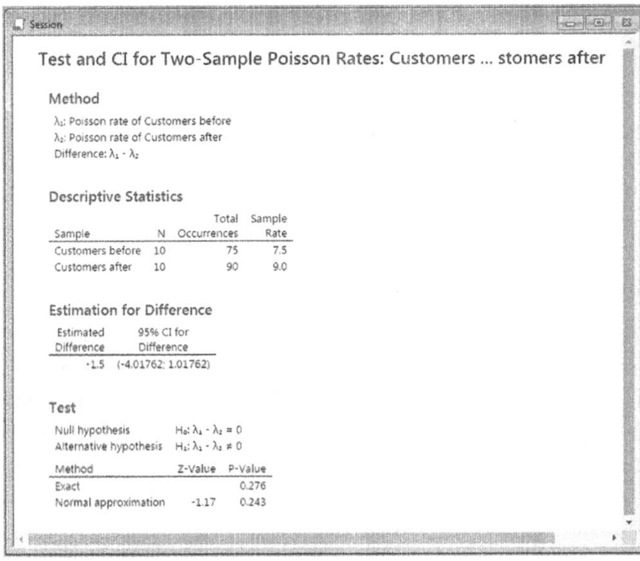

Session

Test and CI for Two-Sample Poisson Rates: Customers ... stomers after

Method

λ_1: Poisson rate of Customers before
λ_2: Poisson rate of Customers after
Difference: $\lambda_1 - \lambda_2$

Descriptive Statistics

Sample	N	Total Occurrences	Sample Rate
Customers before	10	75	7.5
Customers after	10	90	9.0

Estimation for Difference

Estimated Difference	95% CI for Difference
-1.5	(-4.01762; 1.01762)

Test

Null hypothesis $H_0: \lambda_1 - \lambda_2 = 0$
Alternative hypothesis $H_1: \lambda_1 - \lambda_2 \neq 0$

Method	Z-Value	P-Value
Exact		0.276
Normal approximation	-1.17	0.243

Figure 5.30 Session window output for 2-sample rate test.

In Minitab: **Go to Stat > Basic Statistics > 2-Sample Poisson Rate**. In the drop-down, select **Each sample in its own column** and complete the dialog as shown in Figure 5.29. Click **OK.**

The session window output is shown in Figure 5.30.

The output displays two p-values. The first is an exact calculation and is preferred, but it is only available when the difference of interest is 0. It equals 0.276. The second is based on a normal approximation and is 0.243. Because the p-value is greater than the desired alpha of 0.05, the null hypothesis is not rejected and the manager cannot conclude that the new device results in a different number of customers served in a 15-minute interval.

QUESTIONS AND ANSWERS

Q: What do I do if I want to perform a statistical test that requires the data to be from a stable process, but the process is not stable?

A: Processes that are not stable cannot be relied on to produce similar data in the near future, and statistical tests based on them cannot be used to make inference about the process as a whole. It is important to establish stability before using data from that process.

Q: Can I perform a 2-sample t test if my data are not normally distributed?

A: Yes, the t test is robust to non-normality. However, it is still important to check for stability of the data and consider what distribution might be expected to establish whether the non-normality is naturally part of the process or due to an out-of-control situation.

Q: What is the difference between the *t* distribution and the standard normal distribution?

A: The *t* distribution is very similar to the standard normal distribution, but it has fatter tails to compensate for the uncertainty in the estimate of the standard deviation. As the sample size increases, this uncertainty is reduced and the *t* distribution becomes more similar to the standard normal until they are indistinguishable for practical purposes.

Q: What is the difference between power and p-value?

A: Power is the probability of detecting some true condition given the sample size, alpha level, and true underlying distributions. When performing a statistical test, high power is desirable but comes at the cost of higher sample size or higher risk of falsely rejecting the null hypothesis. A p-value is the probability of observing a particular set of data if the null hypothesis were true. They relate in that power can be thought of as the probability of obtaining a p-value less than the chosen alpha level if the assumed conditions are true.

Q: What is a Z-score?

A: A Z-score represents the number of standard deviations a value from the normal distribution is from the mean.

Q: What is an overpowered study?

A: An overpowered study is one in which the amount of data collected allows for statistical significance to be easily detected, even for differences that may have no practical significance. While high power is desirable, one should always consider whether the effect detected is meaningful in real life.

Q: What is an underpowered study?

A: An underpowered study is one in which not enough data were collected to be likely to see an effect unless that effect was very extreme. Sample size estimation should be done before collecting data if possible, to ensure that resources are not wasted collecting data that are insufficient to detect the effect of interest.

Q: What is alpha (α)?

A: Alpha is the probability of rejecting the null hypothesis in a hypothesis test when it is true; it is otherwise known as the level of significance.

Q: What is beta (β)?

A: Beta is the probability of not rejecting the null hypothesis in a hypothesis test when it is not true.

Q: What is the power of a statistical test?

A: Power is the opposite of beta and is the probability of rejecting the null hypothesis when it is false. It is desirable to have high power.

Q: Why do we use 95% confidence (α=0.05) when performing a statistical test?

A: This is a common value because many practitioners find it provides a good balance of confidence in results and power to detect them. Depending on the application and implications for rejecting or not rejecting the null hypothesis, higher or lower values may be more appropriate.

Q: What is a p-value?

A: The p-value is the probability of seeing data at least as extreme as was used if the null hypothesis were true. When this probability is very low, it is seen as an indication that the null hypothesis is likely not true.

Q: What is the difference between p-value and probability?

A: A p-value specifically refers to a value obtained in hypothesis testing, whereas probability generally refers to the likelihood of something occurring.

Q: What is the probability (risk) that I make a Type I error or Type II error when the critical p-value is 0.05?

A: The answer depends on the desired alpha level.

Chapter 6

Sample Size and Confidence Intervals

The probability of a Type II error is related to the size of a sample (Montgomery and Runger 1999), and the necessary sample size for a given test is based on the desired risk of both Type I and Type II errors. However, for a given test and sample size, Type I and Type II errors are inversely related. Desired levels of each must be used to determine what sample size will accomplish them.

SAMPLE SIZE

The power of a statistical test is affected by the alpha level, size of the difference of the values being compared, variability in the data, and the sample size. Larger sample sizes result in more power. There are formulas that can be used for determining sample size (Boslaugh and Watters 2008).

Sample Size for Estimating the Mean

The sample size needed to obtain a margin of error E for estimating the mean is calculated as:

$$n = \left(Z_{\alpha/2}\frac{\sigma}{E}\right)^2 \text{ for a known } \sigma$$

$$n = \left(t_{\alpha/2,n-1}\frac{s}{E}\right)^2 \text{ for an unknown } \sigma$$

A $(1-\alpha)100\%$ confidence interval (CI) on the mean would be equal to the mean plus or minus the margin of error. Therefore, for a desired confidence interval width, the appropriate margin of error formula can be solved for n to determine the sample size necessary.

Example

An engineer wants to take a sample from a process producing screws that are measured for weight. The weights have a known standard deviation of 7.4 grams and the engineer wants to estimate the true mean to within 2 grams with 95% confidence.

Because 95% confidence is desired, the α is 5% = 0.05 and $\alpha/2 = .025$. From a Z table it can be found that $Z_{.025} = -1.96$. The required sample size would be

$\left(\frac{-1.96*7.4}{2}\right)^2 = 52.6$. When estimating sample size, results are always rounded up to the nearest whole number so that the required sample size is 53.

In Minitab: Go to **Stat > Power and Sample Size > Sample Size for Estimation** and enter the standard deviation and margins of error for the confidence interval (see Figure 6.1). Then select **Options** to place a checkmark next to **Assume population standard deviation is known** (as shown in Figure 6.2) and then click **OK**.

The resulting sample size is 53 (see Figure 6.3).

Sample Size for Estimating a Proportion

Sample sizes for estimating a proportion using the normal approximation can be determined using the formula $n = \left(\frac{Z_{\alpha/2}}{E}\right)^2 \hat{P}(1-\hat{P})$, where E is the desired margin of error and the estimated proportion is represented by $\hat{P}$. This provides good results when both $n*\hat{P}$ and $n*(1-\hat{P})$ are at least 5. Otherwise, the exact method must be used. The sample size estimated using the exact method is the value of n that results in both the lower and upper limits being within the margin of error of the estimated proportion. The limits are calculated as

$$\text{Lower limit } P_L = \frac{2x * F_{\alpha/2,2x,2(n-x+1)}}{2(n-x+1)2x * F_{\alpha/2,2x,2(n-x+1)}}$$

$$\text{Upper limit } P_U = \frac{2(x+1) * F_{1-\alpha/2,2(x+1),2(n-x)}}{2(n-x)2(x+1) * F_{1-\alpha/2,2(x+1),2(n-x)}},$$

where x is the number of events.

Example

A quality engineer would like to determine the proportion of parts that are defective. The quality engineer is willing to accept a margin of error (E) of 4%

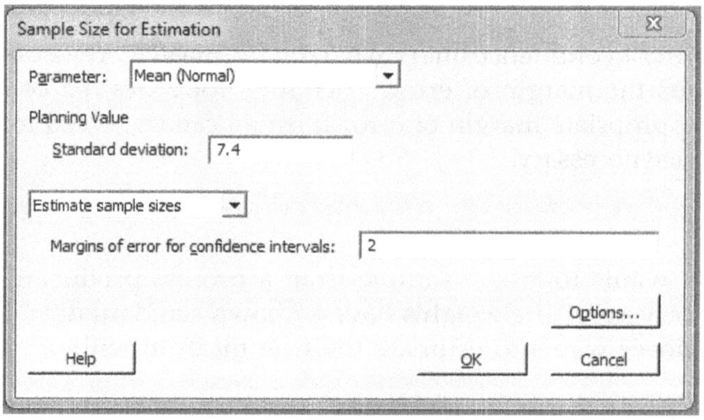

Figure 6.1 Dialog box for sample size determination.

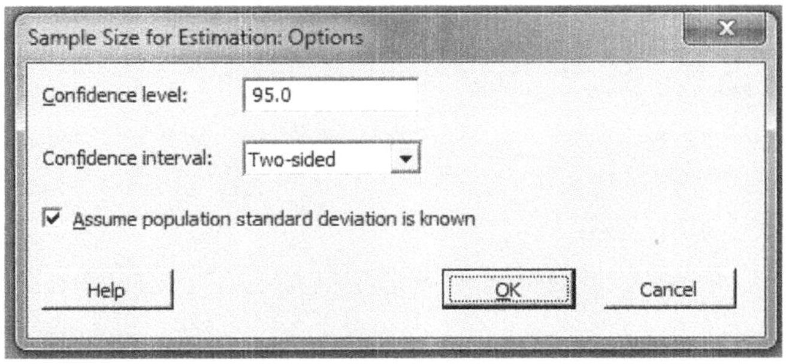

Figure 6.2 Options for sample size determination.

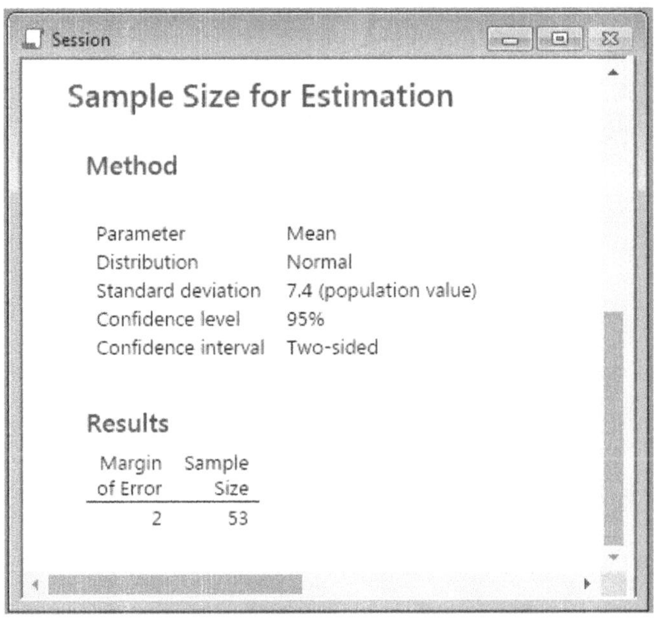

Figure 6.3 Session window for size determination.

and would like 95% confidence (Z = 1.96). The estimated proportion defective is 20%.

$$n = \left(\frac{Z_{\alpha/2}}{E}\right)^2 \hat{P}(1-\hat{P}) = \left(\frac{1.96}{0.04}\right)^2 * 0.20\,(1-0.20) = 384.16$$

In Minitab: Go to **Stat > Power and Sample Size > Sample Size for Estimation** and select **Proportion (Binomial)** as the parameter. Then enter the estimated proportion of 0.20 as the planning value. Enter the margin of error of 0.04 for the confidence interval (see Figure 6.4) and click **OK**.

The resulting sample size is 450 (see Figure 6.5). This differs from the result by hand because Minitab is using the exact method.

In almost all cases it is important to consider the power of a test before you collect data. By making some assumptions and considering how powerful you want the

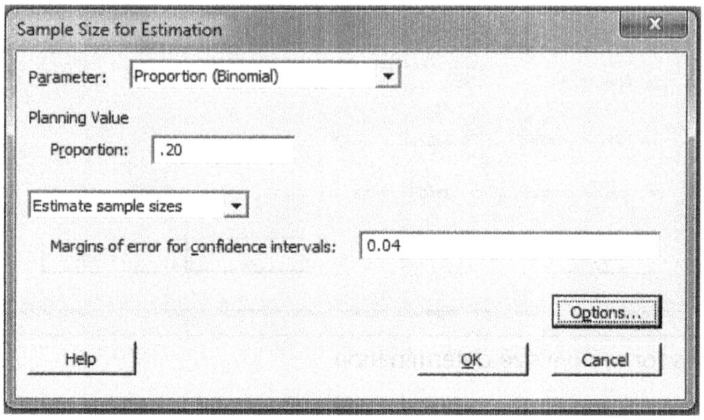

Figure 6.4 Dialog box for size for proportion tests.

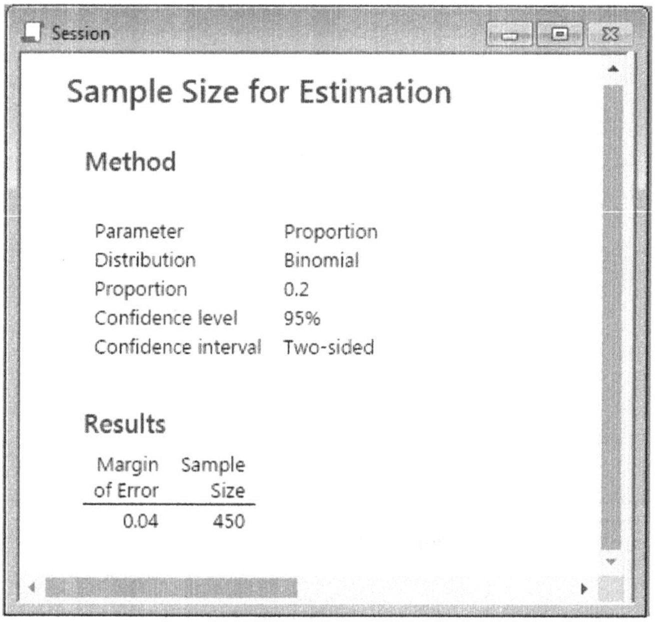

Figure 6.5 Session window for proportion tests.

test to be, a sample size calculation can be performed to ensure that the number of data points taken is sufficiently large. Recall from Chapter 6 that power is the chance that you are able to reject the null hypothesis when it really is not true. Higher power is obviously very desirable, but it comes at a cost in the form of a larger sample.

Because the calculations require software, the sections on sample size for hypothesis tests will focus on the use of Minitab and not the equations themselves.

Sample Size for Z Test

To perform a sample size calculation for a Z test, you must first provide a true value for the standard deviation. For a Z test, this should be a known value already. Additionally, the alternative hypothesis and alpha level must be provided.

Two other pieces of information are required to calculate the necessary sample size. The first is the difference that you are interested in detecting. This should be the value that would have practical meaning and that would cause you to modify a process. Detecting smaller differences requires a larger sample size, so the calculation should only be done for a meaningful difference. The other is the power value that is desired, which is expressed as a proportion that is between 0 and 1. The calculations for sample size for most hypothesis tests are difficult but can be done using Minitab.

Example

A bottle-filling process results in fill volumes with a known standard deviation of 0.12 ounces. You are interested in performing a Z test to determine whether the true fill volume is 16 ounces, and a statistically significant difference of at least 0.05 ounces would require the machine to be recalibrated. You would like to know the required sample size for either 80% or 90% power.

In Minitab, go to **Stat > Power and Sample Size > 1-Sample Z**. By default, the alternative hypothesis is that the mean is not equal to the hypothesized value and the alpha level is 0.05. These can be changed in the **Options** subdialog. Complete the dialog as shown in Figure 6.6 and click **OK**.

The session window output is shown in Figure 6.7.

The session window output shows that detecting the difference of interest with 80% power would require a sample size of 46; with 90% power it would require a sample size of 61. The true values would be a decimal, so Minitab rounds up to the nearest whole number to provide sample size and additionally, provides the true power achieved with that sample size.

Minitab also provides graphical output as shown in Figure 6.8.

The graphical output makes it easier to identify the trade-off between the difference of interest and the power of the test. It can be seen that while these tests provide high power for the specified difference of 0.05, they would be very poor at

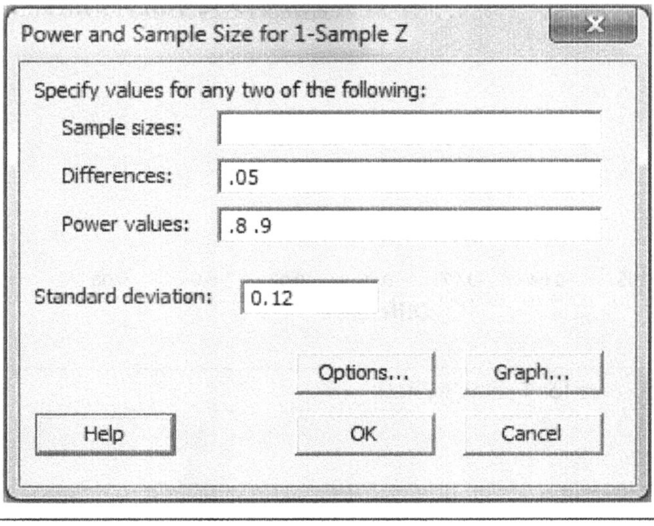

Figure 6.6 Minitab dialog box for sample size for Z test.

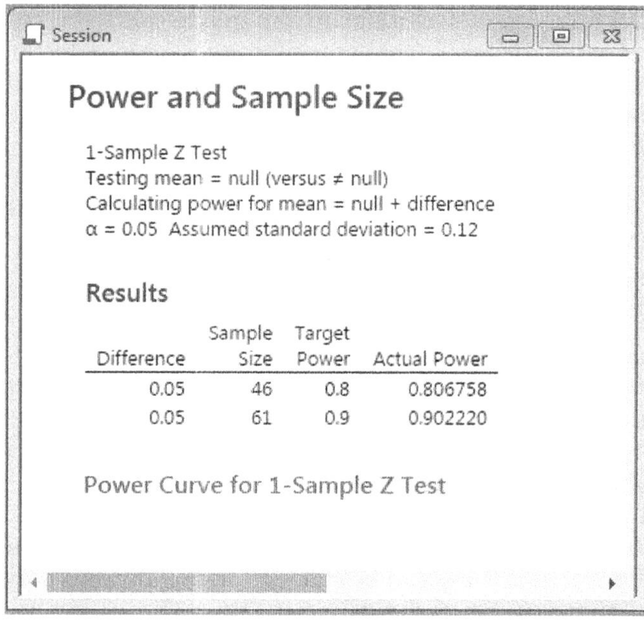
Session

Power and Sample Size

1-Sample Z Test
Testing mean = null (versus ≠ null)
Calculating power for mean = null + difference
α = 0.05 Assumed standard deviation = 0.12

Results

Difference	Sample Size	Target Power	Actual Power
0.05	46	0.8	0.806758
0.05	61	0.9	0.902220

Power Curve for 1-Sample Z Test

Figure 6.7 Session window output for sample size for Z test.

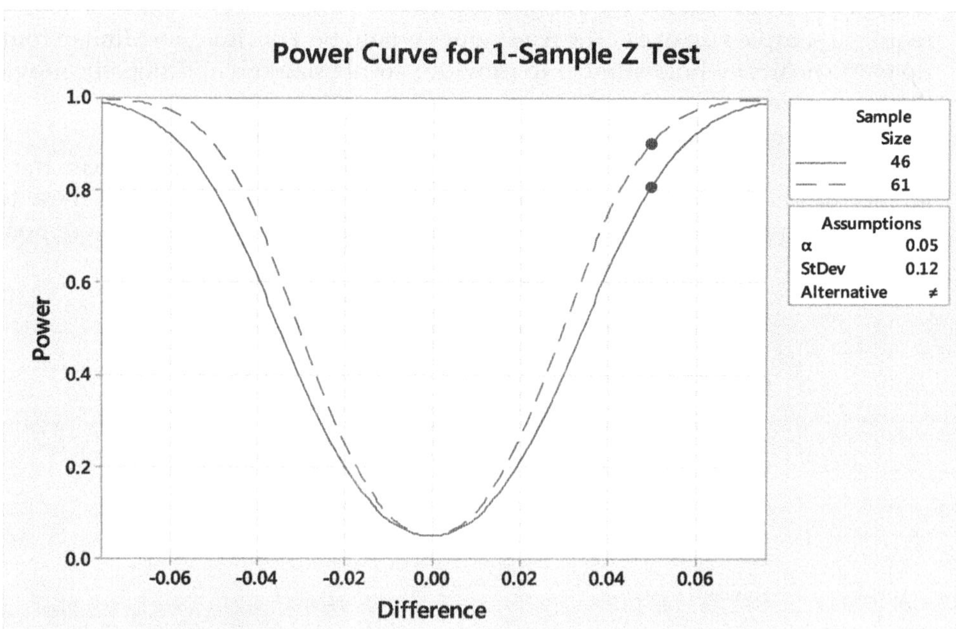

Figure 6.8 Power curve for 1-sample Z test.

detecting a difference of 0.02. Conversely, a difference of 0.07 or higher would be almost certain to result in a significant test.

Sample Size for 1-Sample t Test

Calculating sample size for a 1-sample t test requires the same inputs as the Z test. First, a standard deviation must be assumed. This can be based on historical data or knowledge. If the best guess for the true value falls somewhere in a range, the conservative way to perform the calculation is to use the highest standard deviation in that range. Next, the alternative hypothesis and alpha level must be specified.

One or more differences of interest are considered, and as with the Z test, they should be based on what size difference would have practical meaning. Finally, the desired power levels are required.

Example

The acidity of a new soft drink is measured and you wish to perform a 1-sample t test to determine if the true acidity differs from a target value by more than 0.2. Historical data suggest the standard deviation of acidity measurements across all other drinks is 0.15. You want the test to have 90% power.

In Minitab: Go to **Stat>Power and Sample Size>1-Sample t**. Complete the dialog as shown in Figure 6.9 and click **OK**.

The session window output is shown in Figure 6.10.

The session window output indicates that to achieve the desired power of 90%, only 9 samples are required and that the true power achieved with 9 samples is actually 93.67%. The graphical output is shown in Figure 6.11.

The power curve graph illustrates the high power of the test, and that even at 80% power, the test will still detect a difference of about 0.15.

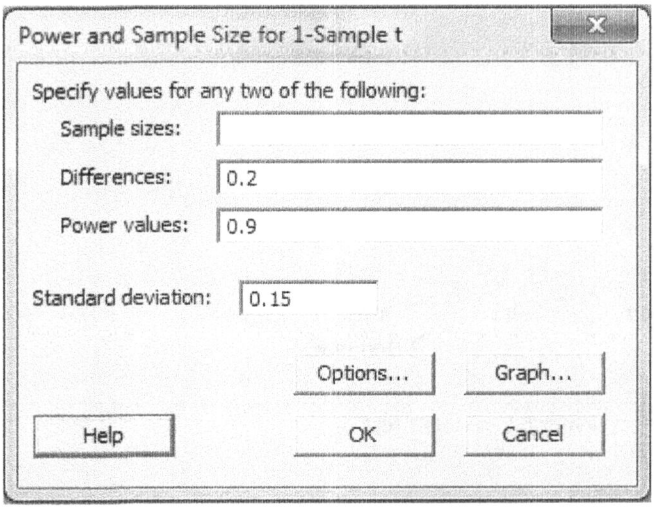

Figure 6.9 Minitab dialog box for sample size for 1-sample t test.

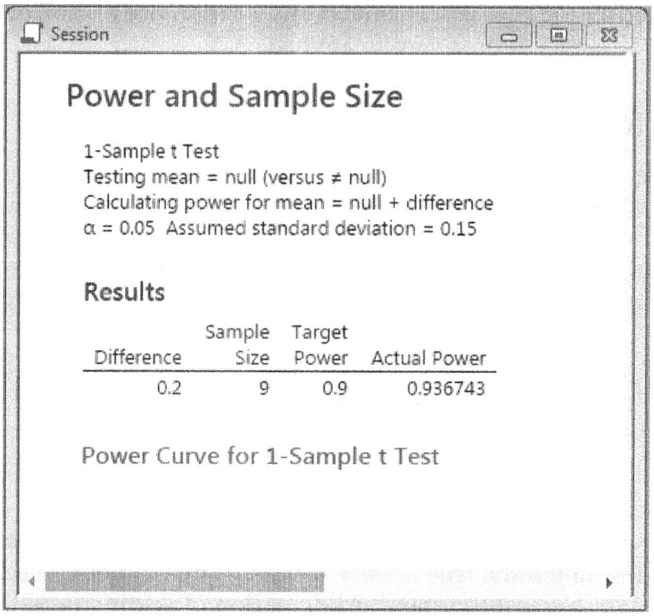

Session

Power and Sample Size

1-Sample t Test
Testing mean = null (versus ≠ null)
Calculating power for mean = null + difference
α = 0.05 Assumed standard deviation = 0.15

Results

Difference	Sample Size	Target Power	Actual Power
0.2	9	0.9	0.936743

Power Curve for 1-Sample t Test

Figure 6.10 Session window output for sample size for 1-sample t test.

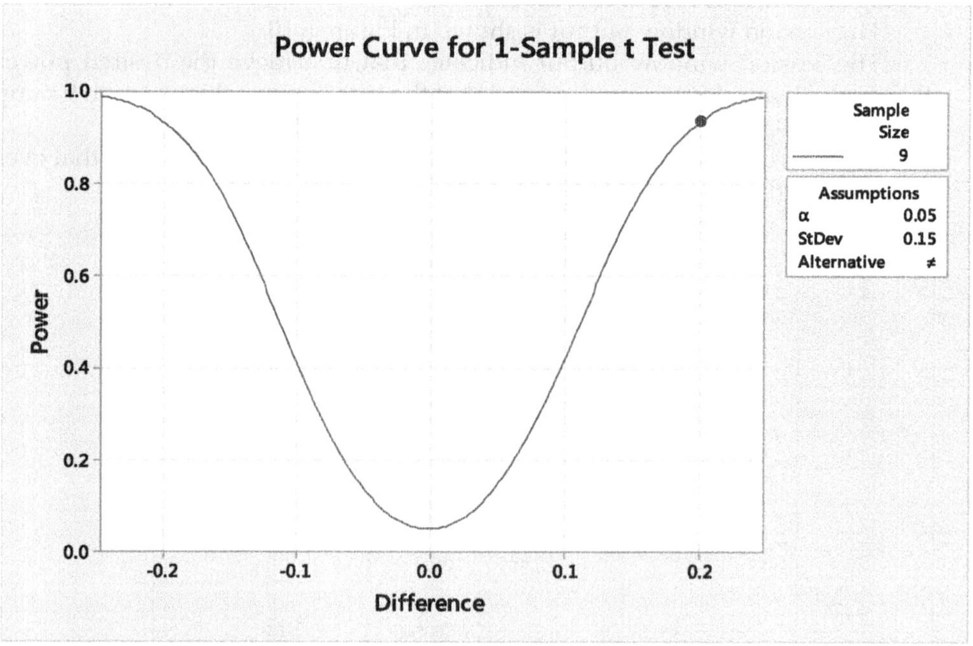

Figure 6.11 Power curve for 1-sample t test.

Sample Size for 2-Sample t Test

Calculating sample size for a 2-sample t test requires an assumed standard deviation, which is treated as equal for both populations. As with the 1-sample t test, the largest suspected standard deviation should be used in order to be conservative in the calculation. The alternative hypothesis and alpha level are also required.

Then one or more differences of interest are needed, and as with the 1-sample version, this difference should be the minimum difference that has practical significance if detected. Smaller values are harder to detect and require a larger sample size. Finally, one or more power values of interest are needed.

Example

The compression of tennis balls is measured in millimeters (mm), and you wish to compare the compression of balls for two production methods. Compression differences of 2 mm or more are noticeable by the customer. Historically the standard deviation of compression has been about 4.4 mm for both methods. Because switching from one method to the other would be very expensive, you desire 97% power for the test.

In Minitab: Go to **Stat > Power and Sample Size > 2-Sample t**. Complete the dialog as shown in Figure 6.12 and click **OK**.

The session window output is shown in Figure 6.13.

The session window output states that a sample size of 144 is needed. The stated sample size is for each group, so a total of 288 balls would be necessary. The graphical output is shown in Figure 6.14.

The power curve demonstrates that while the power is very high for the difference of interest, it quickly becomes much lower for smaller differences.

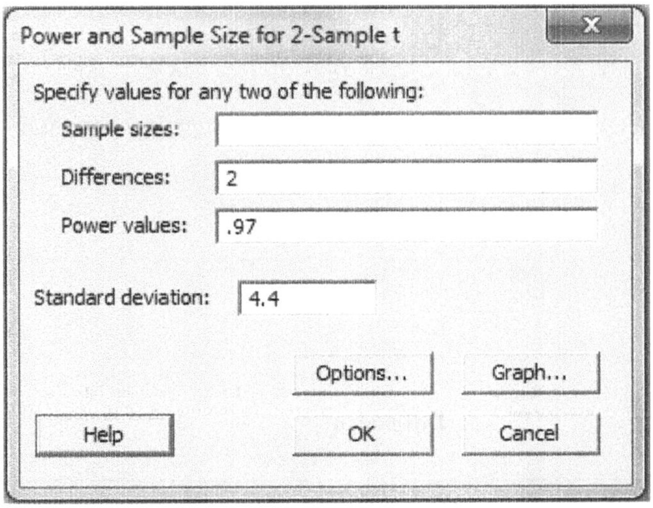

Figure 6.12 Minitab dialog box for sample size for 2-sample t test.

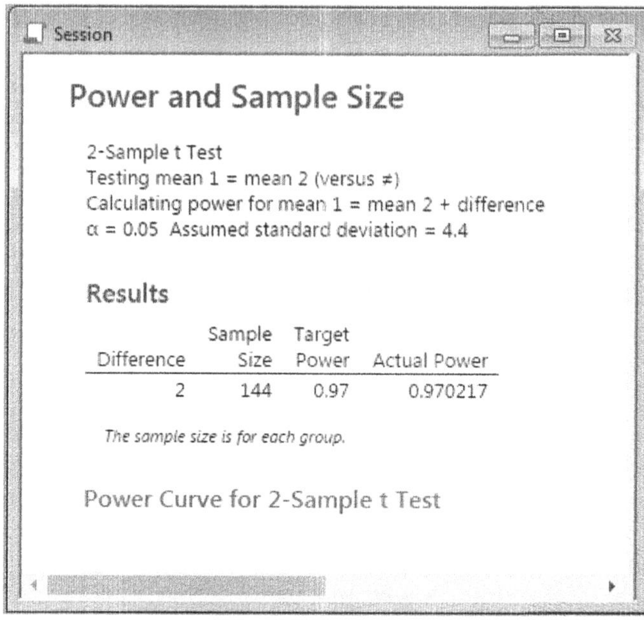

Session

Power and Sample Size

2-Sample t Test
Testing mean 1 = mean 2 (versus ≠)
Calculating power for mean 1 = mean 2 + difference
α = 0.05 Assumed standard deviation = 4.4

Results

Difference	Sample Size	Target Power	Actual Power
2	144	0.97	0.970217

The sample size is for each group.

Power Curve for 2-Sample t Test

Figure 6.13 Session window output for sample size for 2-sample t test.

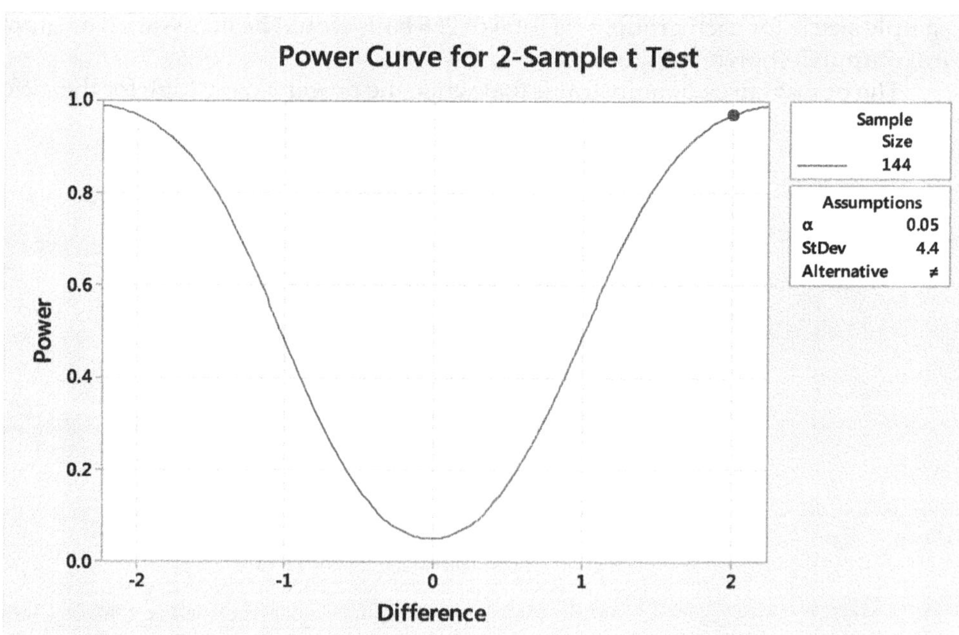

Figure 6.14 Power curve for 2-sample t test.

Sample Size for Paired t Test

The sample size calculation for the paired t test is the same as that of the 1-sample t test. It requires the alternative hypothesis, alpha level, difference(s) of interest, desired power value(s), and the assumed standard deviation of the paired differences. It is important to understand that the standard deviation is that of the differences between the two samples and not the two samples themselves. For example, if the hardness of plastic parts is measured before and after some treatment, the difference in hardness for each piece measured must first be calculated. Then the standard deviation would be on those differences.

Example

You manufacture toy building blocks and are considering chemically treating each piece after molding to increase the hardness. The chemical supplier claims the hardness will increase by 5, but a difference of 3 would be sufficient for you to proceed with the new treatment. You want the test to have 80% power. Based on similar tests in the past, your engineers believe the standard deviation of the differences will be somewhere between 2 and 5.

In Minitab: Go to **Stat > Power and Sample Size > Paired t** and complete the dialog as shown in Figure 6.15. Then click on **Options** and complete the dialog as shown in Figure 6.16. Click **OK** in each dialog.

The standard deviation of 5 was used to provide the most conservative estimate of sample size needed based on the range of values provided by the engineers. The session window output is shown in Figure 6.17.

The output shows that if the true difference is 3, 19 samples are needed and would achieve 80.8% power. The graphical output is shown in Figure 6.18.

From the power curve, it can be seen that while a difference of 3 would be detected with just over 80% power, if the manufacturer's claim of 5 is true, then there is nearly a 100% chance of detection.

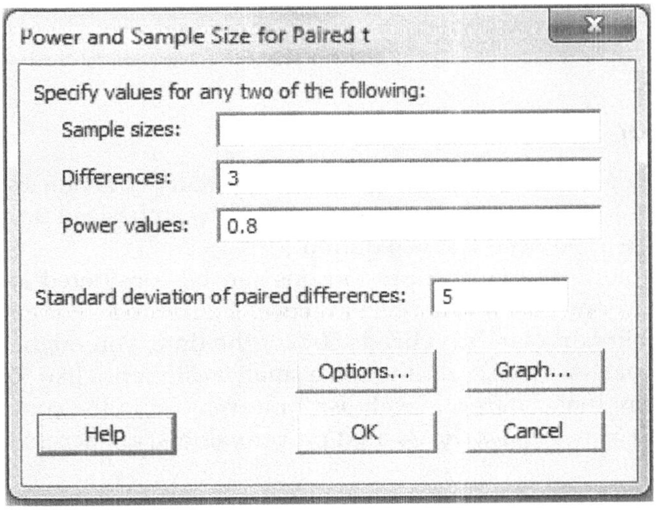

Figure 6.15 Minitab dialog box for sample size for paired t.

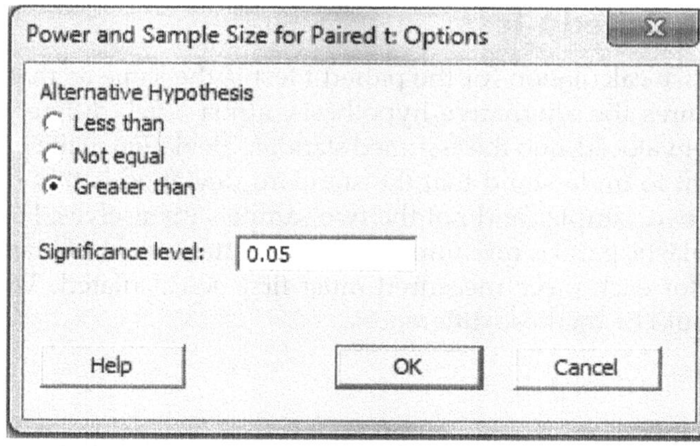

Figure 6.16 Subdialog box for sample size for paired t.

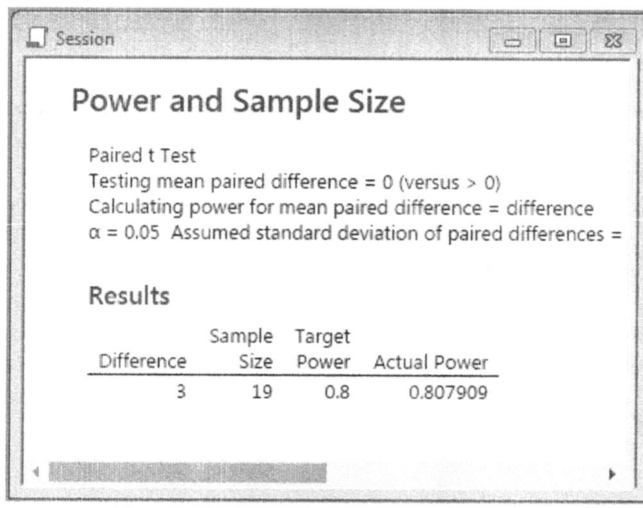

Figure 6.17 Session window output for sample size for paired t.

Sample Size for 1 Proportion Test

As with other sample size calculations, the alternative hypothesis and alpha level are necessary for the 1 proportion test. Then the hypothesized proportion that will be used in the hypothesis test is required.

One or more comparison proportions can be considered and these should be the proportions that would have practical significance. For example, if testing whether a coin fairly lands on heads 50% of the time, you might use comparison proportions of 48% and 52%. Although a smaller difference like 50.1% might show an unfair coin, that difference likely isn't meaningful in the context of most uses for flipping a coin. Finally, the desired power value(s) are necessary.

Example

A technical support call center wants to determine whether the proportion of calls that must be put on hold differs from the target value of 5%. The desired alpha level

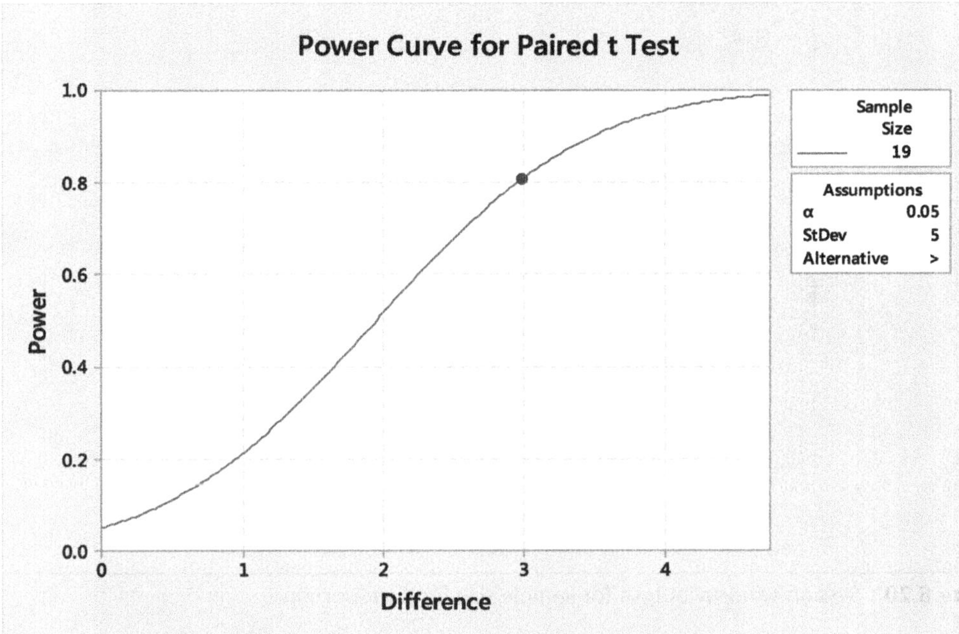

Figure 6.18 Power curve for sample size for paired t test.

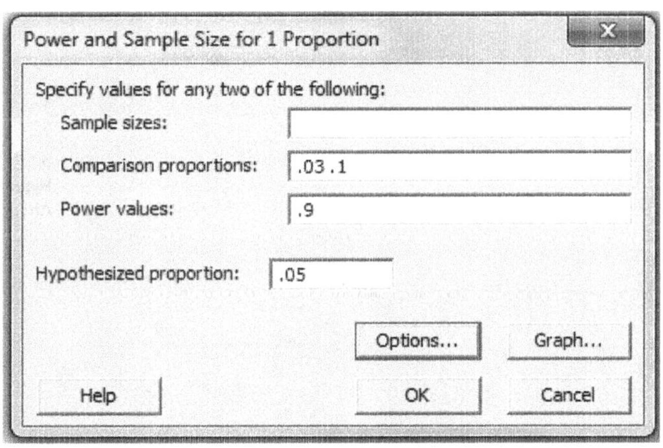

Figure 6.19 Minitab dialog box for sample size for 1 proportion test.

for the test is 0.05. A proportion of less than 0.03 indicates overstaffing, while one greater than 0.1 results in poor service ratings. You desire 90% power for the test.

In Minitab: Go to **Stat > Power and Sample Size > 1 Proportion**. Complete the dialog as shown in Figure 6.19 and click **OK**.

The session window output is shown in Figure 6.20.

The output demonstrates that if the true proportion was 0.03, a sample of 1043 calls would be required to obtain the desired power. If it was 0.1 a sample of 264 would be required. Because a single sample will be taken, the larger value of 1043 should be used. The graphical output is shown in Figure 6.21.

Session

Power and Sample Size

Test for One Proportion
Testing p = 0.05 (versus ≠ 0.05)
α = 0.05

Results

Comparison p	Sample Size	Target Power	Actual Power
0.03	1043	0.9	0.900133
0.10	264	0.9	0.900470

Power Curve for One Proportion

Figure 6.20 Session window output for sample size for 1 proportion.

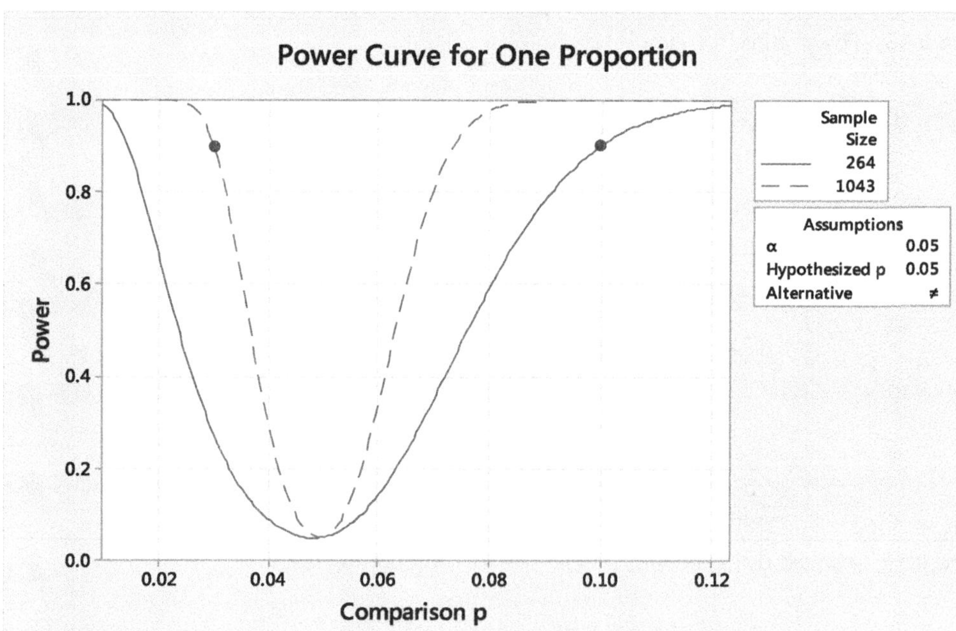

Figure 6.21 Power curve for sample size for 1 proportion test.

As can be seen from the two curves, the larger sample size of 1043 will obtain at least 90% power for both comparison proportions. If the smaller size of 264 were used, the power would be a little above 20% for detecting a change to 0.03 and not sufficient.

Sample Size for 2 Proportion Test

Sample size calculations for the 2 proportions test require the alternative hypothesis and alpha level to be specified. Then, one proportion is specified as a baseline proportion.

One or more comparison proportions must be specified as values that would have practical meaning relative to the baseline proportion. Finally, the desired power must be indicated.

Example

Two shipping companies are available to deliver product. Because you have a history of shipping with one of them, you estimate the true proportion of late shipments to be 11%. You wish to determine how many shipments you need to test with each to have 80% probability of detecting the other company to have either a 5% or 15% rate.

In Minitab: Go to **Stat > Power and Sample Size > 2 Proportions**. Complete the dialog as shown in Figure 6.22 and click **OK**.

The session window output is shown in Figure 6.23.

The output shows that to detect a difference of 5% and 11% with 80% power would require a sample size of 320 in each group. To detect a difference between 11% and 15% would require a much larger sample size of 1109 in each group. The graphical output is shown in Figure 6.24.

The power curve illustrates the needed sample sizes. The larger sample size indicated would result in the desired 80% power for the comparison proportion of 15% and nearly 100% power for the comparison proportion of 5%.

Sample Size for 1 Variance Test

Calculating sample size for the 1 variance test requires the alternative hypothesis and alpha level to be provided. The desired power value(s) must also be chosen and specified. Instead of specifying the hypothesized value and alternative value,

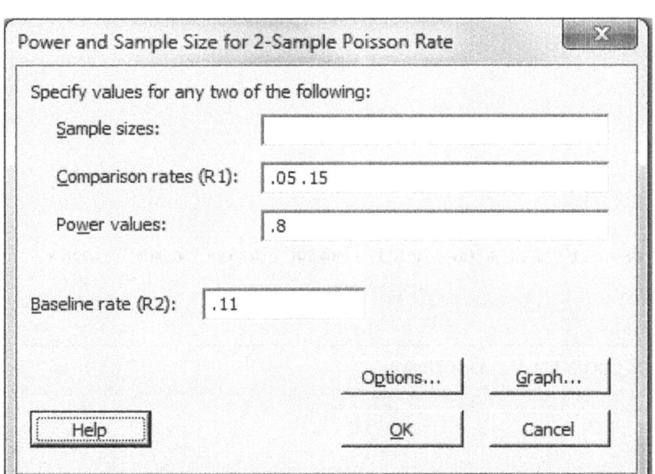

Figure 6.22 Minitab dialog box for sample size for 2 proportions.

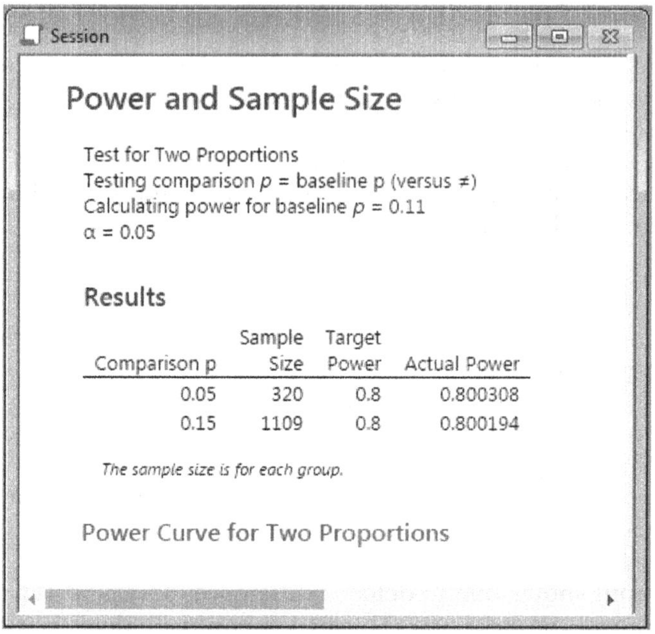
Session

Power and Sample Size

Test for Two Proportions
Testing comparison p = baseline p (versus ≠)
Calculating power for baseline p = 0.11
α = 0.05

Results

Comparison p	Sample Size	Target Power	Actual Power
0.05	320	0.8	0.800308
0.15	1109	0.8	0.800194

The sample size is for each group.

Power Curve for Two Proportions

Figure 6.23 Session window output for sample size for 2 proportions.

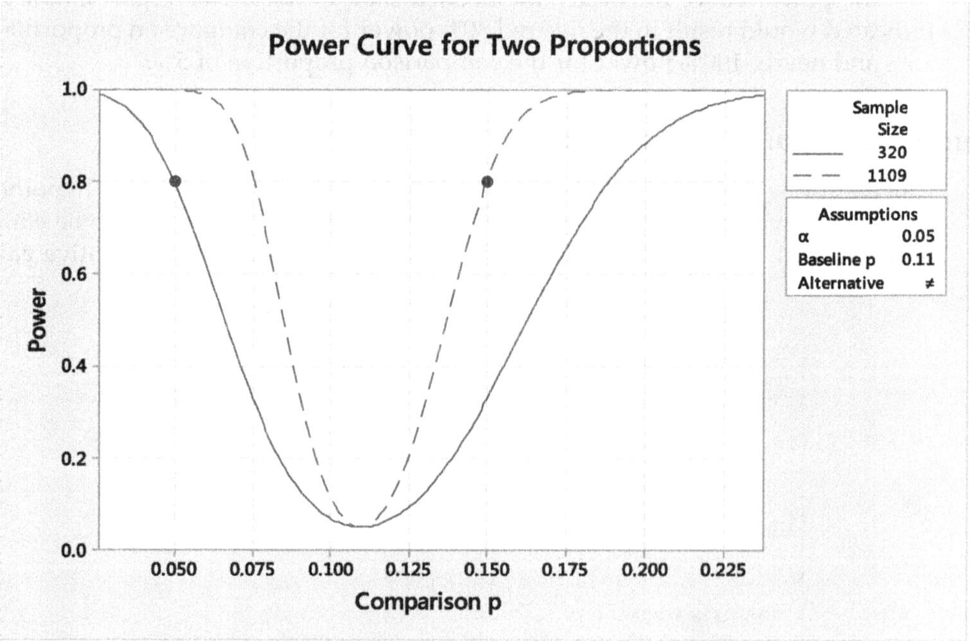

Figure 6.24 Power curve for 2 proportions.

only the ratio of the two must be specified. The calculation may use either the ratio of standard deviations, or variances. For example, you might specify a ratio of 2 to test whether the true standard deviation is at least twice the hypothesized value.

Example

You are considering a new supplier for purified water for a chemical process and suspect that although the total amount of residual material in the water is the same, the new supplier may provide a more consistent product. A reduction of 50% in the standard deviation of residual material would be sufficiently large to warrant a change in supplier. You desire a power value of 90% for the test.

In Minitab: Go to **Stat > Power and Sample Size > 1 Variance**. Complete the dialog as shown in Figure 6.25 and click **OK**.

The session window output is shown in Figure 6.26.

The session window output shows that a sample size of 14 units of water will be needed from the new supplier to achieve the desired power. The graphical output is shown in Figure 6.27.

The power curve illustrates that the calculated sample size has the desired power for a ratio of 0.5, but it has much lower power for ratios closer to 1. For example, even a slightly less extreme ratio of 0.6 only has about 60% power.

Sample Size for 2 Variance Test

The calculation of sample size for the two variance test requires the alternative hypothesis and desired alpha level. Additionally, the desired power value(s) are necessary. As with the 1 variance test, either standard deviations or variances may be used, and the ratio of standard deviations is used instead of raw values for each. The calculated sample size applies to each of the two groups.

Example

A hospital wishes to determine whether the standard deviation of wait times for emergency room admission is different for either of two different check-in

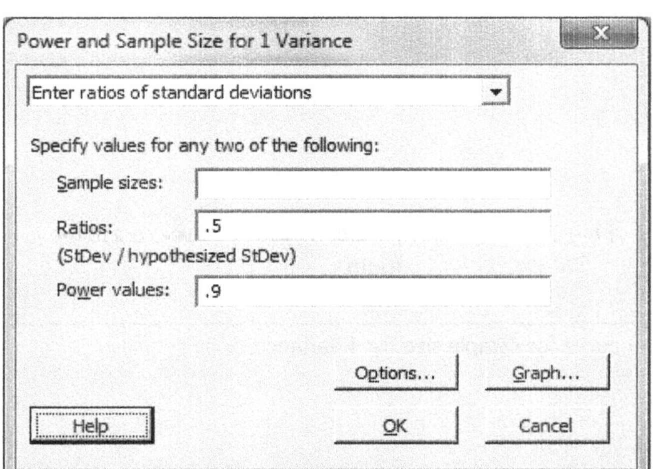

Figure 6.25 Minitab dialog box for sample size for 1 variance test.

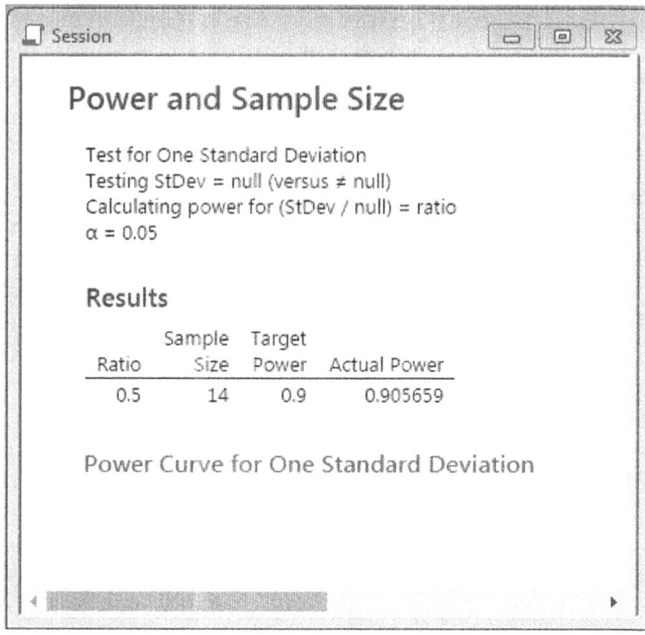

Figure 6.26 Session window output for sample size for 1 variance.

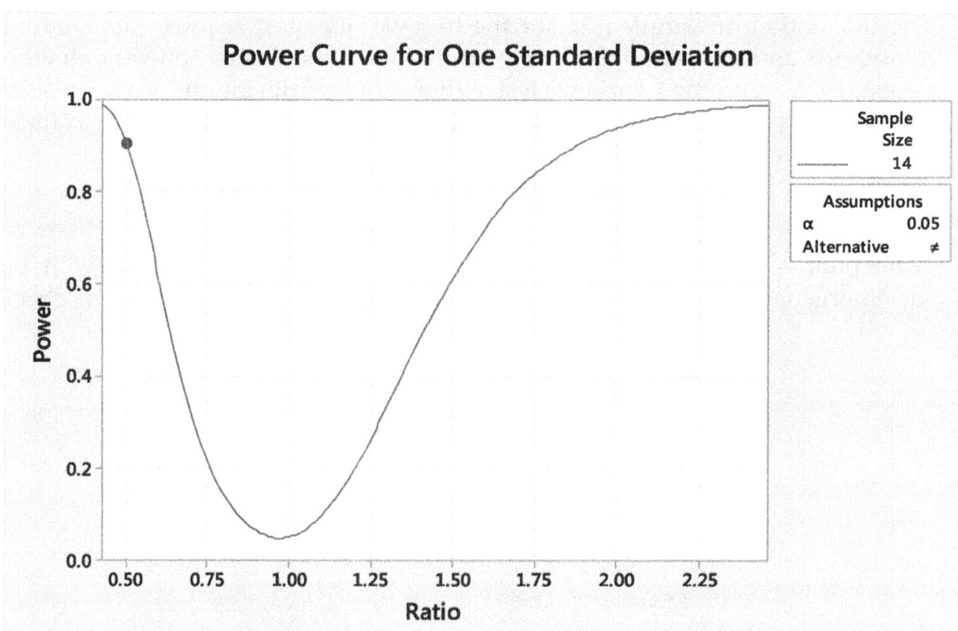

Figure 6.27 Power curve for sample size for 1 variance.

procedures. The first procedure involves patients checking themselves in at a kiosk and the second involves checking in at a desk where there is often a line or an attendant is busy. If either standard deviation is at least 25% larger than the other, the difference is noticeable by the patients. Power of at least 80% is needed, but if the sample size is reasonable, then power of 90% is more desirable.

First, the ratios of interest for the first standard deviation to the second must be calculated. If the second standard deviation is 25% more than the first, then the ratio is 1/1.25 = 0.8. If the first is greater than the second by 25%, the ratio is 1.25/1 = 1.25.

In Minitab: Go to **Stat > Power and Sample Size > 2 Variances**. Complete the dialog as shown in Figure 6.28 and click **OK**.

The session window output is shown in Figure 6.29.

The session window output shows that for the required power of 80%, 185 patients will be needed to utilize each of the two procedures. To achieve 90%, a modest increase to 247 patients for each procedure is necessary. The graphical output is shown in Figure 6.30.

The power curve illustrates the improved power of the larger sample size across most of the ratio values, including those indicated. It would be reasonable to consider taking the larger sample to gain higher power when performing the test.

CONFIDENCE INTERVALS

A sample statistic may provide the best estimate of the true population parameter, but does not provide information on how precise that estimate is. Because of this, it is often preferable to use a confidence interval, which provides a range of values and the likelihood that the true parameter falls within that range. The confidence interval tells us "the degree of uncertainty associated with a point estimate" (Montgomery and Runger 1999, 289). For example, a sample of ball bearings may have a mean weight of 12.34 grams (g), but a confidence interval on the mean may state that we are 95% confident that the true population mean of bearing weight is between 12.21 and 12.47 g.

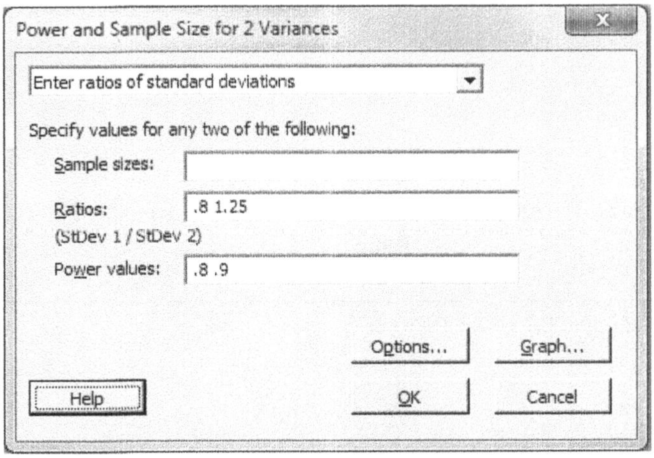

Figure 6.28 Minitab dialog box for sample size for two variances.

Session

Power and Sample Size

Test for Two Standard Deviations
Testing (StDev 1 / StDev 2) = 1 (versus ≠)
Calculating power for (StDev 1 / StDev 2) = ratio
α = 0.05
Method: Levene's Test

Results

Ratio	Sample Size	Target Power	Actual Power
0.80	185	0.8	0.800992
0.80	247	0.9	0.900420
1.25	185	0.8	0.800992
1.25	247	0.9	0.900420

The sample size is for each group.

Figure 6.29 Session window output for sample size for two variances.

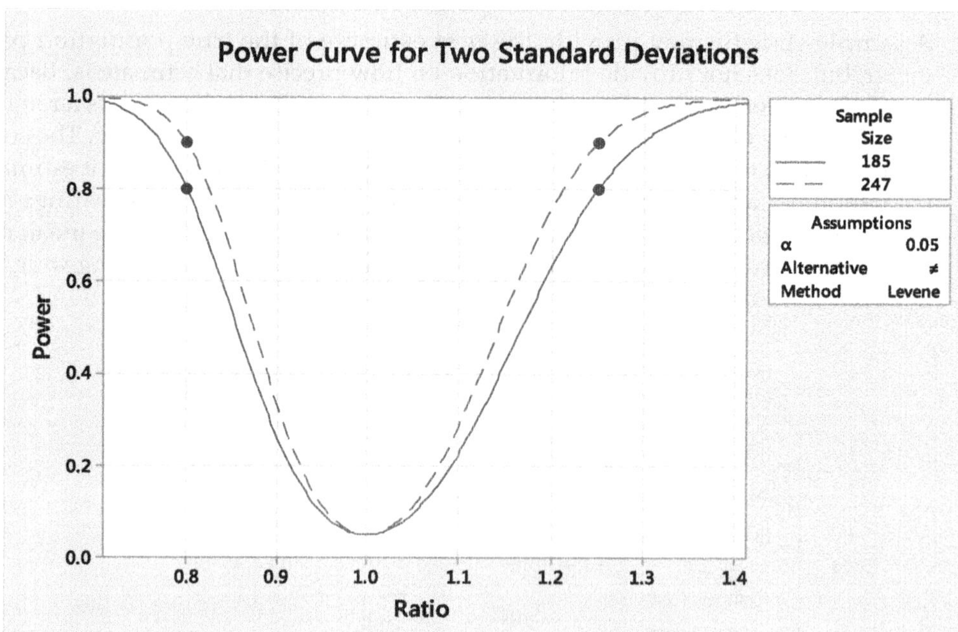

Figure 6.30 Power curve for two variances.

An interval plot illustrates confidence intervals for the mean of one or more samples of data and can be used to indicate whether a statistical test for a difference in means may be needed. When intervals heavily overlap, it is unlikely that there is a statistically significant difference in means. Clearly separated intervals indicate that there is likely a difference in true means.

Confidence Interval for the Population Mean with Known Standard Deviation

The confidence interval for the mean of a population with a known standard deviation can be estimated using the following formula:

$$(1-\alpha)100\%\ CI = \bar{x} \pm Z_{\alpha/2} * \frac{\alpha}{\sqrt{n}}$$

This is generally accurate for sample sizes of at least 30 or for smaller samples with data that is not highly skewed.

Example

A sample of 40 parts has a mean of 17.2 and a standard deviation of 0.7. The value of $Z_{\alpha/2} = -1.96$ can be found in Table 6.1, which depicts Table C.1 in Appendix C. The 95% confidence interval for the mean is:

$$\bar{x} \pm Z_{\alpha/2} * \frac{\sigma}{\sqrt{n}} = 17.2 \pm 1.96 * \frac{0.7}{\sqrt{40}} = (16.98, 17.42)$$

In Minitab: Go to **Stat > Basic Statistics > 1-Sample Z** and select **Summarized data**. Then enter the sample size, sample mean, and known standard deviation (see Figure 6.31) and click **OK**.

The resulting 95% confidence interval is (16.983, 17.417) (see Figure 6.32).

Confidence Interval for a Mean with Unknown Standard Deviation

The confidence interval for the population mean with unknown standard deviation uses the *t* distribution. The test statistic is:

$$(1-\alpha)\,100\%\ CI = \bar{x} \pm t_{\alpha/2,n-1} * \frac{s}{\sqrt{n}}$$

With degrees of freedom = n − 1, this formula is accurate for samples of at least 30 and for smaller samples with distributions that are not highly skewed.

Example

A sample of 14 is taken and found to have a mean of 52.5 and standard deviation of 0.6. What is the 95% confidence interval for the mean? The t value corresponding

Table 6.1 Lower-tail Z table for $\alpha/2 = .025$.

Z	0	0.01	0.02	0.03	0.04	0.05	0.06	0.07
0	0.5000	0.4960	0.4920	0.4880	0.4840	0.4801	0.4761	0.4721
–0.1	0.4602	0.4562	0.4522	0.4483	0.4443	0.4404	0.4364	0.4325
–0.2	0.4207	0.4168	0.4129	0.4090	0.4052	0.4013	0.3974	0.3936
–0.3	0.3821	0.3783	0.3745	0.3707	0.3669	0.3632	0.3594	0.3557
–0.4	0.3446	0.3409	0.3372	0.3336	0.3300	0.3264	0.3228	0.3192
–0.5	0.3085	0.3050	0.3015	0.2981	0.2946	0.2912	0.2877	0.2843
–0.6	0.2743	0.2709	0.2676	0.2643	0.2611	0.2578	0.2546	0.2514
–0.7	0.2420	0.2389	0.2358	0.2327	0.2296	0.2266	0.2236	0.2206
–0.8	0.2119	0.2090	0.2061	0.2033	0.2005	0.1977	0.1949	0.1922
–0.9	0.1841	0.1814	0.1788	0.1762	0.1736	0.1711	0.1685	0.1660
–1	0.1587	0.1562	0.1539	0.1515	0.1492	0.1469	0.1446	0.1423
–1.1	0.1357	0.1335	0.1314	0.1292	0.1271	0.1251	0.1230	0.1210
–1.2	0.1151	0.1131	0.1112	0.1093	0.1075	0.1056	0.1038	0.1020
–1.3	0.0968	0.0951	0.0934	0.0918	0.0901	0.0885	0.0869	0.0853
–1.4	0.0808	0.0793	0.0778	0.0764	0.0749	0.0735	0.0721	0.0708
1.5	0.0668	0.0655	0.0643	0.0630	0.0618	0.0606	0.0594	0.0582
–1.5	0.0548	0.0537	0.0526	0.0516	0.0505	0.0495	0.0485	0.0475
–1.7	0.0446	0.0436	0.0427	0.0418	0.0409	0.0401	0.0392	0.0384
–1.8	0.0359	0.0351	0.0344	0.0336	0.0329	0.0322	0.0314	0.0307
–1.9	0.0287	0.0281	0.0274	0.0268	0.0262	0.0256	0.0250	0.0244
–2.0	0.0228	0.0222	0.0217	0.0212	0.0207	0.0202	0.0197	0.0192

to a cumulative probability of $\alpha/2 = .025$ and 13 degrees of freedom can be found in in Table C.6 in Appendix C as –2.160. The confidence interval is then equal to:

$$\bar{x} \pm t_{\alpha/2n-1} * \frac{s}{\sqrt{n}} = 52.5 \pm 2.160 * \frac{0.6}{\sqrt{14}} = (52.154, 52.846)$$

In Minitab: Go to **Stat > Basic Statistics > 1-Sample t** and select **Summarized data**. Then enter the sample size, sample mean, and known standard deviation (see Figure 6.33) and click **OK**.

The resulting confidence interval is (52.154, 52.846) (see Figure 6.34).

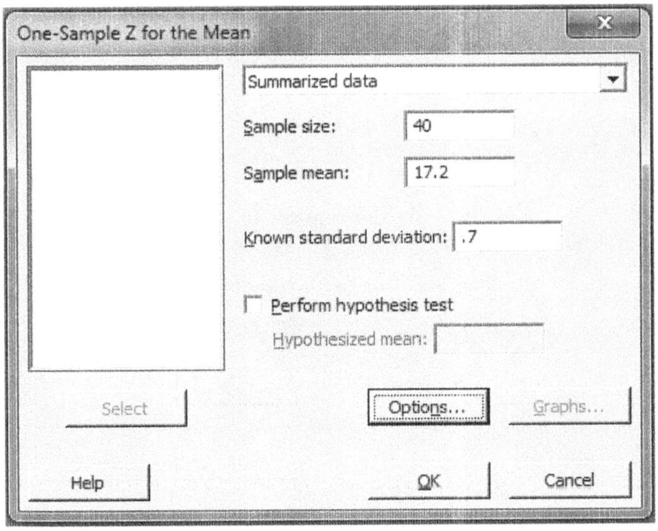

Figure 6.31 Dialog box for a confidence interval for the population mean with known standard deviation.

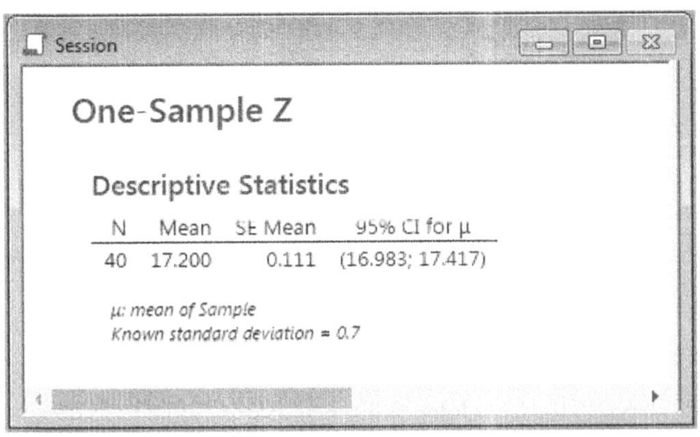

Figure 6.32 Session window for a confidence interval for the population mean with known standard deviation.

Confidence Interval for Proportions

The exact confidence interval for a proportion of x events in n opportunities can be determined using the following formulas:

$$\text{Lower limit } \frac{\nu_1 * F_{\alpha/2,\nu_1,\nu_2}}{\nu_2 + \nu_1 * F_{1-\alpha/2,\nu_1,\nu_2}}, \text{ where: } \nu_1 = 2x \text{ and } \nu_2 = 2(n - x + 1)$$

$$\text{Upper Limit } \frac{\nu_1 * F_{\alpha/2,\nu_1,\nu_2}}{\nu_2 + \nu_1 * F_{1-\alpha/2,\nu_1,\nu_2}}, \text{ where: } \nu_1 = 2(x + 1) \text{ and } \nu_2 = 2(n - x)$$

When the count of events and nonevents are both greater than or equal to 5, a normal approximation can be used:

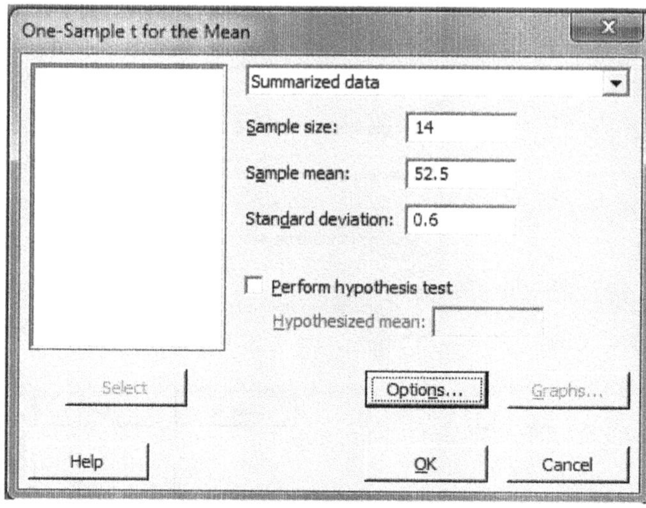

Figure 6.33 Dialog box for a confidence interval for a mean with unknown standard deviation.

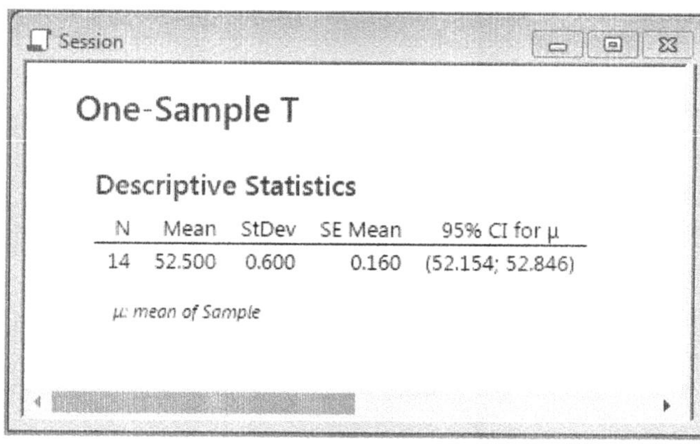

Figure 6.34 Session window for a confidence interval for the population mean with unknown standard deviation.

$$(1-\alpha)\ 100\%\ CI = \hat{p} \pm Z_{\alpha/2} * \sqrt{\frac{\hat{p}(1-\hat{p})}{n}}$$

Example

A quality inspector checks 720 parts and finds 68 are defective for a proportion of (68/720) = 0.094. Determine the 95% confidence interval for the proportion. Because the number of events (68) and nonevents (652) are large, the normal approximation can be used:

$$\hat{p} \pm Z_{\alpha/2} * \sqrt{\frac{\hat{p}(1-\hat{p})}{n}} = 0.094 \pm 1.96 * \sqrt{\frac{0.094(1-0.094)}{720}} = (0.073,\ 0.116)$$

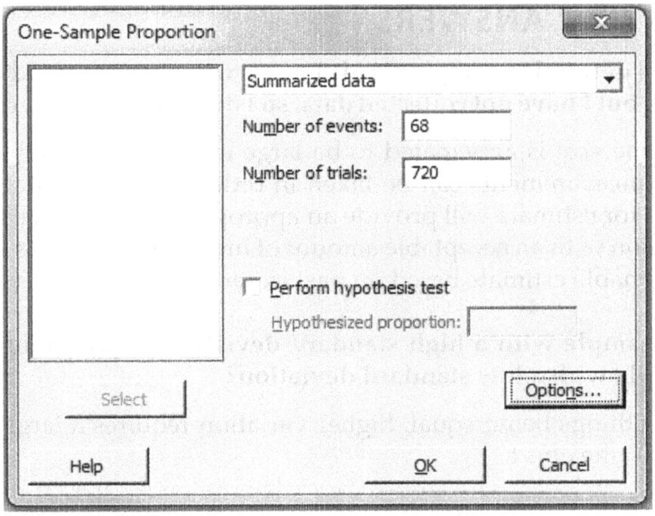

Figure 6.35 Dialog box for a confidence interval for proportions.

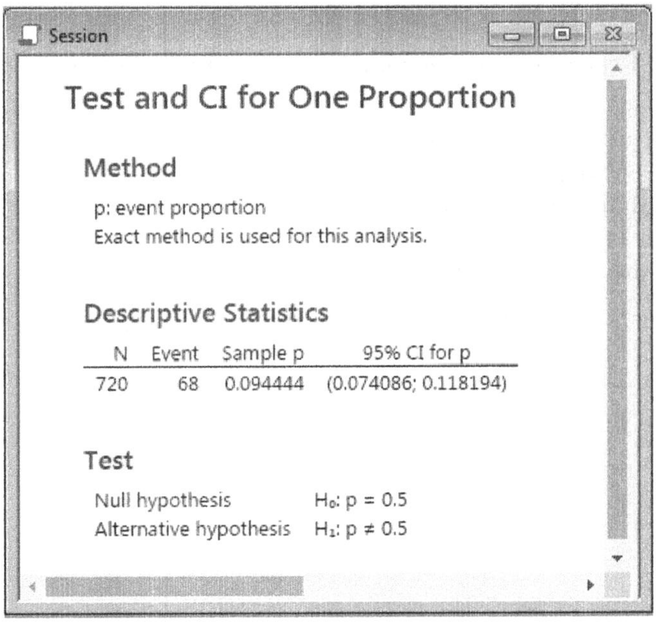

Figure 6.36 Session window for a confidence interval for proportions.

In Minitab: Go to **Stat > Basic Statistics > 1-Proportion** and select **Summarized data**. Then enter the number of events and number of trials (see Figure 6.35) and click **OK**. If desired, the method can be changed to the normal approximation in the **Options** subdialog.

The resulting 95% confidence interval is (0.074, 0.118), as shown in Figure 6.36.

QUESTIONS AND ANSWERS

Q: How do I determine sample size if I need to use the standard deviation in the calculation, but I have not collected data, so I don't know the standard deviation?

A: If the same size is anticipated to be large and it is relatively cheap to collect data, some measurements can be taken in order to estimate standard deviation. Sample size for estimate will provide an appropriate sample size to estimate standard deviation with an acceptable amount of uncertainty. Process experts may also have a reasonable estimate based on past experience.

Q: Does a sample with a high standard deviation require more or less values than a sample with a low standard deviation?

A: All other things being equal, higher variation requires a larger sample size to observe the same effect.

Q: What is the power function?

A: Power function describes the likelihood of rejecting the null hypothesis when it is in fact not true.

Q: How do I calculate sample size to get a representative sample for selecting the parts?

A: Obtaining a representative sample is dependent on the sampling scheme and not the sample size. However, more samples will allow for more precise estimation of population parameters.

Q: What is the margin of error for confidence intervals?

A: It represents the probability of the interval containing the true population value.

Chapter 7
Test of Normality and Distribution Transformations

TEST OF NORMALITY

Probability Plot for a Test of Normality

Using a probability plot it can be determined whether it is reasonable to assume that data follows the normal distribution. The probability plot graphs the data values (x) against the expected percentiles (y) with the y-axis spaced proportionately with the percentiles of the normal distribution (Montgomery, Runger, and Hubele 2001). This permits us to compare the position of actual values to where they would be expected to be if the data follow the normal distribution. The points plotted for the expected percentiles are calculated using the Median Rank method, which is defined as

$$\text{Plotted point} = \frac{i - 0.3}{n + 0.4},$$

where i is the rank of that value in the data set and n is the sample size. Data that fits the normal distribution will fall in a straight line. Suppose there is a sample of 6 items with the following values: 25.26, 26.53, 25.16, 25.87, 25.95, and 27.66. To determine the points to be plotted, a table can be created such as the one in Table 7.1. The rank order i is listed followed by a column containing the individual values. The y-values of i minus 0.3 divided by n plus 0.4 is then listed in the last column. The results are then plotted on a normal probability plot as shown in Figure 7.1.

In Minitab: Go to **Graph > Probability Plot** and a gallery will appear with an option to select **Single** or **Multiple** probability plots (Figure 7.2). The **Multiple** probability plot option permits the evaluation of multiple sets of data in one graphical output. Select **Single** if there is only one column of data and then select the column for the graph variable as shown in Figure 7.3. Click **OK**.

The resulting probability plot shown in Figure 7.4 graphically depicts the data with a fitted line through the plotted points and 95% prediction limits. Additionally, a legend provides the mean, standard deviation and sample size of the data as well as an Anderson-Darling statistic and p-value for a statistical test of normality. Unlike statistics such as t and F, the Anderson-Darling statistic itself has no useful interpretation; it is simply used to calculate the p-value as described below. P-values less than .05 are interpreted as indicating that the data are not reasonably fit by the normal distribution, and larger p-values are interpreted as indicating there is not enough evidence to determine the data are not fit well.

Table 7.1 Probability plot calculations.

i	X_j	$Y_i = \frac{i-0.3}{6+0.4}$
1	25.19	0.11
2	25.26	0.27
3	25.87	0.42
4	25.95	0.58
5	26.53	0.73
6	27.66	0.89

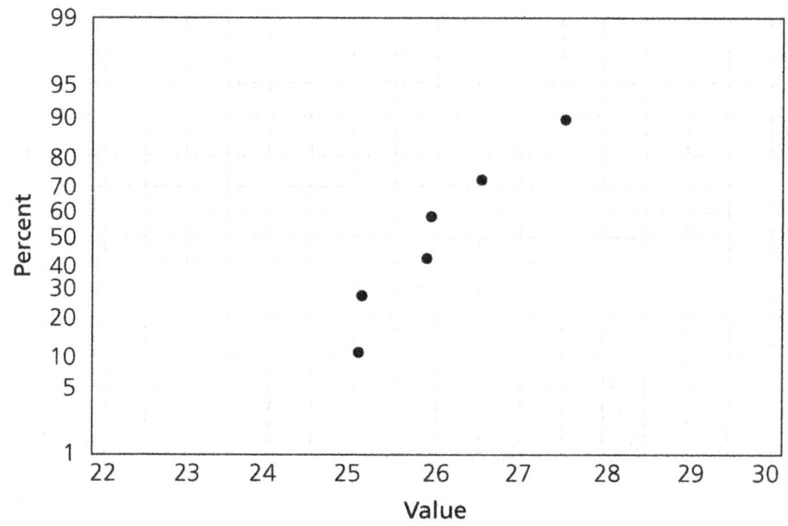

Figure 7.1 Probability plot.

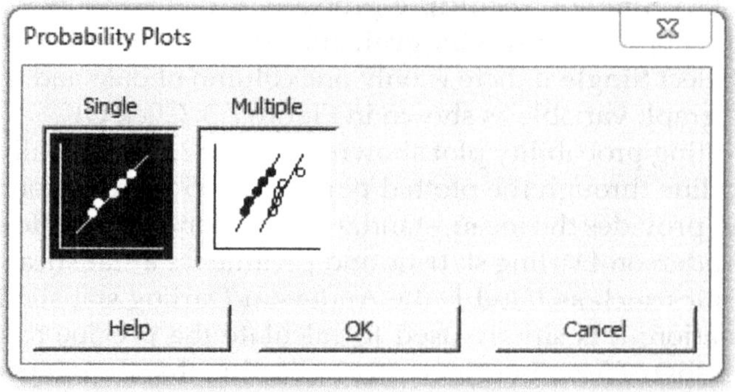

Figure 7.2 Probability plot menu.

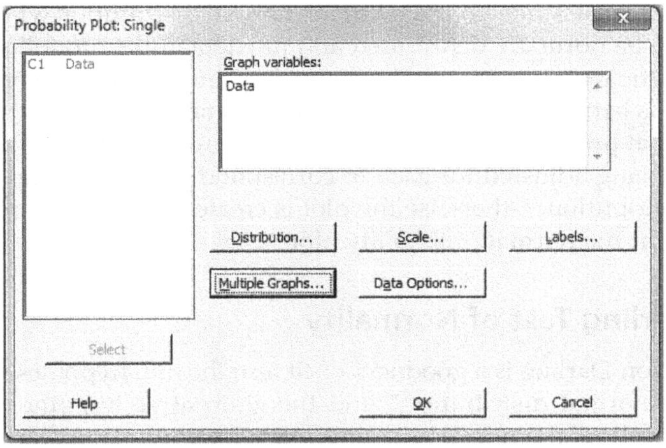

Figure 7.3 Minitab dialog box for a probability plot

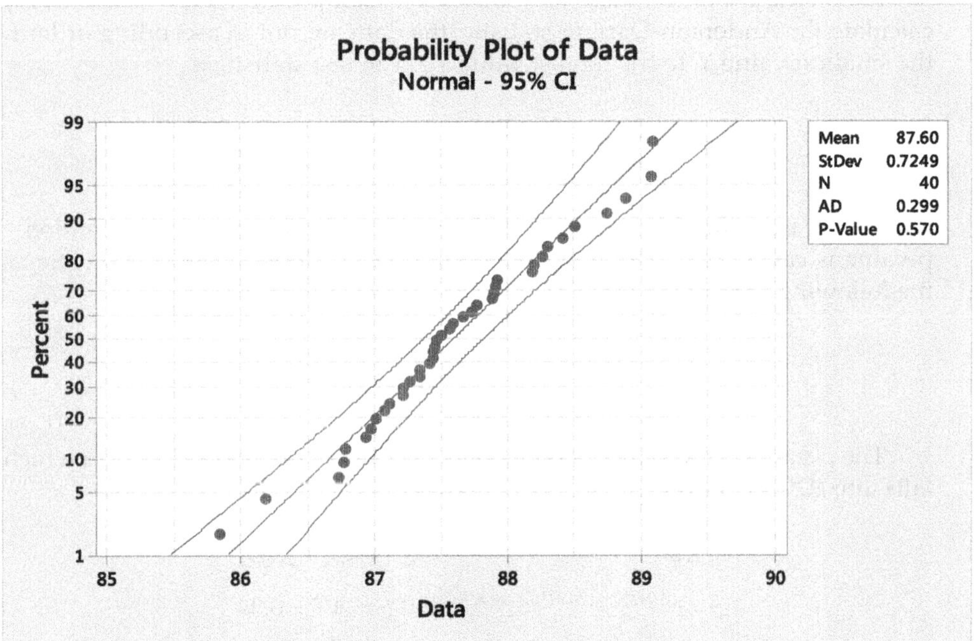

Figure 7.4 Minitab probability plot with normally distributed data.

Using the p-value as a formal statistical test of whether the data follow the normal distribution is generally not advisable. With small samples, the test is not powerful enough to detect the data are non-normal unless the difference is very severe. With large samples, the test becomes too powerful, and even tiny departures for normality that have no practical impact on further analysis are detected with small p-values. A better check for normality would be to consider whether it makes sense that the data would follow the normal distribution. Also look at the normal probability plot and ask, "Do the points reasonably fall along a straight line on the normal probability plot with no obvious nonlinear pattern such as a curve?" You can assume your data fit the normal distribution when the data can be expected to be normally distributed and the data appear to follow a straight line. A test statistic can supplement the visual interpretation of the normal probability plot, but it is better to trust your visual interpretation.

Note that probability plots can be created for other distributions as well, with the y-axis being adjusted for each to correspond to the spacing of the percentiles for that distribution. Otherwise the plot is created in the same manner and interpreted just as the normal probability plot.

Anderson-Darling Test of Normality

The Anderson-Darling is a goodness-of-fit test; the null hypothesis (H_0) is "the data follow the normal distribution" and the alternative hypothesis is (H_a) is "the data don't follow the normal distribution." The Anderson-Darling test will provide a value for assessing the normality of data; however, it is still advisable to create a normal probability plot to see whether the data truly appear to follow the normal distribution. It is also not robust when sample size is low (Kahn 2013). To calculate the Anderson-Darling statistic, the data are put in ascending order from the smallest value Y_i to the largest value Y_N. The test statistic is

$$A^2 = -N - \Sigma_{i=1}^{N} \frac{(2i-1)}{N} \left[\ln F(Y_i) + \ln (1 - F(Y_{N+1-i})) \right],$$

where F is the cumulative distribution function of the normal distribution. The p-value is calculated by first standardizing the Anderson-Darling statistic using the following formula:

$$A'^2 = A^2 \left(1 + \frac{0.75}{N} + \frac{2.25}{N^2} \right)$$

The p-value is then calculated using a formula based on the range in which A'^2 falls into (D'Agostino and Stephens 1986):

$$p = e^{1.2937-5.709A'^2 + 0.0186(A'^2)^2} \text{ if } 13 > A'^2 > 0.6$$

$$p = e^{0.9177-4.279A'^2 - 1.38(A'^2)^2} \text{ if } 0.6 > A'^2 > 0.34$$

$$p = 1 - e^{-8.318 + 42.796A'^2 - 59.938(A'^2)^2} \text{ if } .34 > A'^2 > 0.2$$

$$p = 1 - e^{-13.436 + 101.14A'^2 - 223.73(A'^2)^2} \text{ if } 0.2 > A'^2$$

The null hypothesis is rejected when the resulting p-value is less than the significance level, generally 0.05. Lower Anderson-Darling values correspond to a better fit to a given distribution than a higher Anderson-Darling value.

In Minitab: An Anderson-Darling test of normality is performed with the same function as a probability plot. Go to **Stat > Graph > Probability Plot**, as shown in Figures 7.5 and 7.6, and use the p-value to interpret the results; then click **OK**. The resulting p-value of <0.005 is less than the alpha of 0.05, so we can conclude the data do not fit the distribution (see Figure 7.7).

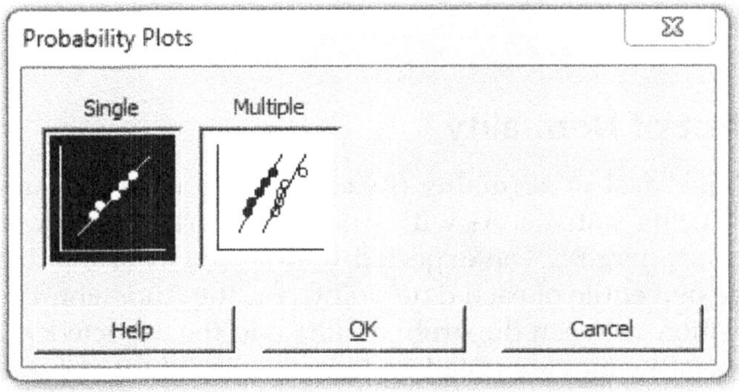

Figure 7.5 Probability plot menu.

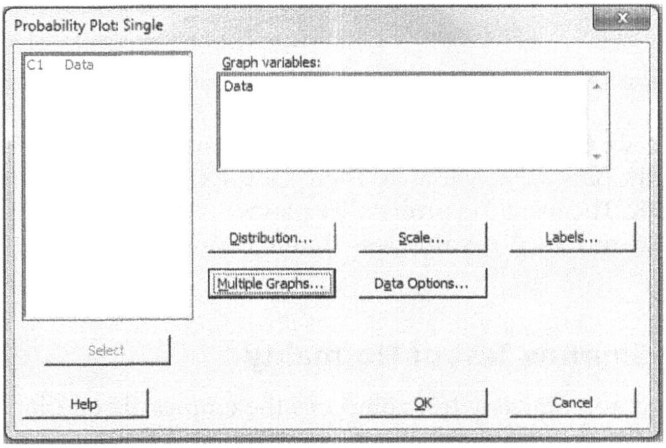

Figure 7.6 Minitab dialog box for a probability plot.

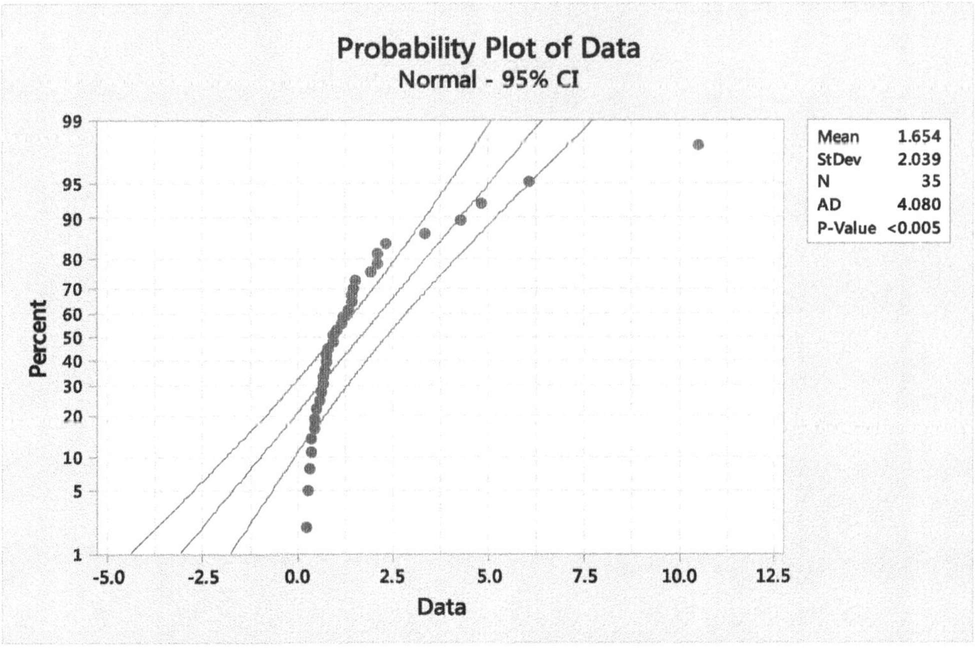

Figure 7.7 Minitab normal probability plot with non-normally distributed data.

Ryan-Joiner Test of Normality

The Ryan-Joiner test of normality (Ryan and Joiner 1976) is an alternative to the Anderson-Darling statistic. As with Anderson-Darling, the data are first ordered from smallest to largest. The expected normal scores of the data are then found based on the percentile of each data point, and the Anderson-Darling test statistic is the correlation between the original data and the expected values based on the normal scores. P-values are obtained from a series of fitted equations provided in Ryan and Joiner (1976). The Ryan-Joiner test of normality is a complicated calculation, so it is best performed using a statistical software program. The formula is

$$R_p = \frac{\Sigma Y_i b_i}{\sqrt{s^2 (n-1) \Sigma b_i^2}},$$

where Y_i represents ordered observations and b_i is the normal score of the ordered data.

In Minitab: Go to **Stat > Basic Statistics > Normality test** and enter the column containing the data as the variable. Then select **Ryan-Joiner** (as shown in Figure 7.8) and click **OK**. The resulting probability plot will have a p-value for the Ryan-Joiner test (see Figure 7.9). A p-value less than 0.05 indicates the data is not normally distributed.

Kolmogorov-Smirnov Test of Normality

The Kolmogorov-Smirnov test compares the empirical cumulative distribution of the data to the distribution that can be expected if the data are normally distributed (Daniel 1990). For symmetric distributions like normal, it is a more powerful test than Anderson-Darling (Yap and Sim 2011). The test statistic is the larger of the two values:

$$D^{+} = \max_i \left(\frac{i}{n - F(X_i)} \right)$$

$$D^{-} = \max_i \left(F(X_i) - \frac{i-1}{n} \right)$$

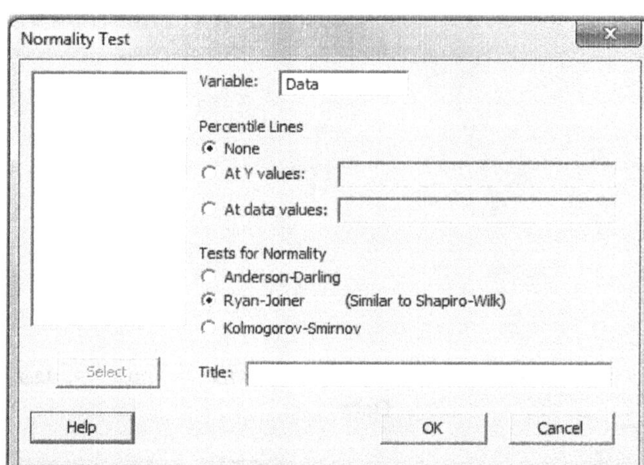

Figure 7.8 Minitab dialog box for a Ryan-Joiner test.

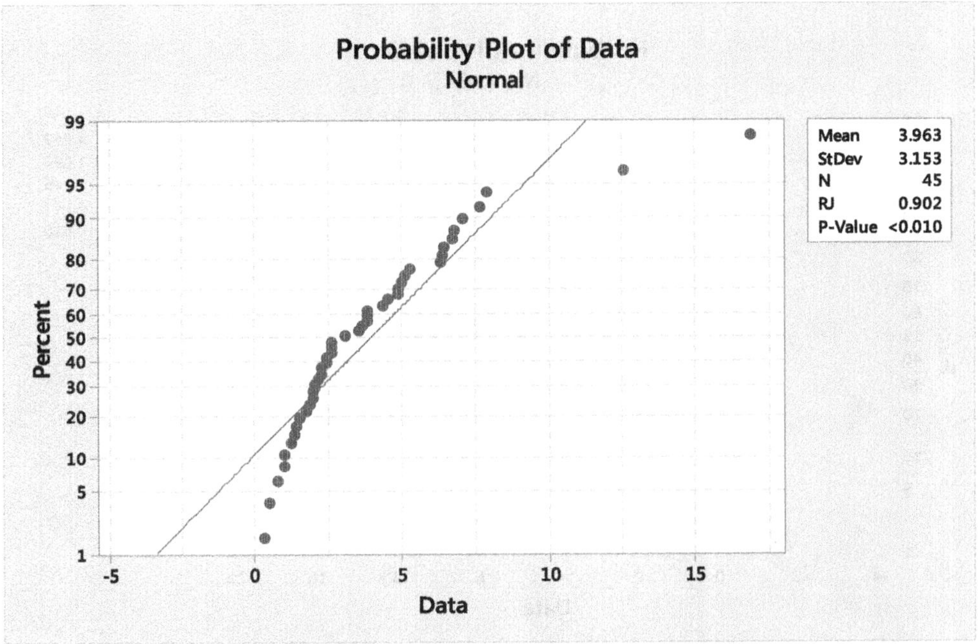

Figure 7.9 Minitab probability plot for a Ryan-Joiner test with non-normally distributed data.

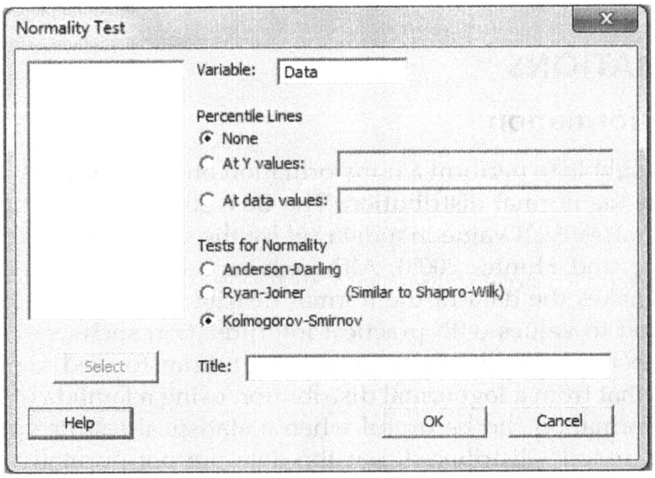

Figure 7.10 Minitab dialog box for a Kolmogorov-Smirnov test.

X_i is observation i when the data are put in order, and $F(X_i)$ is the cumulative distribution function of the fitted normal distribution at observation i.

In Minitab: Go to **Stat > Basic Statistics > Normality Test** and enter the column containing the data as the variable as shown in Figure 7.10. Then select **Kolmogorov-Smirnov** and click **OK**. The resulting probability plot will have a p-value for the Kolmogorov-Smirnov test as shown in Figure 7.11. A p-value less than 0.05 indicates the data is not normally distributed.

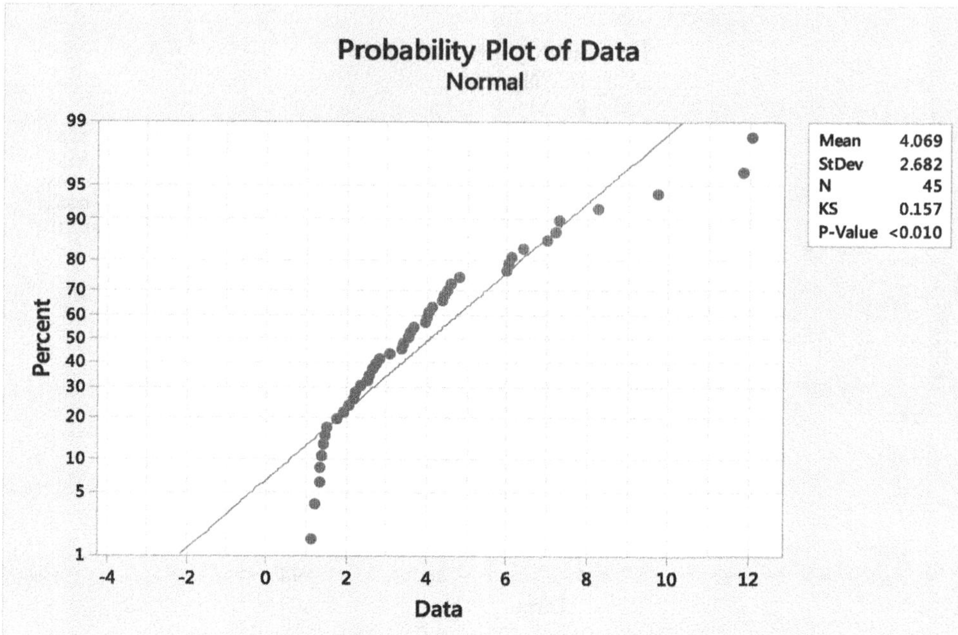

Figure 7.11 Minitab probability plot for a Kolmogorov-Smirnov test with non-normally distributed data.

TRANSFORMATIONS

Box-Cox Transformation

It may be desirable to perform a transformation on non-normally distributed data so that it fits the normal distribution. The Box-Cox transformation is often used and simply raises each value in a data set by the same power, known as lambda (Box, Hunter, and Hunter 2005). Although a precise "optimal" lambda can be found that makes the data fit the normal distribution most closely, this value is often rounded to values with practical interpretation such as –2, –1, –0.5, 0, 0.5, and 2. One example of data that can be directly transformed to normally distributed data is that from a lognormal distribution using a lambda of 0.

A transformation can be useful when a statistical test or operation requires data to be normally distributed, but the data are not normally distributed and do not fit any distribution such as heavily skewed data (Breyfogle 2003). However, Hoerl and Snee warn that "transformations add complexity to our analysis; therefore, we generally do not apply them unless there is a particular reason to do so" (2012, 407). Additionally, before simply applying a transformation it is critical to consider why the data are non-normal. Data on cycle times, offsets, and many other common measurements are expected to be skewed and therefore non-normal. However, many of these are well-fit by other distributions, and when possible, it is better to utilize the data in their natural distribution rather than transforming the data.

Other than considering whether the data are likely to be skewed and follow some other natural distribution, the data should be checked for stability before applying a transformation. Stability can be checked using a control chart such as an I-MR chart for data without subgroups or an Xbar-R or Xbar-S chart for data with subgroups. Control charts are discussed in Chapter 13. A lack of stability in data that are naturally normally distributed can cause the data to be poorly fit by the normal distribution. In this case, transforming the data may improve the fit to the normal distribution, but the data cannot be used to make inference about the process or future performance. Therefore it is critical that the process be made stable before performing further analysis regardless of whether transformation causes the data to appear normal.

To determine the optimal lambda, first the geometric mean of the data is calculated as G. Then the value of λ is found that minimizes the within-subgroup standard deviation of the variable W_i, which is calculated as

$$W_i = \frac{Y_i^{\lambda} - 1}{\lambda G^{\lambda-1}} \text{ when } \lambda \neq 0$$

$$W_i = G * \ln(Y_i) \text{ when } \lambda = 0,$$

where Y_i is the ith data point. Typically, values of λ ranging from −5 to 5 are searched to find the value that minimizes the within-subgroup standard deviation of W_i, which is defined as

$$S_{\text{Within}} = \frac{\sqrt{\dfrac{\Sigma\Sigma(x_{ij} - \bar{x}_i)^2}{\Sigma(n_i - 1)}}}{c_4(d+1)} \text{ when subgroup size} > 1,$$

where x_{ij} = the jth observation from the ith subgroup, x_i = the mean of the ith subgroup, n_i = the size of the ith subgroup $d = \Sigma(n_i - 1)$, and c_4 = an unbiasing constant. or defined as

$$S_{\text{Within}} = \frac{\dfrac{\Sigma R_i}{n-1}}{d_2(2)} \text{ when subgroup size} = 1,$$

where R_i = the range of the ith pair of consecutive points and d_2 = an unbiasing constant.

In Minitab: Go to **Stats > Control Charts > Box-Cox Transformation** and select the column containing the data. A subgroup size must also be entered; however, if the data lack natural subgroups a subgroup size of 1 may be used (Figure 7.12). Select the **Options** button and enter a column to store the transformed data (as shown in Figure 7.13) and then click **OK**.

The transformed values will then appear in the indicated column in the Minitab worksheet and Minitab will create a Box-Cox plot of the data as shown in Figure 7.14.

Although an optimal lambda value of −0.09 was estimated from the data, a confidence interval on the true optimal value includes the "natural" value of 0, and by default Minitab uses this rounded value to perform the transformation.

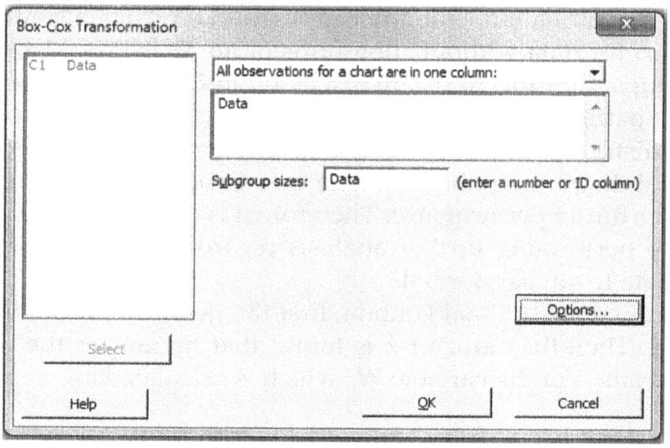

Figure 7.12 Minitab dialog box for a Box-Cox transformation.

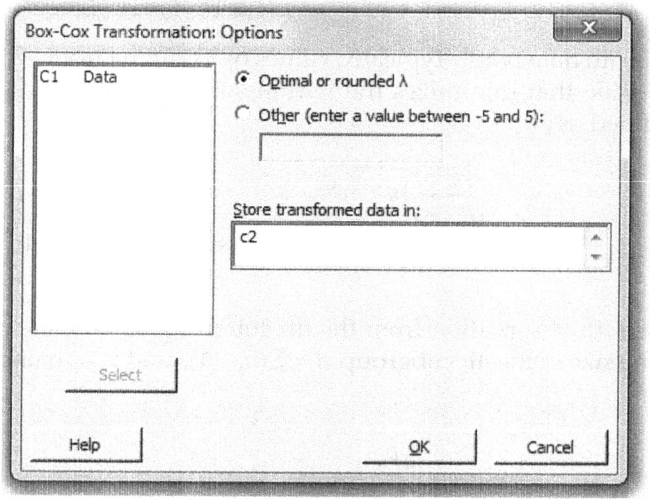

Figure 7.13 Options for a Box-Cox transformation.

Johnson Transformation

The Johnson transformation transforms non-normally distributed data so that it fits the normal distribution as closely as possible using a family of equations (Yeo and Johnson 2000). This may be necessary when the data do not follow a common distribution but tools requiring a known distribution will be used. As with the Box-Cox transformation, it is important to understand the reason the data are not normally distributed before simply transforming the data. Whenever possible, the data should be checked for stability and if stable for a natural distribution that is logical for the type of data. Processes that are not stable should be made stable before further analysis is done on data from those processes.

Unlike the Box-Cox transformation, the Johnson transformation does not take subgroups into account when finding the optimal transformation. However, it

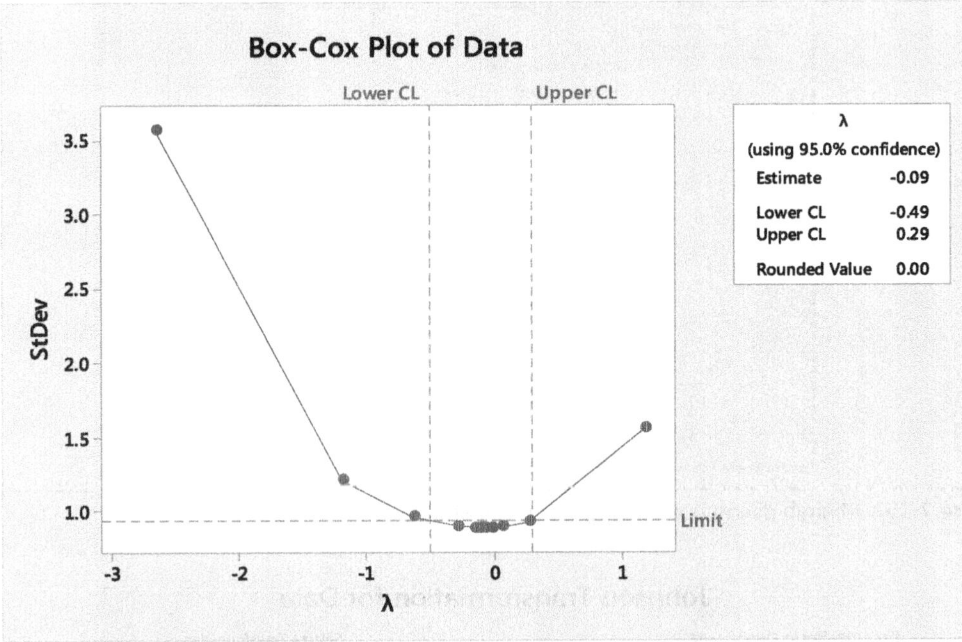

Figure 7.14 Box-Cox plot of the standard deviation of W_i for various lambda values.

often successfully finds a reasonable transformation of the data when the simpler Box-Cox transformation is unable to. However, a distribution that is unexpectedly non-normal and not transformed by Box-Cox is often a sign that the process is not stable and should be examined before proceeding with statistical tools. Even if a transformation such as the Johnson transformation is able to make the data appear normally distributed, the underlying process is still unstable in this case and the true problem to be solved is the lack of stability, not the lack of normality. A prediction based on data from an unstable process is very risky, as the next sample taken from the unstable process might not resemble the previous sample that was used for prediction purposes.

The Johnson transformation uses three different families of functions to attempt to transform the data. The three families are:

$$\text{Family 1: } S_{\text{Bounded}} : \gamma + \eta \ln\left(\frac{x-\varepsilon}{\lambda+\varepsilon-x}\right)$$

$$\text{Family 2: } S_{\text{Lognormal}} : \gamma + \eta \ln(x-\varepsilon)$$

$$\text{Family 3: } S_{\text{Unbounded}} : \gamma + \eta \ln\left(\frac{x-\varepsilon}{\lambda} + \sqrt{1+\left(\frac{x-\varepsilon}{\lambda}\right)^2}\right)$$

An algorithm finds the parameters of the family that results in the highest p-value for the Anderson-Darling test for normality.

In Minitab: **Go to Stat > Quality Tools > Johnson Transformation** and enter the column containing the data. Then select a column to contain the transformed data as shown in Figure 7.15 and click **OK.**

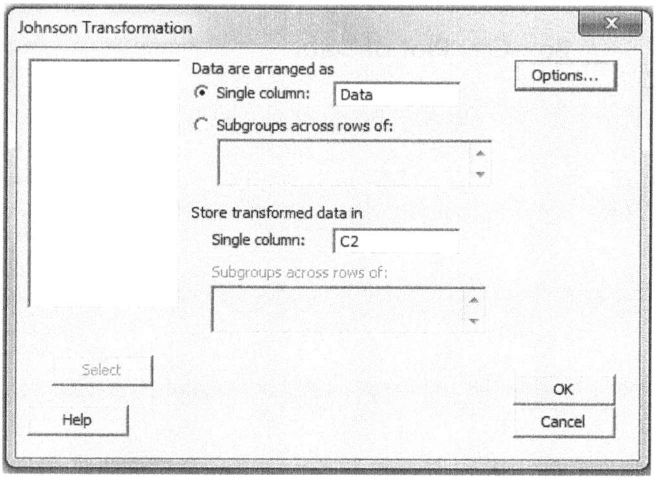

Figure 7.15 Minitab dialog box for a Johnson transformation.

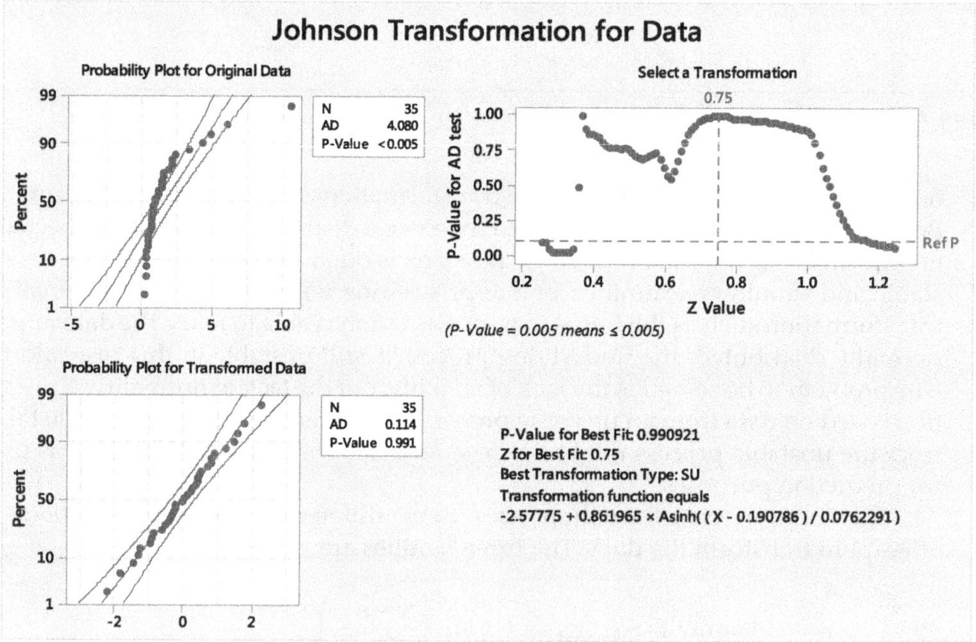

Figure 7.16 Johnson transformation results.

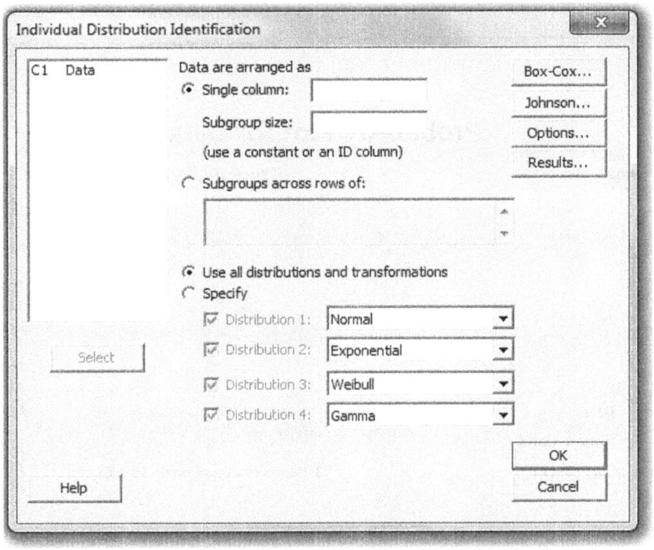

Figure 7.17 Minitab dialog box of an individual distribution identification.

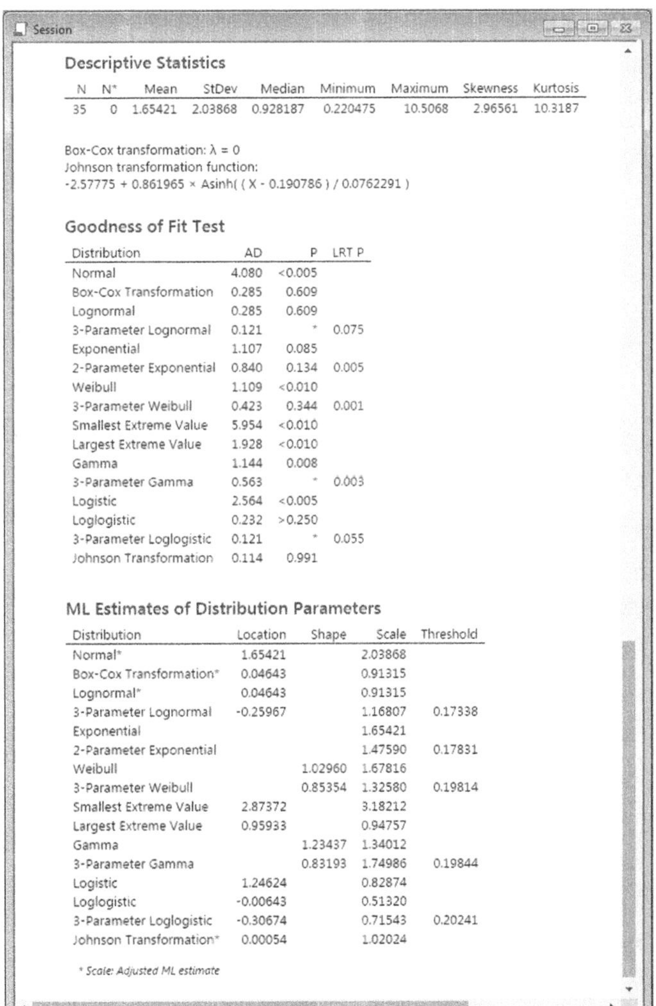
Session

Descriptive Statistics

N	N*	Mean	StDev	Median	Minimum	Maximum	Skewness	Kurtosis
35	0	1.65421	2.03868	0.928187	0.220475	10.5068	2.96561	10.3187

Box-Cox transformation: λ = 0
Johnson transformation function:
-2.57775 + 0.861965 × Asinh((X - 0.190786) / 0.0762291)

Goodness of Fit Test

Distribution	AD	P	LRT P
Normal	4.080	<0.005	
Box-Cox Transformation	0.285	0.609	
Lognormal	0.285	0.609	
3-Parameter Lognormal	0.121	*	0.075
Exponential	1.107	0.085	
2-Parameter Exponential	0.840	0.134	0.005
Weibull	1.109	<0.010	
3-Parameter Weibull	0.423	0.344	0.001
Smallest Extreme Value	5.954	<0.010	
Largest Extreme Value	1.928	<0.010	
Gamma	1.144	0.008	
3-Parameter Gamma	0.563	*	0.003
Logistic	2.564	<0.005	
Loglogistic	0.232	>0.250	
3-Parameter Loglogistic	0.121	*	0.055
Johnson Transformation	0.114	0.991	

ML Estimates of Distribution Parameters

Distribution	Location	Shape	Scale	Threshold
Normal*	1.65421		2.03868	
Box-Cox Transformation*	0.04643		0.91315	
Lognormal*	0.04643		0.91315	
3-Parameter Lognormal	-0.25967		1.16807	0.17338
Exponential			1.65421	
2-Parameter Exponential			1.47590	0.17831
Weibull		1.02960	1.67816	
3-Parameter Weibull		0.85354	1.32580	0.19814
Smallest Extreme Value	2.87372		3.18212	
Largest Extreme Value	0.95933		0.94757	
Gamma		1.23437	1.34012	
3-Parameter Gamma		0.83193	1.74986	0.19844
Logistic	1.24624		0.82874	
Loglogistic	-0.00643		0.51320	
3-Parameter Loglogistic	-0.30674		0.71543	0.20241
Johnson Transformation*	0.00054		1.02024	

* Scale: Adjusted ML estimate

Figure 7.18 Minitab session window for individual distribution identification.

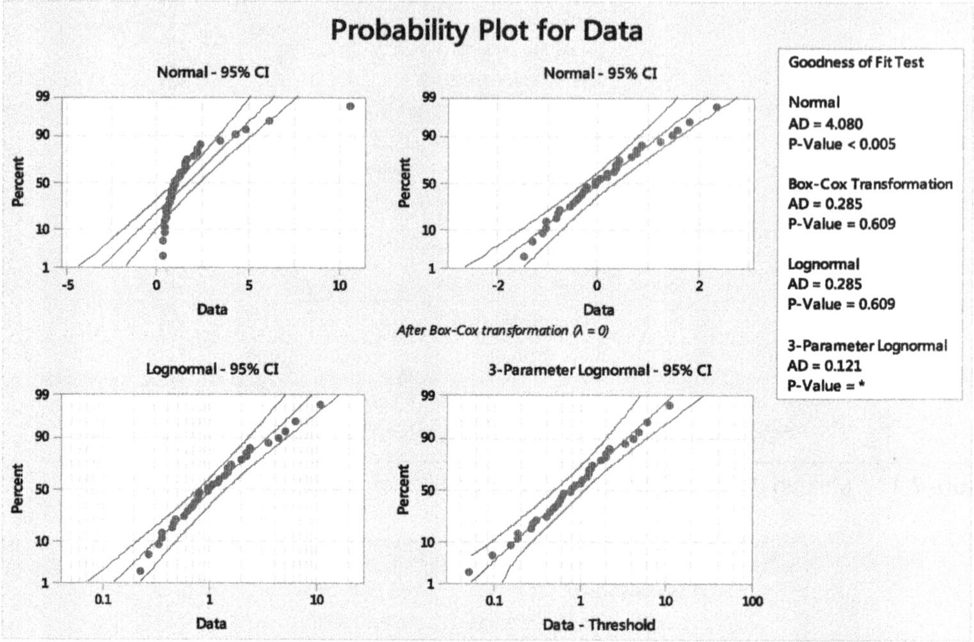

Figure 7.19 Minitab probability plots for normal, normal after a Box-Cox transformation, lognormal, and 3-parameter lognormal distributions.

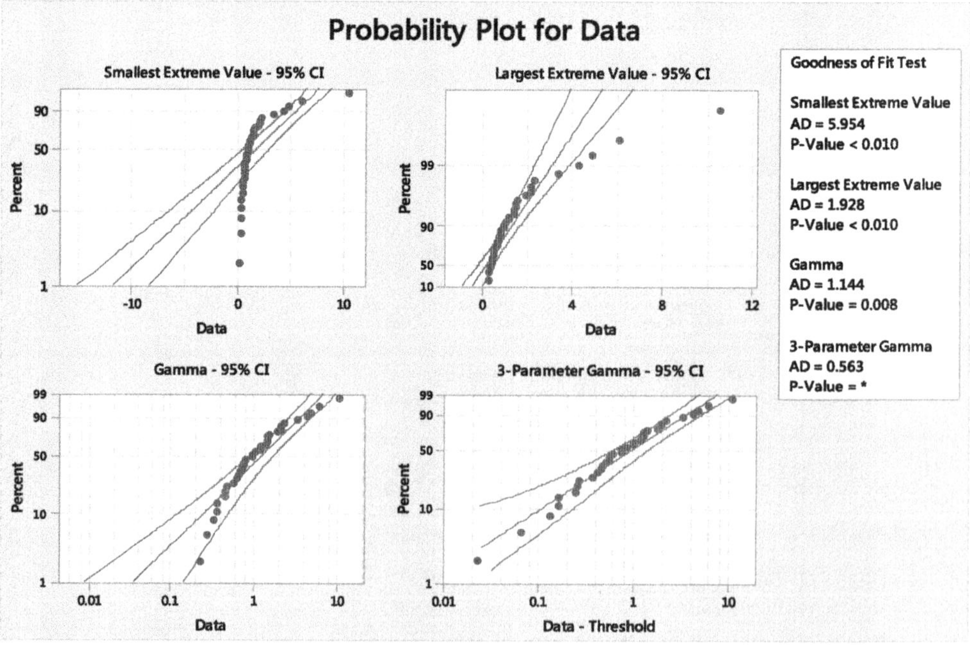

Figure 7.20 Minitab probability plots for smallest extreme value, largest extreme value, gamma, and 3-parameter gamma distributions.

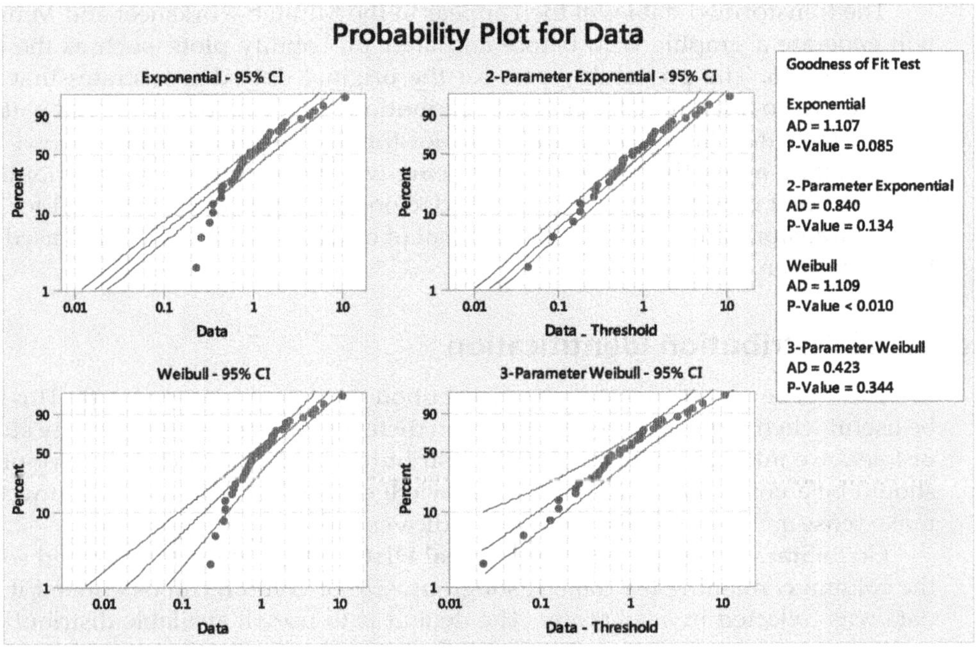

Figure 7.21 Minitab probability plots for exponential, 2-parameter exponential, Weibull, and 3-parameter Weibull distributions.

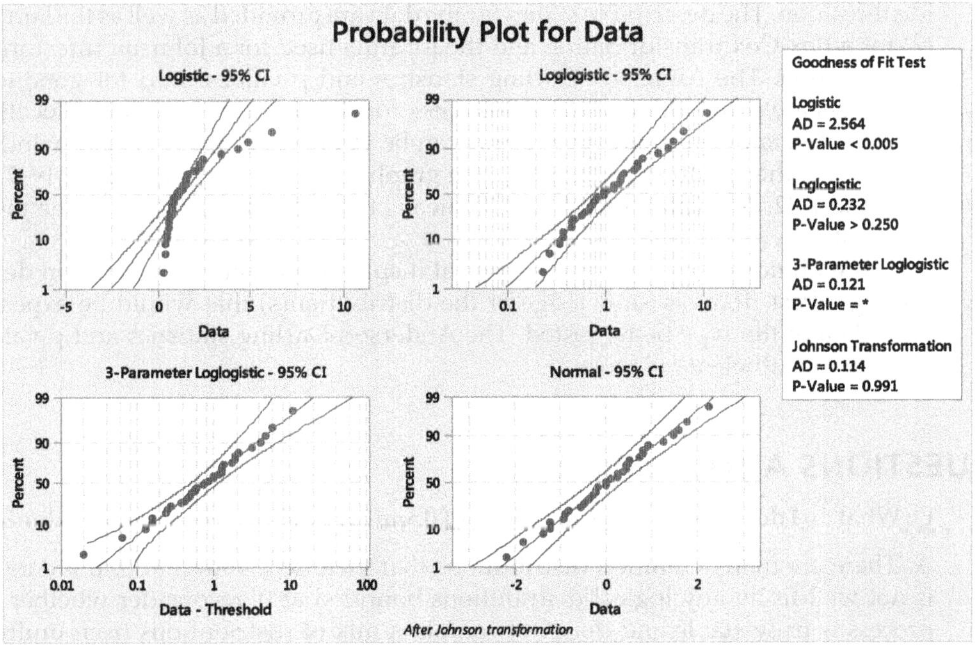

Figure 7.22 Minitab probability plots for logistic, loglogistic, 3-parameter loglogistic, and the normal distribution after a Johnson transformation.

The transformed data will then appear in the Minitab worksheet and Minitab will generate a graphic with before and after probability plots, such as the one in Figure 7.16. The probability plot for the original data demonstrates that the data clearly do not follow the normal distribution as they display a curved pattern around the fitted line. The probability plot for the transformed data shows the points falling along the fitted line and being well-fit by the normal distribution. The third plot demonstrates that the Anderson-Darling p-value achieved for the chosen transformation was higher than that of other values that were not used for transformation.

Individual Distribution Identification

Minitab can be used to identify the distribution that best fits a set of data. This can be useful when requiring tools that rely on distribution, such as a capability study or tolerance interval on non-normally distributed data. Selecting the distribution should be a combination of process knowledge about which distribution would make sense and a probability plot that indicates the data are a good fit.

Go to **Stat > Quality Tools > Individual Distribution Identification** and select the column containing the data. A subgroup size or column can be entered if the data was collected in a subgroup. The default is to use all available distributions and both transformations; however, up to four specific distributions can be selected, which is advisable if only certain distributions are logical for the data of interest (see Figure 7.17).

Figure 7.18 shows the Minitab session window for the individual distribution identification. The descriptive statics for the data are provided as well as the lambda (λ) for a Box-Cox transformation and the formula used for a Johnson transformation function. The Anderson-Darling statistics and p-values used for goodness-of-fit tests are provided as well as estimates for distribution parameters location, shape, scale, and threshold. The results can be interpreted in the session window or in one of the resulting probability plot graphics, as shown in Figures 7.19, 7.20, 7.21, and 7.22. A p-value less than 0.05 means the data likely do not fit the given distribution.

As with normality tests, the graphical display is of high importance in determining a best fit, as is knowledge of the distribution(s) that would be expected to fit data of the type being tested. The Anderson-Darling statistics and p-values alone are insufficient evidence.

QUESTIONS AND ANSWERS

Q: What do I do if there is a natural limit of 0 and I can't fit the data to a distribution?

A: There are many common distributions that include a bound at 0. If a data set is not well-fit by any logical distributions bounded at 0, reconsider whether the process is truly stable and does not contain a mix of distributions from multiple processes or conditions. If the data are likely stable, consider whether the statistical tool you are using requires a distribution or can be done for most distributions. If a distribution is required, consider using a transformation or nonparametric test.

Q: I want to perform a test that requires normally distributed data, but the p-value for the normal probability plot is less than 0.05, although most values fall close to the line, with a few exceptions. Can I still perform the test?

A: If the data fall randomly along the fitted line on a probability plot with no obvious patterns, then the normal distribution is a good fit for the data and you can proceed with the test. If, however, patterns exist, then the data likely do not follow the normal distribution. You should reconsider whether the data are stable, whether a non-normal distribution might be expected for that process, and whether there is an alternate test that can be done for that non-normal distribution or that does not assume a distribution.

Q: What is the difference between the Johnson transformation and the Box Cox transformation? When should I use one or the other?

A: The Box-Cox transformation is a simpler transformation that may have a theoretical rationale and can use subgrouping information if data were collected that way. The Johnson transformation, while more complicated and without a straightforward theoretical rationale for most data, is often more successful at transforming the data to an approximately normal distribution. Because of its simplicity and straightforward explanation, the Box-Cox transformation is more desirable in situations needing an effective transformation. It also allows for data collected in subgroups. Before using a transformation, consider whether the data are expected to be normal and whether the process is stable.

Q: What do I do if a test of normality and the resulting p-value indicates the data are not normally distributed, but the values are close to the line in the probability plot?

A: If the data fall randomly along the fitted line on a probability plot with no obvious patterns, then the distribution is a good fit for the data.

Q: When should I use a Box-Cox transformation? When should I use a Johnson transformation?

A: These are useful when the distribution was expected to be non-normal but a tool requiring normality is desired. Care should be taken to ensure the non-normality is natural and not due to an unstable process.

Q: My data failed a test of normality with p=0.39, but the probability plot showed most values fell close to the line. I did a capability study using the normal distribution and the Weibull distribution. The resulting capability values were both high, but differed. Which should I use?

A: If the data were expected to be normally distributed and there were not obvious patterns in the points on the probability plot, then the normal distribution can be assumed.

Q: What's the difference between Anderson-Darling, Ryan-Joiner, Shapiro-Wilk, and Kolmogorov-Smirnov tests of normality?

A: The Ryan-Joiner test is a variation of the Shapiro-Wilk test. Both the Ryan-Joiner and the Anderson-Darling test are more sensitive to departures from normality than the Kolmogorov-Smirnov test, although statistical significance in a normality test may not be meaningful given that many tools work well even with fairly large departures from normality.

Chapter 8
Nonparametric Tests

Nonparametric tests, also known as distribution-free tests, are statistical tests performed without using or assuming any parameters of a particular distribution (Conover 1999). Whereas a parametric test such as the one-way ANOVA tests whether means are different, a nonparametric test is used to test whether the location of the populations or samples differ (Stephens 2004). For example, such tests often compare central values such as the median, which is not a parameter of a particular distribution. Parametric tests are preferable to nonparametric because they generally tell us more about the data by using the mean or standard deviation as compared to a test of medians.

Making an inference on the median is usually not particularly valuable because the median is not the expected value of the population and you cannot use your results to make a prognosis such as the number of defective parts to anticipate when making cost-savings estimates. Nonparametric tests have often been used as an alternative for when the assumptions required for a parametric test have not been met. As discussed in Chapter 5 on hypothesis testing, those parametric tests are often robust to assumptions that have not been met; therefore, parametric tests can often be used even when those assumptions are not met instead of switching to nonparametric tests.

1-SAMPLE SIGN

The 1-sample sign test compares the true median to a hypothesized median. Similar to the 1-sample t test, the available alternative hypotheses are H_a: $m \neq m_0$, H_a: $m < m_0$, and H_a: $m > m_0$.

To calculate the p-value, the data are first divided into the number of points falling below the hypothesized median (or in the case of H_a: $m > m_0$, above) and those that do not. The p-value is calculated by finding the odds that at least that number of points out of the total occurs on that side of the median, using the binomial distribution with $p = 0.5$ and multiplying by 2 for H_a: $m \neq m_0$. More information on the binomial distribution can be found in Chapter 4.

Example

A light bulb manufacturer claims the median bulb life is at least 2000 hours, and you want to test whether the bulbs are performing as claimed. To do the test, 20 bulbs are burned until failure and the life of each bulb is recorded as shown in Table 8.1.

Table 8.1 Failure times for 20 light bulbs.

Hours	
1858	1662
1988	1969
1733	2332
2226	1892
1642	1772
1893	1609
1958	2146
2014	1734
2473	1589
1870	1827

Because we are only interested in whether the median burn time is below the claimed value, the null hypothesis is H_0: $m \geq 2000$ and the alternative hypothesis is H_a: $m < 2000$. The number of bulbs with failure times less than 2000 was 15. The p-value is then determined by using the binomial distribution to calculate the probability of at least 15 events out of 20 occurring if the true proportion is 0.5. This probability is found to be 0.0207. Because this is less than the alpha level of 0.05, the null hypothesis is rejected and we conclude that the median burn time is likely less than the claimed value of 2000 hours.

In Minitab: **Go to Stat > Nonparametrics > 1-Sample Sign** and complete the dialog as shown in Figure 8.1 and click **OK**.

The session window output is shown in Figure 8.2.

The p-value of 0.021 is less than the alpha level of 0.05, so the null hypothesis is rejected and we conclude that the median burn time is less than 2000 hours.

1-SAMPLE WILCOXON

An alternative to the 1-sample sign test that is accurate for symmetric distributions is the Wilcoxon test, which uses the same possible hypotheses and also tests the median against a hypothesized value (Conover 1999). The test uses the statistic W, which requires a lengthy calculation that is performed easily by statistical software.

Example

You wish to test the median light bulb burn time from the 1-sample sign test example using the Wilcoxon test.

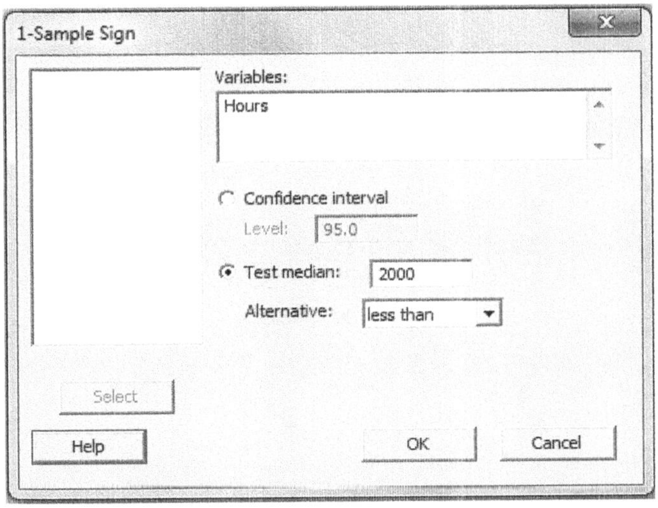

Figure 8.1 Minitab dialog box for 1-sample sign test.

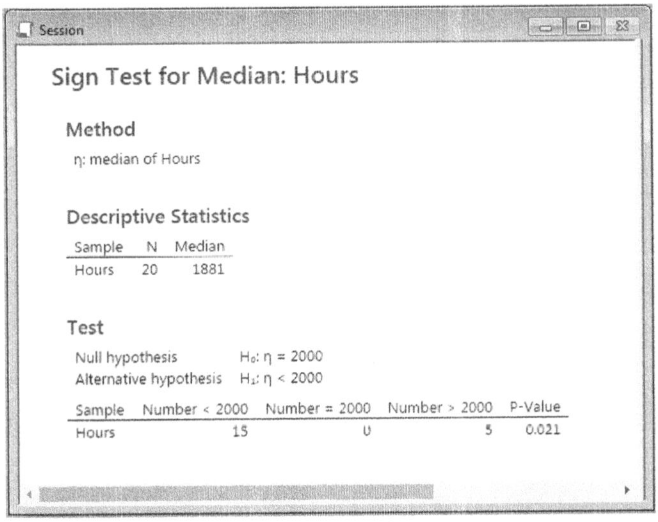

Figure 8.2 Session window output for 1-sample sign test.

In Minitab: Go to **Stat > Nonparametrics > 1-Sample Wilcoxon** and complete the dialog as shown in Figure 8.3 and then click **OK**.

The session window output is shown in Figure 8.4.

The p-value is calculated to be 0.038, which is less than the alpha of 0.05, so the null hypothesis is rejected and we conclude that the median burn time is less than 2000 hours.

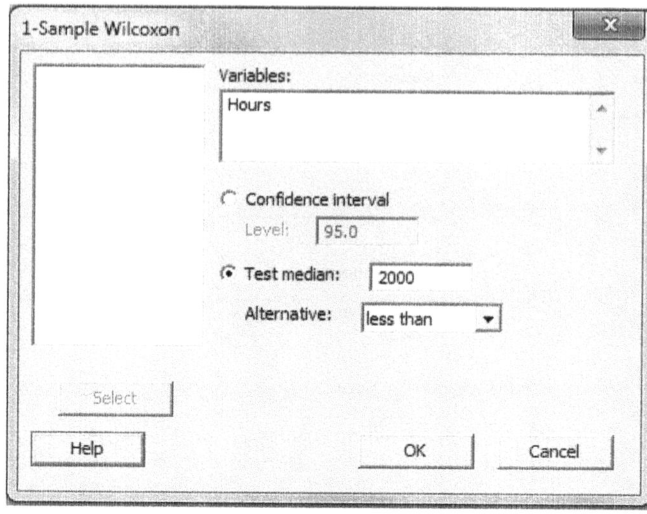

Figure 8.3 Minitab dialog box for 1-sample Wilcoxon test.

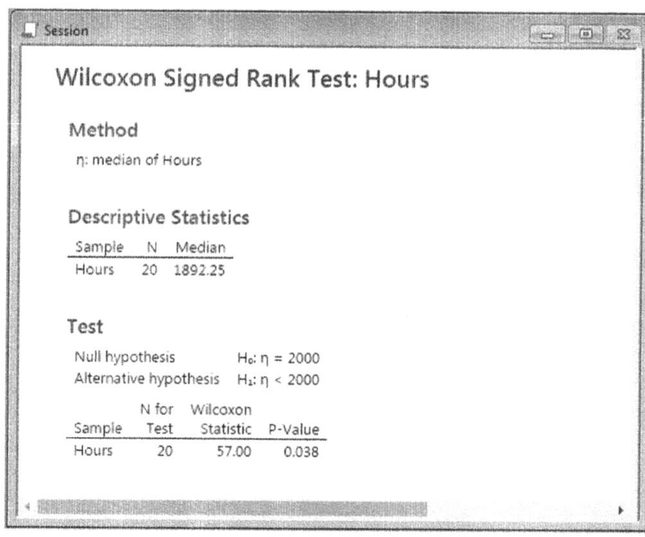
Session

Wilcoxon Signed Rank Test: Hours

Method

η: median of Hours

Descriptive Statistics

Sample	N	Median
Hours	20	1892.25

Test

Null hypothesis	H_0: η = 2000
Alternative hypothesis	H_1: η < 2000

Sample	N for Test	Wilcoxon Statistic	P-Value
Hours	20	57.00	0.038

Figure 8.4 Session window output for 1-sample Wilcoxon test.

MANN-WHITNEY TEST

The Mann-Whitney test is performed to compare the medians of two populations that have the same shape (Conover 1999). The possible alternative hypotheses are:

1. H_a: $m_1 \neq m_2$
2. H_a: $m_1 < m_2$
3. H_a: $m_1 > m_2$

To calculate the test statistic W, the two samples are combined and then ranked from smallest (rank 1) to largest. If any observations are tied, they are given the average rank of the points that are tied at that value. The test statistic W is then the sum of the ranks from the first sample. The p-value is based on a normal approximation using Z and is calculated for each alternative hypothesis as follows:

For H_a: $m_1 \neq m_2$, $p = 2 * P\left(|Z| > \dfrac{\left| W - \dfrac{n(m+n+1)}{2} \right| - 0.5}{\sqrt{\dfrac{mn(m+n+1)}{12}}} \right) = 2$ times the upper-tail probability of the absolute value of the calculated statistic as found in Table C.5 in Appendix C.

For H_a: $m_1 < m_2$, $p = P\left(Z < \dfrac{W - n(m+n+1) - \dfrac{n(m+n+1)}{2} + 0.5}{\sqrt{\dfrac{mn(m+n+1)}{12}}} \right) =$ the lower-tail probability of the calculated statistic as found in Tables C.1 and C.2 in Appendix C.

For H_a: $m_1 > m_2$, $p = P\left(Z > \dfrac{W - \dfrac{n(m+n+1)}{2} - 0.5}{\sqrt{\dfrac{mn(m+n+1)}{12}}} \right) =$ the upper-tail probability of the calculated statistic as found in Table C.5 in Appendix C.

In the formulas, n is the sample size of the first sample and m is the sample size of the second sample.

Example

Two fill heads are used in the production of bottles of lubricating oil and you suspect that the median fill volumes for each are not equal. You decide to sample 16 bottles randomly from each fill head and measure the volume of lubricating oil in each, recording the data as shown in Table 8.2.

Because a difference in either direction would be of interest, the null hypothesis is H_0: $m_1 = m_2$ and the alternative hypothesis is H_a: $m_1 \neq m_2$. The test statistic W is calculated by first ranking all points from smallest to largest, as depicted in Table 8.3.

The sum of the ranks from fill head 1 is calculated as $W = 22 + 18 + \ldots + 3.5 = 262$. The test statistic is calculated as:

$$\frac{\left| W - \dfrac{n(m+n+1)}{2} \right| - 0.5}{\sqrt{\dfrac{mn(m+n+1)}{12}}} = \frac{\left| 262 - \dfrac{16(16+16+1)}{2} \right| - 0.5}{\sqrt{\dfrac{16*16(16+16+1)}{12}}} = 0.057$$

The p-value is calculated as $p = 2 * P(|Z| > 0.057)$ or 2 times the upper-tail probability of the Z distribution corresponding to 0.057. Using Table C.5 in Appendix C for Z = 0.06 (the closest value to 0.057 found in the table), this probability is found to be 0.4761. The p-value is then 2 * 0.4761 = 0.9522. This value is greater than the alpha level of 0.05 and therefore the null hypothesis is not rejected and we do not have sufficient evidence that the median fill volumes are not equal.

Table 8.2 Values for head 1 and head 2.

Head 1	Head 2
16.03	16.05
16.02	15.98
16.07	15.98
15.97	15.88
16.04	16.05
16.02	15.93
16.12	16.07
15.8	15.92
16	15.98
16.1	16.1
16.02	16.03
15.95	16.03
15.85	16.02
16.1	16.01
15.9	15.98
15.88	16.02

In Minitab: Go to **Stat > Nonparametrics > Mann-Whitney** and complete the dialog as shown in Figure 8.5 and click **OK**.

The session window output is shown in Figure 8.6.

The output provides the median volume for each fill head as well as the difference in medians and a confidence interval on the true difference in population median. It also provides a p-value and a second p-value that is adjusted for ties in the data. The p-value of 0.955 is larger than the alpha level of 0.05, so the null hypothesis is not rejected and we cannot conclude that the median volumes are different for the two fill heads.

KRUSKAL-WALLIS TEST

The Kruskal-Wallis test is a generalization of the Mann-Whitney test to compare the medians of two or more populations with the same shape (Boslaugh and Watters 2008). The null hypothesis is that all groups have the same median and the alternative hypothesis is that at least one median is different. As with Mann-Whitney, the data are combined and ranked and the p-value is derived from the average rank of each group by first calculating the test statistic H using the formula

$$H = \frac{12\sum n_j(\bar{R}_j - \bar{R})^2}{N(N+1)},$$

where n_j is the sample size of the jth group, R_j is the average rank for that group, R is the overall average rank, and N is the total sample size. If the data set contains ties, there is also an adjustment that can be made resulting in a test statistic H(adj). The p-value is calculated as $P(\chi^2_{k-1} > H)$, where k is the number of groups.

Example

Three suppliers provide cutting heads for a machine lathe and you want to know whether the median hours of use until replacement is needed differs among the suppliers. Data is collected on the time each of 5 cutting heads is in use until replacement and is found in Table 8.4, along with the rank of each data point in the data set.

Table 8.3 Values and rank order for head 1 and head 2.

Head 1	Rank	Head 2	Rank
16.03	22	16.05	25.5
16.02	18	15.98	11.5
16.07	27.5	15.98	11.5
15.97	9	15.88	3.5
16.04	24	16.05	25.5
16.02	18	15.93	7
16.12	32	16.07	27.5
15.8	1	15.92	6
16	14	15.98	11.5
16.1	30	16.1	30
16.02	18	16.03	22
15.95	8	16.03	22
15.85	2	16.02	18
16.1	30	16.01	15
15.9	5	15.98	11.5
15.88	3.5	16.02	18

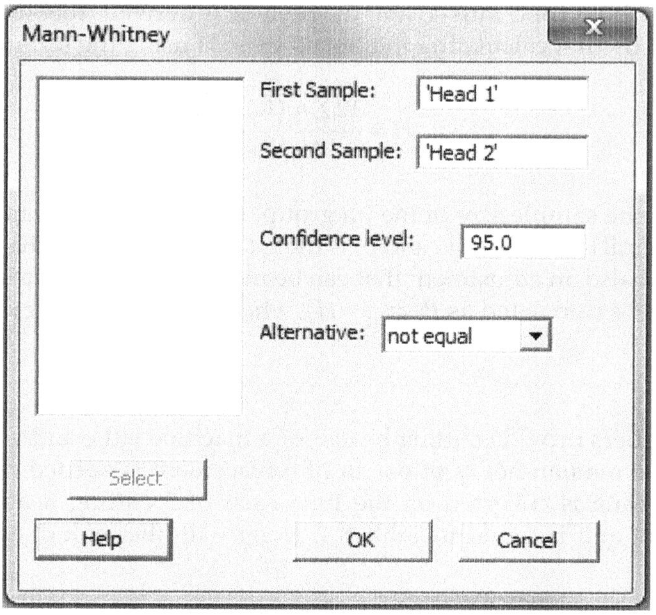

Figure 8.5 Minitab dialog box for Mann-Whitney test.

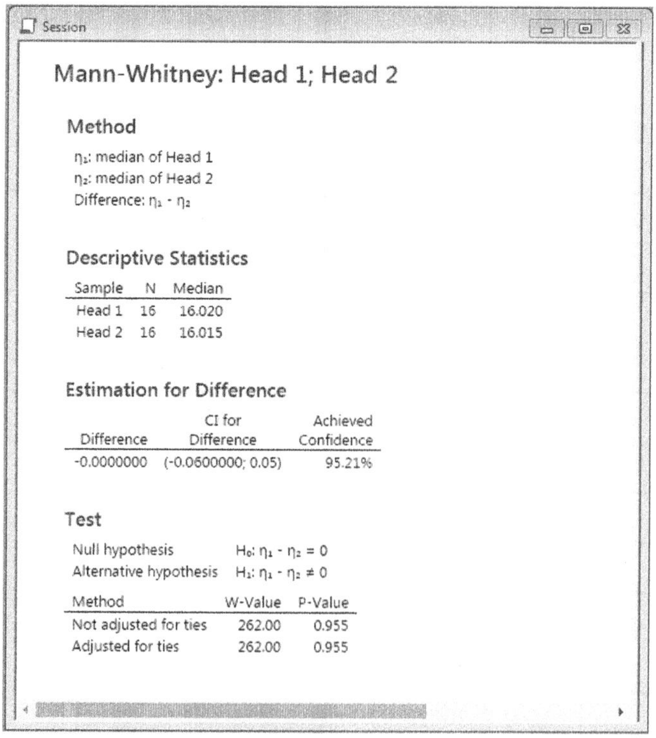

Figure 8.6 Session window output for Mann-Whitney test.

Table 8.4 Cutting head replacement data.

Hours	Supplier	Rank
34	A	6
35	A	7.5
33	A	5
39	A	11
36	A	9
45	B	14
43	B	13
35	B	7.5
41	B	12
50	B	15
31	C	4
24	C	2
25	C	3
22	C	1
38	C	10

Table 8.5 Average rank by supplier.

Supplier	Avg. Rank
A	7.7
B	12.3
C	4

The overall average rank $\bar{R}$ is 8, and the average rank $\bar{R}_j$ for each supplier is found in Table 8.5.

The test statistic H is calculated as:

$$H = \frac{12\sum n_j(\bar{R}_j - \bar{R})^2}{N(N+1)} = \frac{12*\left[5(7.7-8)^2 + 5(12.3-8)^2 + 5(4-8)^2\right]}{15(15+1)} = 8.645$$

The critical value is the value of χ_2^2 corresponding to an upper-tail probability of 0.05, as can be found in Table C.13 in Appendix C. This value is 5.99. Because

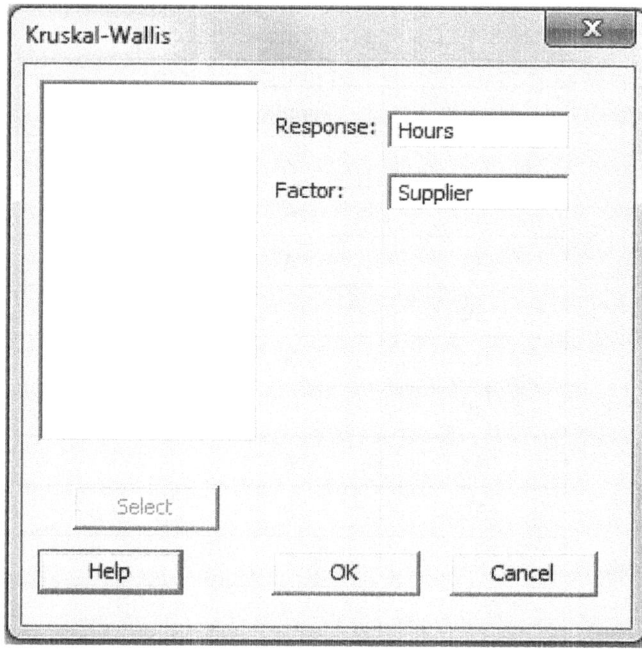

Figure 8.7 Minitab dialog box for Kruskal-Wallis test.

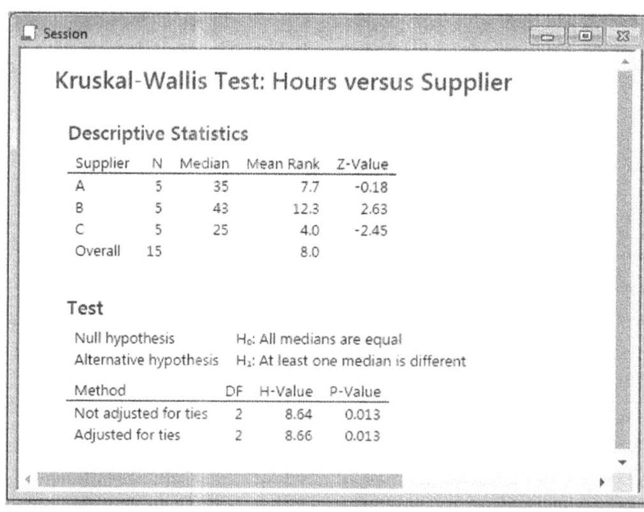

Session

Kruskal-Wallis Test: Hours versus Supplier

Descriptive Statistics

Supplier	N	Median	Mean Rank	Z-Value
A	5	35	7.7	-0.18
B	5	43	12.3	2.63
C	5	25	4.0	-2.45
Overall	15		8.0	

Test

Null hypothesis H_0: All medians are equal
Alternative hypothesis H_1: At least one median is different

Method	DF	H-Value	P-Value
Not adjusted for ties	2	8.64	0.013
Adjusted for ties	2	8.66	0.013

Figure 8.8 Session window output for Kruskal-Wallis test.

the test statistic H=8.645 is greater than the critical value, the null hypothesis is rejected and we conclude the median time until replacement of cutting heads provided by the three suppliers is not equal. At least one supplier's cutting heads have a median time until replacement that is different.

In Minitab: Go to **Stat**>**Nonparametrics**>**Kruskal-Wallis** and complete the dialog as shown in Figure 8.7 and click **OK**.

The session window output is shown in Figure 8.8.

The output shows that the median time until replacement for the cutting heads supplied by supplier C is the lowest and that of supplier B is the highest. The p-value of 0.013 is smaller than the alpha level of 0.05, so the null hypothesis can be rejected and we conclude that the true population median replacement time for the three suppliers is not equal.

MOOD'S MEDIAN TEST

Mood's median test is used to determine if two or more population medians are equal, but unlike Kruskal-Wallis, it does not assume each population has the same shape. The null hypothesis is that all groups have the same median and the alternative hypothesis is that at least one median is different. To calculate Mood's median, the overall median of the data is first calculated and then for each group the number of observations less than or equal to the overall median are counted and the greater than are counted. Then a Chi-square statistic is calculated as

$$\chi^2 = \frac{\Sigma(O_{ij} - E_{ij})^2}{E_{ij}},$$

where O_{ij} is the observed count of observations less than or equal to the median or above the median for group j and E_{ij} is the expected number of observations for the combination, calculated as the number of total observations in that group times the total observations on that side of the median, divided by the total number of observations. The critical value is the value of χ^2_{k-1} corresponding to an upper-tail probability of alpha. The p-value is calculated as:

$$P(\chi^2_{k-1} > \chi^2)$$

Example

Using the cutting head replacement time data from the Kruskal-Wallis test, you wish to perform a Mood's median test as you are unsure whether each population has the same shape.

The overall median of the data is found to be 35. A table is first formed to tabulate the number of observations for each supplier that are less than or equal to 35, and then the number greater than 35. Additionally, the total number of

Table 8.6 Counts relative to the median.

	# ≤ 35	# > 35	Sum
Supplier A	3	2	5
Supplier B	1	4	5
Supplier C	4	1	5
Sum	8	7	15

Table 8.7 Expected counts.

	# ≤ 35	# > 35
Supplier A	2.67	2.33
Supplier B	2.67	2.33
Supplier C	2.67	2.33

observations for each row and column in the table are calculated. The results are shown in Table 8.6.

Then the expected counts for each cell of the table are calculated. For example, the expected count of observations less than or equal to 35 for supplier B is calculated by taking the number of observations for that row (5) times the number of observations for that column (8) divided by the total number of observations (15). The expected count is 5 * 8/15 = 2.67. The expected count for each cell is shown in Table 8.7.

The test statistic is calculated as:

$$X^2 = \sum \frac{(O_{ij} - E_{ij})^2}{E_{ij}} = \frac{(3-2.67)^2}{2.67} + \frac{(1-2.67)^2}{2.67} + \frac{(4-2.67)^2}{2.67} + \frac{(2-2.33)^2}{2.33}$$

$$+ \frac{(4-2.33)^2}{2.33} + \frac{(1-2.33)^2}{2.33} = 3.75$$

The critical value is the value of χ_2^2 corresponding to an upper-tail probability of 0.05 and 2 degrees of freedom and is found in Table C.9 in Appendix C to be 5.99. Because the test statistic of 3.75 is smaller than this value, we fail to reject the null hypothesis and cannot conclude the true population median replacement times are different. This result differs from that of the Kruskal-Wallis test because the Mood's median test is less powerful in detecting differences.

In Minitab: Go to **Stat > Nonparametrics > Mood's Median Test** and complete the dialog as shown in Figure 8.9 and click **OK**.

The session window output is shown in Figure 8.10.

The output provides a p-value of 0.153, which is greater than the alpha of 0.05, so we fail to reject the null hypothesis. The output also includes confidence intervals on the true population medians for each supplier but warns that because there were less than six observations for each supplier the true confidence is less than the stated 95%.

CHI-SQUARE GOODNESS-OF-FIT TEST

In some cases a variable can take on any one of two or more values and a count can be taken of how often the variable takes each value. One example would be a packaging line in which a single conveyor splits into three different chutes for boxing the items. In this case, it may be desirable to assess whether the variable takes each value the same proportion of the time. Or you might compare to some other

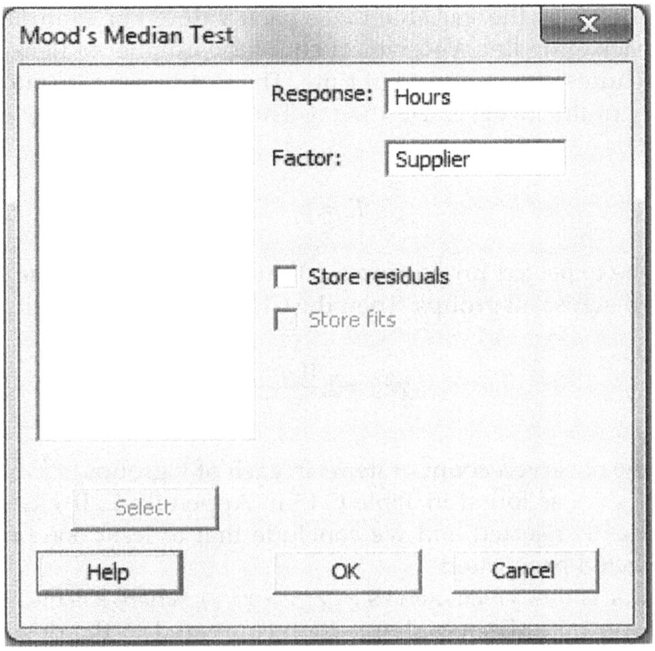

Figure 8.9 Minitab dialog box for Mood's median test.

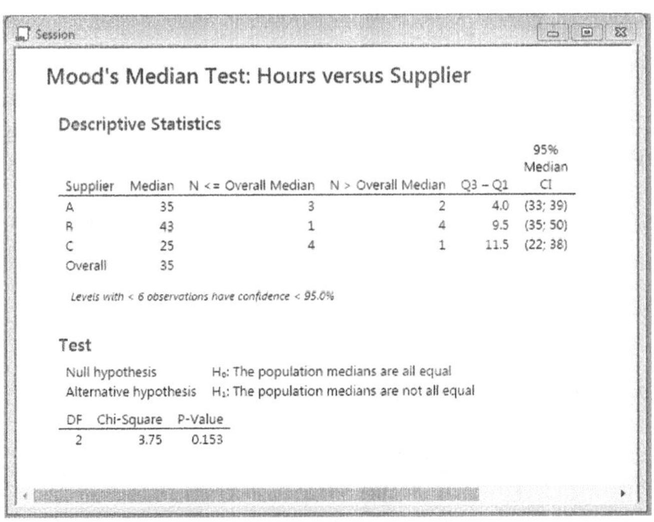

Session

Mood's Median Test: Hours versus Supplier

Descriptive Statistics

Supplier	Median	N <= Overall Median	N > Overall Median	Q3 – Q1	95% Median CI
A	35	3	2	4.0	(33; 39)
B	43	1	4	9.5	(35; 50)
C	25	4	1	11.5	(22; 38)
Overall	35				

Levels with < 6 observations have confidence < 95.0%

Test

Null hypothesis H_0: The population medians are all equal
Alternative hypothesis H_1: The population medians are not all equal

DF	Chi-Square	P-Value
2	3.75	0.153

Figure 8.10 Session window output for Mood's median test.

proportion of interest. In either case, a one-way Chi-square goodness-of-fit test can be performed to determine if the proportions likely do not equal the expected proportions. The one-way Chi-square goodness-of-fit test can be used when the underlying distribution is unknown (Montgomery, Runger, and Hubele 2001).

The null hypothesis for the test is H_0: The true proportions equal the expected proportions, and the alternative hypothesis is H_a: At least one proportion does not equal the expected proportion. Data must be collected by counting the number of times the variable takes each value. For example, on the aforementioned packaging line, you would count the number of items that enter each of the three chutes over a period of time. The test is less accurate if the expected count for any of the levels is less than 5. The expected count E_i for each group is calculated as

$$E_i = p_i * N,$$

where p_i is the expected proportion for that group and N is the total number of items counted across all groups. Then the Chi-square test statistic is calculated as

$$\chi^2_{test} = \sum \frac{(O_i - E_i)^2}{E_i},$$

where O_i is the observed count of items in each of k groups. χ^2_{test} can be compared to $\chi^2_{critical} = \chi^2_{k-1,1-\alpha}$ as found in Table C.13 in Appendix C. If $\chi^2_{test} > \chi^2_{critical}$ then the null hypothesis is rejected and we conclude that at least one proportion differs from the expected proportion.

The p-value is then calculated as $P(\chi^2_{test} > \chi^2_{k-1})$, where k is the number of values the variable can take. The p-value is then compared to the desired alpha and if smaller, the null hypothesis is rejected in favor of the alternative hypothesis.

Example

A tile supplier provides boxes of tiles and claims there are equal proportions of blue, green, and yellow tile for creating random mosaic floor patterns. After the last order was installed, the customer thought there were a disproportionate number of blue tiles. A few boxes of tiles are opened and the number of each tile color is counted as shown in Table 8.8.

The expected proportion for each color under the null hypothesis is $\frac{1}{3}$ and the total number of tiles is 270. The expected count of tiles for each color is:

$$E_i = p_i * N = \frac{1}{3} * 270 = 90$$

Then the Chi-square test statistic is calculated as:

$$\chi^2_{test} = \sum \frac{(O_i - E_i)^2}{E_i} = \frac{(94-90)^2}{90} + \frac{(92-90)^2}{90} + \frac{(84-90)^2}{90} = 0.622$$

Table 8.8 Tile colors and counts.

Color	Count
Blue	94
Green	92
Yellow	84

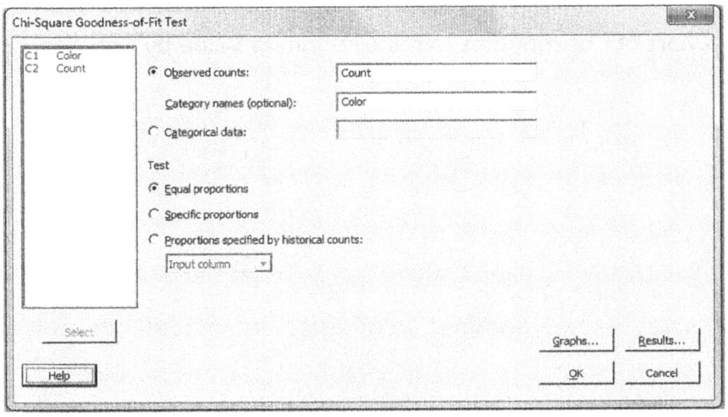

Figure 8.11 Minitab dialog for Chi-square test for one variable.

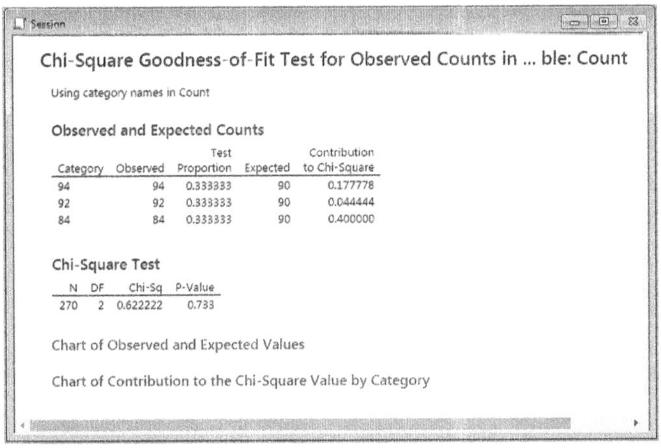

Figure 8.12 Session window output for Chi-square test for one variable.

Using the Chi-square critical values found in Table C.13 in Appendix C, it can be seen that the critical value for a Chi-square with 2 degrees of freedom and an alpha level of 0.05 is 5.99. Because the test statistic of 0.622 is less than this value, the null hypothesis is not rejected and we cannot conclude that the proportion of any color is different than $\frac{1}{3}$.

In Minitab: Go to **Stat > Tables > Chi-square Goodness-of-Fit test (One variable)**, complete the dialog as shown in Figure 8.11, and click **OK**.

The session window output is shown in Figure 8.12.

The session window output shows a p-value of 0.733. Because this value is larger than the alpha level of 0.05, the null hypothesis is not rejected.

Minitab also provides two graphs that are useful in determining which of the values deviated most from its expected value and therefore contributed the most to the Chi-square test statistic. These graphs are shown in Figure 8.13 and 8.14 and should only be interpreted if the test was statistically significant. The larger the bar in Figure 8.13, the more that level contributed to the significance of the test. Figure 8.14 further helps understand why a test is significant by showing how much the observed count exceeded or fell short of the expected count.

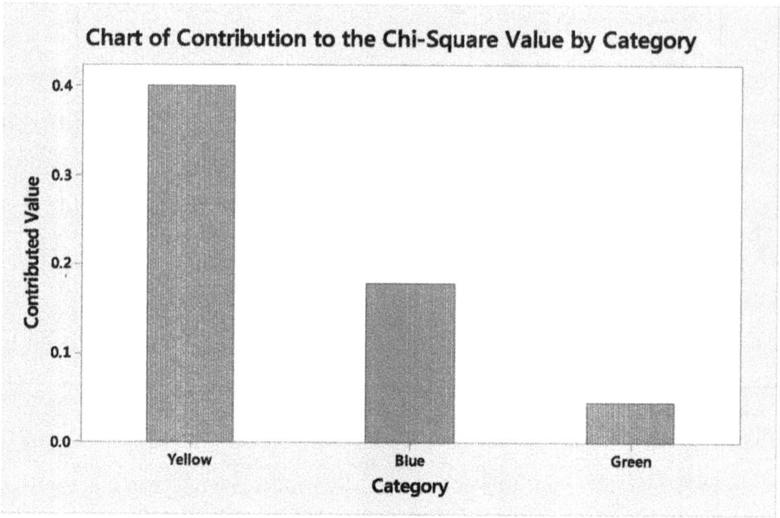

Figure 8.13 Chart of Chi-square contribution.

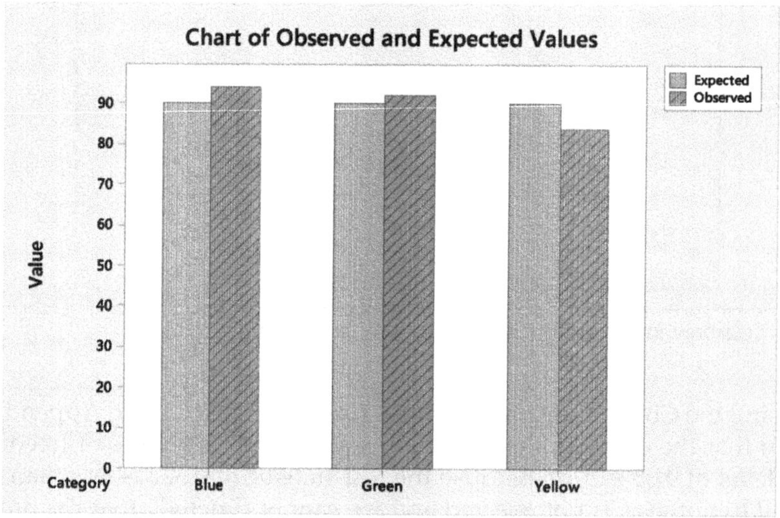

Figure 8.14 Chart of observed and expected values.

CHI-SQUARE TEST (TWO VARIABLES)

The Chi-square test for two variables is used for situations in which there are two categorical variables and you wish to test whether they are independent of one another. If the variables are independent, then the likelihood of any particular level of one of the variables is the same regardless of the level of the other. For example, if a card in a deck of cards is drawn at random, the odds of that card being a 5 is independent of what suit it is. Knowing the suit is spades does not affect the likelihood of it being a 5 or any other value.

The structure of the test is the same as the Chi-square test for one variable, but the expected count is calculated differently. The formula is

$$E_{ij} = \frac{n_{i*}n_{*j}}{N},$$

where n_{i*} is the count of observations at that level of the first variable and n_{*j} is the count of observations at that level for the second variable and N is the total number of observations.

The Chi-square test statistic is then calculated as

$$\chi^2_{test} = \Sigma \frac{(O_{ij} - E_{ij})^2}{E_{ij}},$$

where O_{ij} is the observed count of items in each combination of groups from the two variables. χ^2_{test} can be compared to $\chi^2_{critical} = \chi^2_{(m-1)(n-1),\,1-\alpha}$ as found in Table C.13 in Appendix C, where m is the number of groups for the first variable and n is the number of groups for the second variable. If $\chi^2_{test} > \chi^2_{critical}$ then the null hypothesis is rejected and we conclude that at least one proportion differs from the expected proportion.

Alternatively, the p-value is calculated as $P(\chi^2_{test} > \chi^2_{(m-1)(n-1)})$. The p-value is then compared to the desired alpha and if smaller, the null hypothesis is rejected in favor of the alternative hypothesis.

Example

Customers rate their satisfaction with each of three different manufacturing plants and you want to determine whether the plants perform equally. The number of customers rating each plant at each level are found in Table 8.9.

First, the totals for each row and column are calculated as well as the overall total, as seen in Table 8.10.

For each cell in the table the expected count is calculated as well as that cell's contribution to the Chi-square statistic using the following formulas:

$$E_{ij} = \frac{n_{i*}n_{*j}}{N}$$

$$\chi^2_{test} = \Sigma \frac{(O_{ij} - E_{ij})^2}{E_{ij}}$$

The calculations are shown in Table 8.11.

Table 8.9 Satisfaction by plant.

		Satisfaction		
		Poor	**Okay**	**Good**
Plant	Chicago	6	23	10
	Monterrey	4	17	16
	Madrid	8	20	9

Table 8.10 Satisfaction by plant with totals.

		Satisfaction			
		Poor	**Okay**	**Good**	**Total**
Plant	Chicago	6	23	10	39
	Monterrey	4	17	16	37
	Madrid	8	20	9	37
	Total	18	60	35	113

Table 8.11 Satisfaction by plant with totals and contributions.

		Satisfaction			
		Poor	**Okay**	**Good**	**Total**
Plant	**Chicago**	6 $E_{11} = \frac{39*18}{113} = 6.21$ $X^2_{cell} = \frac{(6-6.21)^2}{6.21} = .007$	23 $E_{12} = \frac{39*60}{113} = 20.71$ $X^2_{cell} = \frac{(23-20.71)^2}{20.71} = .254$	10 $E_{13} = \frac{39*18}{113} = 12.08$ $X^2_{cell} = \frac{(10-12.08)^2}{12.08} = .358$	39
	Monterrey	4 $E_{21} = \frac{37*18}{113} = 5.89$ $X^2_{cell} = \frac{(4-5.89)^2}{5.89} = .609$	17 $E_{22} = \frac{37*60}{113} = 19.65$ $X^2_{cell} = \frac{(17-19.65)^2}{19.65} = .356$	16 $E_{23} = \frac{37*35}{113} = 11.46$ $X^2_{cell} = \frac{(16-11.46)^2}{11.46} = 1.798$	37
	Madrid	8 $E_{31} = \frac{37*18}{113} = 5.89$ $X^2_{cell} = \frac{(8-5.89)^2}{5.89} = .753$	20 $E_{32} = \frac{37*60}{113} = 19.65$ $X^2_{cell} = \frac{(20-19.65)^2}{19.65} = .006$	9 $E_{33} = \frac{37*35}{113} = 11.46$ $X^2_{cell} = \frac{(9-11.46)^2}{11.46} = .528$	37
	Total	18	60	35	113

Table 8.12 Chi-square table for 4 degrees of freedom and an alpha of 0.05.

df	0.995	0.990	0.975	0.900	0.750	0.500	0.250	0.100	0.050	0.025
1	0.00004	0.0002	0.001	0.02	0.10	0.45	1.32	2.71	3.84	5.02
2	0.01	0.02	0.05	0.21	0.58	1.39	2.77	4.61	5.99	7.38
3	0.07	0.11	0.22	0.58	1.21	2.37	4.11	6.25	7.81	9.35
4	0.21	0.30	0.48	1.06	1.92	3.36	5.39	7.78	9.49	11.14
5	0.41	0.55	0.83	1.61	2.67	4.35	6.63	9.24	11.07	12.83

Worksheet 1 ***

↓	C1-T	C2	C3	C4	C5
	Plant	Poor	Okay	Good	
1	Chicago	6	23	10	
2	Monterrey	4	17	16	
3	Madrid	8	20	9	
4					
5					

Figure 8.15 Data for a Chi-square test of two variables.

The test statistic is formed by summing the individual contributions as follows:

$$\chi^2_{test} = \Sigma \frac{(O_{ij} - E_{ij})^2}{E_{ij}} = .007 + .254 + .358 + .609 + .356 + 1.798 + .753 + .006 + .528$$
$$= 4.669$$

Using Table C.9, $\chi^2_{critical}$ is found by looking up the row with $(3-1) * (3-1) = 4$ degrees of freedom and an alpha of 0.05 (see Table 8.12).

Because $\chi^2_{test} = 4.669$ is less than $\chi^2_{critical} = 9.49$, we fail to reject the null hypothesis that the plant and ratings are independent and cannot conclude that any plants are rated differently from the others.

In Minitab: Enter the data as shown in Figure 8.15.

Go to **Stat > Tables > Chi-Square Test for Association** and complete the dialog as shown in Figure 8.16. Be sure to type a name such as *Rating* in the column field and then click **OK**.

The session window output is shown in Figure 8.17.

The output shows that the test resulted in a p-value of 0.323, which is greater than the alpha level of 0.05. Therefore the null hypothesis is not rejected and we cannot conclude that the customers' ratings depend on the plant.

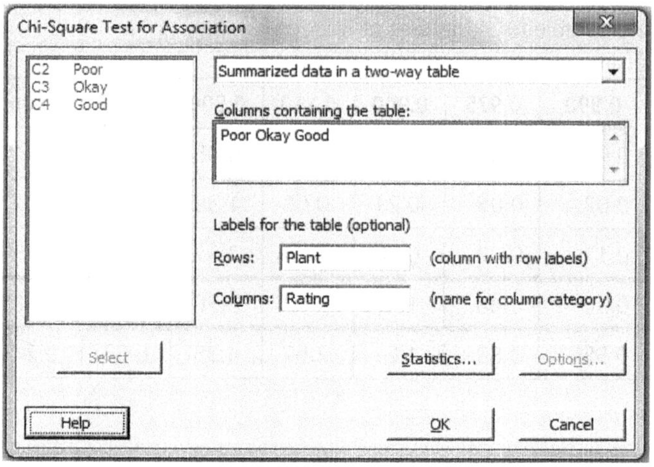

Figure 8.16 Minitab dialog box for Chi-square test for association.

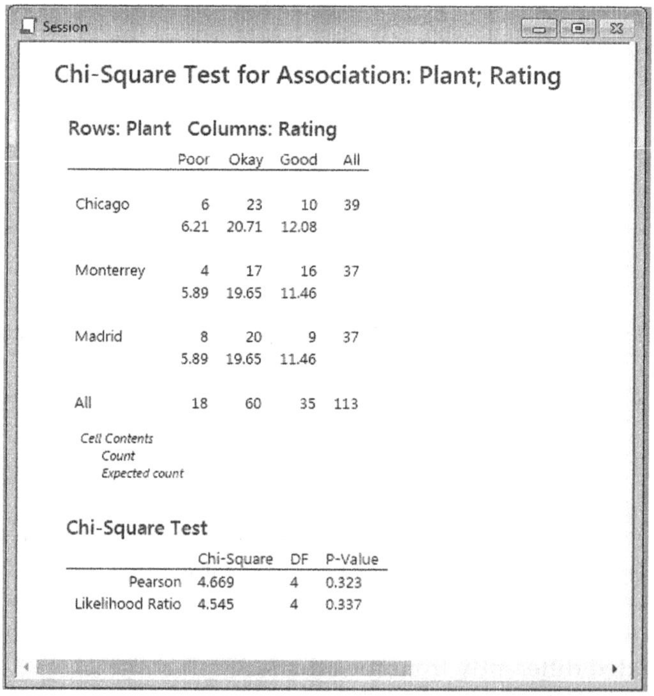

Chi-Square Test for Association: Plant; Rating

Rows: Plant Columns: Rating

	Poor	Okay	Good	All
Chicago	6	23	10	39
	6.21	20.71	12.08	
Monterrey	4	17	16	37
	5.89	19.65	11.46	
Madrid	8	20	9	37
	5.89	19.65	11.46	
All	18	60	35	113

Cell Contents
Count
Expected count

Chi-Square Test

	Chi-Square	DF	P-Value
Pearson	4.669	4	0.323
Likelihood Ratio	4.545	4	0.337

Figure 8.17 Session window output for Chi-square test for association.

QUESTIONS AND ANSWERS

Q: Which test should I use to compare two heavily skewed samples if their shapes are different?

A: The Mood's median test does not assume populations have the same shape.

Q: What is the difference between nonparametric test and parametric tests?

A: Nonparametric tests do not assume a specific distribution when determining a p-value, whereas parametric tests do. However, many parametric tests are robust to that assumption.

Q: Should I always use a nonparametric test if my data fail a test of normality?

A: Not necessarily. Many parametric tests that assume normality are robust to that assumption and even outperform the nonparametric equivalent.

Chapter 9
Analysis of Variance (ANOVA)

CLASSICAL ONE-WAY ANOVA

One-way analysis of variance (ANOVA) tests whether the mean of at least one of three or more populations is different from the others. A one-way ANOVA is much like a 2-sample t test for more than two samples (Witte 1993). An assumption of the classical test is that all populations are normally distributed and have equal variances, and the test does not perform as well if there are even moderate departures from these assumptions. An alternate calculation known as the Welch test, described later in this chapter, performs well even when these assumptions are not met. The null hypothesis in one-way ANOVA is that all populations have equal means, and the alternative hypothesis is that at least one mean is different.

Recall that the formula for variance uses the sum of the squared differences between the individual data points and the mean. One-way ANOVA is an extension of that concept, except the overall variation is separated into variation between the sample means and variation between each point and the mean of its sample. The equivalent "sum of squares" formulas in one-way ANOVA are shown in Table 9.1.

Here $\bar{y}_{i.}$ is the mean of the observations from sample i, $\bar{y}_{..}$ is the overall mean, y_{ij} is the value of observation j in sample i, and n_i is the number of observations in sample i. Each sum of squares then has a corresponding degrees of freedom, and the mean square for each is defined as the sum of squares divided by its degrees of freedom. This information is defined in a one-way ANOVA table as shown in Table 9.2.

In the table, k is the number of populations being tested and n is the total sample size across all samples. The total degrees of freedom are equal to the factor degrees of freedom plus the error degrees of freedom, and the total sum of squares is equal to SS(Factor) + SS(Error). The test statistic is F, defined as

$$F = \frac{MS_{factor}}{MS_{error}},$$

where the F has k-1 numerator degrees of freedom and n-k denominator degrees of freedom. The p-value is equal to $P(F_{k-1,n-k} > F)$. This p-value can tell us that at least one mean is likely different from the others; however, it does not tell us which mean is different.

Example

A production manager splits employees randomly into three groups. One group receives internal training, one group is sent to a consultant for training, and one

Table 9.1 One-way ANOVA explanations.

Source	Formula	Explanation
SS (Factor)	$\sum n_i(\bar{y}_{i\cdot} - \bar{y}_{\cdot\cdot})^2$	The sum of the squared difference between each sample mean and mean and the overall mean, multiplied by the sample size
SS (Error)	$\sum (y_{ij} - \bar{y}_{i\cdot})^2$	The sum of the squared differences between each data point and point and the mean of the sample from which it came
SS (Total)	$\sum (y_{ij} - \bar{y}_{\cdot\cdot})^2 = SS(Factor) + SS(Error)$	The sum of the squared differences between each data point and point and the overall mean

Table 9.2 One-way ANOVA table.

ANOVA table				
Source	**df**	**Sum of squares (SS)**	**Mean square (MS)**	**F**
Factor	$k-1$	SS (Factor)	$\frac{SS(Factor)}{k-1}$	$\frac{MS(Factor)}{MS(Error)}$
Error	$n-k$	SS (Error)	$\frac{SS(Error)}{n-k}$	
Total	$n-1$	SS (Total)		

Table 9.3 Parts produced.

	Trained by consultant	Internal training	Untrained
	74	78	76
	72	75	73
	73	76	75
	76	80	74
	75	79	71
Total	370	388	369

Table 9.4 Calculating sum of squares.

Factor level	Value	Level mean	(Value-mean)^2
Trained by Consultant	74	74	0
	72	74	4
	73	74	1
	76	74	4
	75	74	1
Internal Training	78	77.6	0.16
	75	77.6	6.76
	76	77.6	2.56
	80	77.6	5.76
	79	77.6	1.96
Untrained	76	73.8	4.84
	73	73.8	0.64
	75	73.8	1.44
	74	73.8	0.04
	71	73.8	7.84

group remains untrained. The average number of parts produced per day by each group is recorded for a week as shown in Table 9.3.

A table is created to indicate the mean of each factor level and the squared difference between each data point and the mean of its factor level, as shown in Table 9.4.

The overall mean of the data is found to be 75.133. The following calculations are then performed, with the results entered into the one-way ANOVA table shown in Table 9.5:

$$SS\,(Factor) = \Sigma n_i(\bar{y}_i. - \bar{y}..)^2 = 5(74-75.133)^2 + 5(77.6-75.133)^2 + 5(73.8-75.133)^2 = 45.73$$

$$SS\,(Error) = \Sigma(y_{ij} - \bar{y}_i.)^2 = (74-74)^2 + (72-74)^2 + \cdots + (71-73.8)^2 = 42$$

$$SS\,(Total) = SS(Factor) + SS(Error) = 45.73 + 42 = 87.73$$

$$\text{MS (Factor)} = \frac{SS\ (Factor)}{k-1} = \frac{45.73}{3-1} = 22.865$$

$$\text{MS (Error)} = \frac{SS(Error)}{n-k} = S_P^2 = \frac{42}{15-3} = 3.5$$

$$\text{F} = \frac{\text{MS}_{\text{factor}}}{\text{MS}_{\text{error}}} = \frac{22.865}{3.5} = 6.53$$

Table 9.5 One-way ANOVA table for the example.

ANOVA table				
Source	**df**	**SS**	**MS**	**F**
Factor	3 – 1	45.73	22.865	6.53
Error	15 – 3	42	3.5	
Total	14			

Table 9.6 F distribution values corresponding to a lower-tail probability of 0.95.

Denominator degrees of freedom		**1**	**2**	**3**
	1	161.4	199.5	215.7
	2	18.51	19.00	19.16
	3	10.13	9.55	9.28
	4	7.71	6.94	6.59
	5	6.61	5.79	5.41
	6	5.99	5.14	4.76
	7	5.59	4.74	4.35
	8	5.32	4.46	4.07
	9	5.12	4.26	3.86
	10	4.96	4.10	3.71
	12	4.75	3.89	3.49
	15	4.54	3.68	3.29

The F statistic can be compared to the critical F value found by looking up the F statistic in Table C.17 in Appendix C corresponding to a lower-tail probability of $1-\alpha=0.95$, numerator degrees of freedom of $df_{factor}=2$, and denominator degrees of freedom $df_{error}=12$. As shown in Table 9.6, $F_{critical}=3.89$. The calculated F statistic is greater than the critical F value, so we reject the null hypothesis.

An approximate p-value can be found by finding the closest F values above and below the calculated statistic of $F=6.53$ in the upper-tail probability Tables C.10–C.14 in Appendix C that have 2 numerator degrees of freedom and 12 denominator degrees of freedom. These closest values will correspond to limits on the p-value. For example, the closest value less than $F=6.53$ with 2 numerator and 12 denominator degrees of freedom found in the tables is $F=5.096$, corresponding to an upper-tail probability of 0.025. The closest value greater than $F=6.53$ with

Table 9.7 F distribution values corresponding to an upper-tail probability of 0.025.

Denominator degrees of freedom	1	2	3
1	647.79	799.5	864.16
2	38.506	39	39.165
3	17.443	16.044	15.439
4	12.218	10.649	9.9792
5	10.007	8.4336	7.7636
6	8.8131	7.2599	6.5988
7	8.0727	6.5415	5.8898
8	7.5709	6.0595	5.416
9	7.2093	5.7147	5.0781
10	6.9367	5.4564	4.8256
12	6.5538	5.0959	4.4742
15	6.1995	4.765	4.1528
20	5.8715	4.4613	3.8587
24	5.7166	4.3187	3.7211

2 numerator and 12 denominator degrees of freedom is 6.927, corresponding to an upper-tail probability of 0.01 (see Table 9.7). These encompass the test statistic 6.53, so the p-value is between 0.01 and 0.025. Because this is smaller than the alpha level of 0.05, we reject the null hypothesis and conclude that at least one mean is different.

In Minitab: Enter the data in the worksheet as shown in Figure 9.1. Go to **Stat > ANOVA > One-Way** and select **Response data are in a separate column for each factor level** as shown in Figure 9.2.

The resulting p-value of 0.012 is less than 0.05, so we reject the null hypothesis that all means are equal (see Figure 9.3). The statistic S = 1.87083 is equal to the square root of MS(Error) and is the standard deviation of the residuals. The statistics R-Sq, R-Sq(adj), and R-Sq(pred) give an indication of how much of the variation in production was explained by the training method. These concepts are discussed in more detail in Chapter 10.

The one-way ANOVA does not tell us which mean is different; however, the resulting interval plot gives us an indication (see Figure 9.4). On the interval plot it can be seen that while the confidence intervals for Consultant and Untrained overlap considerably, the interval for Training does not overlap with Untrained. This can be confirmed by reading the confidence limit values for each sample in the session window.

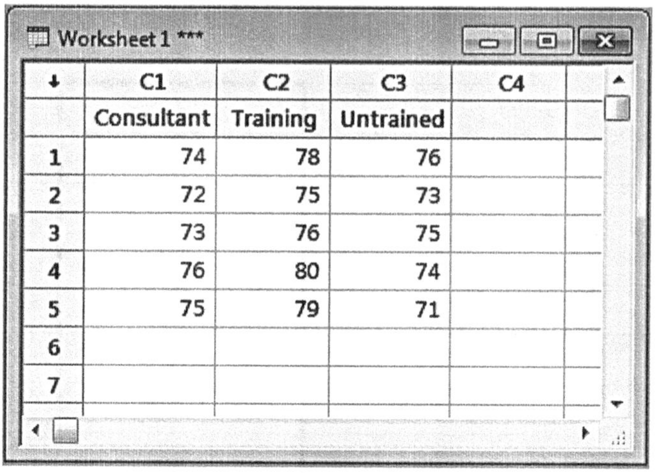

Figure 9.1 Worksheet for regression.

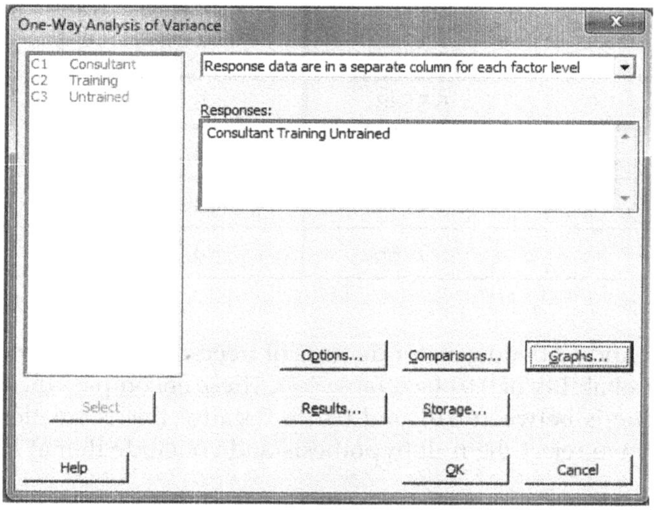

Figure 9.2 Dialog box for regression.

The statistically different mean or means can be quickly identified by using the Comparisons subdialog when performing one-way ANOVA with Minitab. Go to **Stat>ANOVA>One-Way** and click on **Comparisons**. Then select **Tukey** as shown in Figure 9.5 and click **OK**. This method provides a graphic with confidence intervals for the means; confidence intervals of means that do not contain 0.0 are significantly different (see Figure 9.6).

From the interval plot in Figure 9.6 we can see that internal training resulted in higher production rates than consultant training; there was no statistically significant difference in production rates between those untrained and those trained by consultant; and those untrained had lower production rates than those who attended internal training.

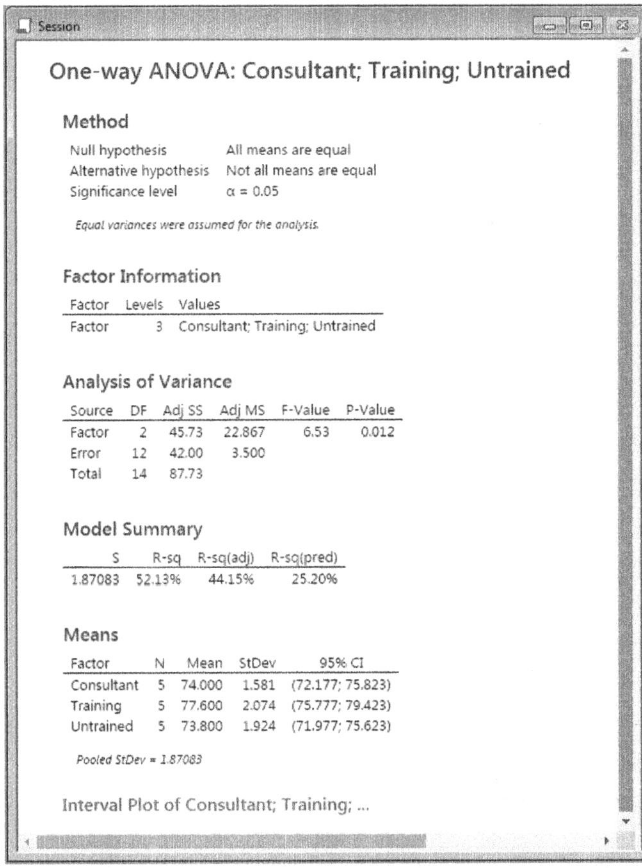

Session

One-way ANOVA: Consultant; Training; Untrained

Method

Null hypothesis	All means are equal
Alternative hypothesis	Not all means are equal
Significance level	α = 0.05

Equal variances were assumed for the analysis.

Factor Information

Factor	Levels	Values
Factor	3	Consultant; Training; Untrained

Analysis of Variance

Source	DF	Adj SS	Adj MS	F-Value	P-Value
Factor	2	45.73	22.867	6.53	0.012
Error	12	42.00	3.500		
Total	14	87.73			

Model Summary

S	R-sq	R-sq(adj)	R-sq(pred)
1.87083	52.13%	44.15%	25.20%

Means

Factor	N	Mean	StDev	95% CI
Consultant	5	74.000	1.581	(72.177; 75.823)
Training	5	77.600	2.074	(75.777; 79.423)
Untrained	5	73.800	1.924	(71.977; 75.623)

Pooled StDev = 1.87083

Interval Plot of Consultant; Training; ...

Figure 9.3 Session window for one-way ANOVA.

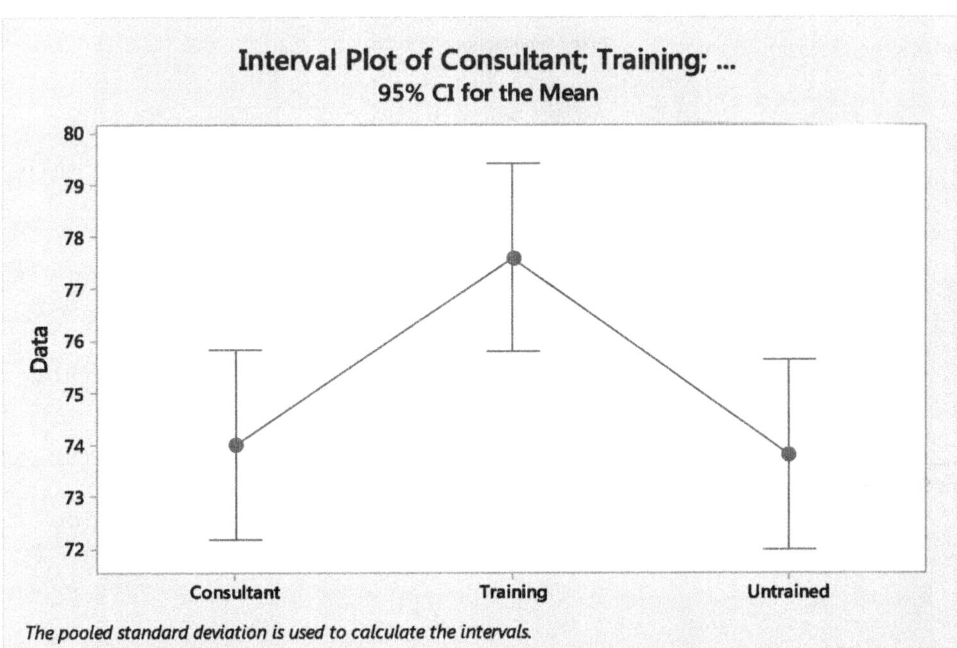

Figure 9.4 Interval plot for one-way ANOVA.

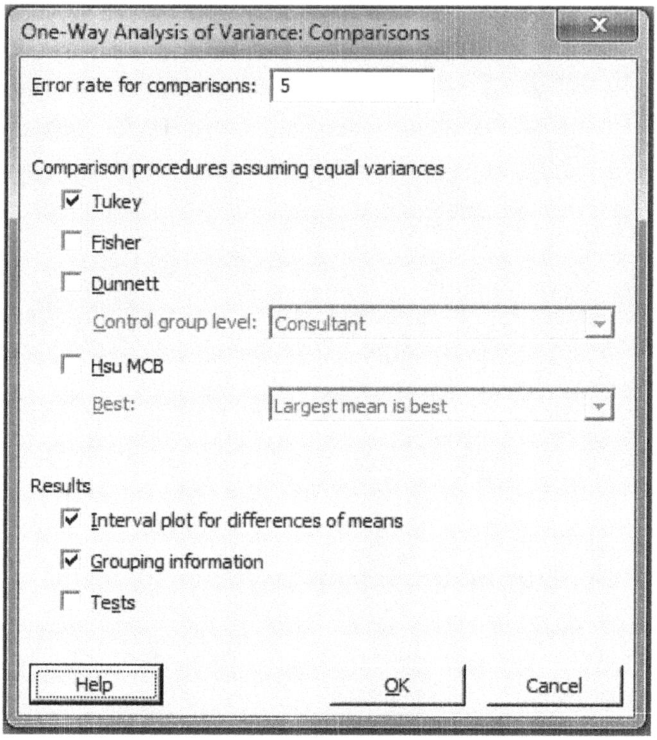

Figure 9.5 Comparisons subdialog box for selecting the Tukey method.

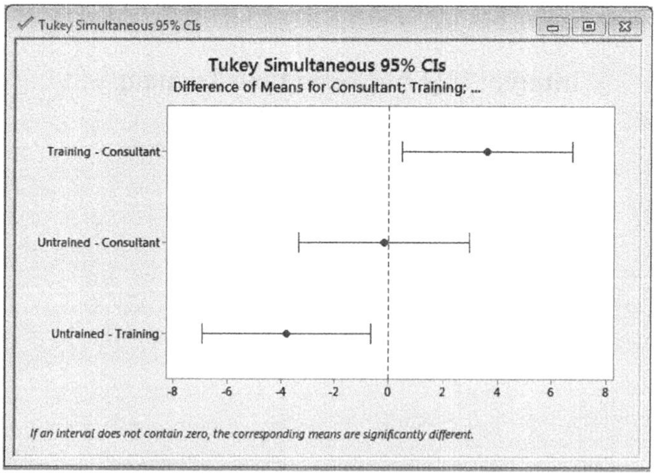

Figure 9.6 Interval plot for Tukey method.

WELCH'S METHOD

Welch's method provides an alternate calculation for the p-value in one-way ANOVA and performs well even when the underlying populations are not normal or variances are not equal. The test statistic W is calculated as

$$W^* = \frac{\Sigma w_i\left[(\bar{x}_{i\cdot} - \hat{\mu})^2 / k - 1\right]}{1 + \left[2(k-2)/k^2 - 1\right]\Sigma h_i},$$

where

$$w_i = \frac{n_i}{s_i^2}$$

$$W = \Sigma w_i$$

$$\hat{\mu} = \frac{\Sigma w_i \bar{x}_i}{W}$$

$$h_i = \frac{\left(1 - \frac{w_i}{W}\right)^2}{n_i - 1}.$$

The p-value is calculated using an F statistic in the same manner as the classical test, with numerator degrees of freedom k-1 and denominator degrees of freedom f, which is calculated as:

$$f = \frac{k^2 - 1}{3\Sigma h_i}$$

Therefore, the p-value is equal to $P(F_{k-1,f} > W^*)$.

In Minitab, the Welch method can be chosen by clicking on the **Options** subdialog in one-way ANOVA and unchecking **Assume equal variances**.

GENERAL LINEAR MODEL

General Linear Model (GLM) is used to create models with a continuous response and multiple factors, including at least one categorical factor. If all factors are continuous, then regression should be used. General Linear Model determines the statistical significance of each term in the model and estimates the linear effect that the term has on the response.

General linear models can include almost any type of factor, including fixed effects, random effects, covariate, and interactions. A *fixed effect* is a categorical factor that has levels of specific interest. For example, you are likely interested in the specific difference between males and females if gender is included in a model. A *random effect* is a categorical factor consisting of levels that are not of specific interest, often because they were chosen from a larger group of potential levels. With a random effect the variation in the impacts of different levels is of interest. For example, if income is the response and you want to examine zip code as a factor, you might randomly select 10 zip codes to collect data on. By treating zip code as a random effect, your model will estimate how much income varies generally between zip codes rather than specifically comparing the 10 randomly chosen zip codes. A *covariate* is a continuous predictor to include in the model, and an *interaction* is a term to account for when the impact of one factor or covariate in the model changes depending on another factor. For example, a model looking at income that includes gender and age might require an interaction between the two. While younger respondents may show little difference between males and females, older respondents may show one gender making much more than the other. Therefore the impact of age on income is dependent on another factor—gender.

General linear models include metrics on the quality of fit such as R^2, R^2(adjusted), and R^2(predicted). These are discussed in more detail in Chapter 10 on regression. Generally, values closer to 100% indicate the factors included in the model explained most of the variation in the response and large differences between R^2 and R^2(predicted) indicate that the factors may correlate with the response but are not likely to be the true driving forces.

Additionally, coefficients for the intercept and each covariate are calculated. For covariates, the coefficient indicates the change in the response estimated to occur for each one unit increase in that covariate. For categorical factors, a reference level is selected and the coefficient for each other level represents the change in the response for that level compared to the reference level. The coefficient for the reference level is then the negative sum of all other levels so that the sum of all levels equals zero. For example, if suppliers A, B, and C are compared and C is the reference level, a coefficient of –1 for supplier A and –3 for supplier B indicates the coefficient for supplier C is +4. An equation to predict the response would be:

$$y = intercept - 1\ (if\ Supplier\ A) - 3\ (if\ Supplier\ B) + 4\ (if\ Supplier\ C) + other\ factors$$

Example

You manufacture crayons and want to better understand the differences in breaking force as customers often complain that the crayons break when being used. You randomly sample crayons across the two work shifts (AM and PM), the three manufacturing lines (1, 2, and 3), and three randomly selected colors (Red, Yellow, and Blue). The data are shown in Table 9.8.

The factors shift and line represent all available levels, and those levels are of specific interest so the factors are treated as fixed. Color, however, contains randomly selected levels among a larger pool of available levels. Rather than being interested in the breaking force for these three specific colors, you are interested in how much color impacts breaking force. Therefore, color is a random effect. There are no covariates in the data and no interactions will be tested.

In Minitab: Go to **Stat > ANOVA > General Linear Model > Fit General Linear Model** and complete the dialog as shown in Figure 9.7 and click **OK**.

Click on **Random/Nest** and complete the dialog as shown in Figure 9.8.

Click **OK** to return to the main dialog. Click on **Graphs** and complete the dialog as shown in Figure 9.9 and then click **OK** in each dialog.

The resulting residuals plot is shown in Figure 9.10.

The residuals plot indicates that the data do not show any extreme non-normality, outliers, or patterns across fits or observation order. Therefore the results of the model can be analyzed. The session window output is shown in Figure 9.11.

Table 9.8 Crayon breaking force.

Shift	Line	Color	Breaking force
AM	1	Red	44.8
AM	1	Yellow	34.2
AM	1	Blue	43.7
AM	2	Red	45.5
AM	2	Yellow	36.0
AM	2	Blue	46.0
AM	3	Red	36.6
AM	3	Yellow	24.5
AM	3	Blue	41.2
PM	1	Red	50.1
PM	1	Yellow	36.3
PM	1	Blue	46.3
PM	2	Red	52.6
PM	2	Red	50.4
PM	2	Yellow	35.6
PM	2	Blue	52.8
PM	2	Blue	47.1
PM	3	Red	43.2
PM	3	Yellow	27.7
PM	3	Yellow	26.0
PM	3	Blue	38.3

The p-values for each term in the one-way ANOVA table in Figure 9.11 indicate that each factor has a statistically significant relationship with breaking force at an alpha level of 0.05. All three R^2 values are very high, indicating that the terms included in the model not only explain over 90% of the variation in breaking forces, but that the model is predictive as well.

The p-values provided in the *coefficients* table indicate whether that level of that factor had a statistically significant different impact on the response than the reference level. The reference level is the one not listed in the coefficients table. For example, shift only lists "AM," so "PM" is the reference level. By examining the

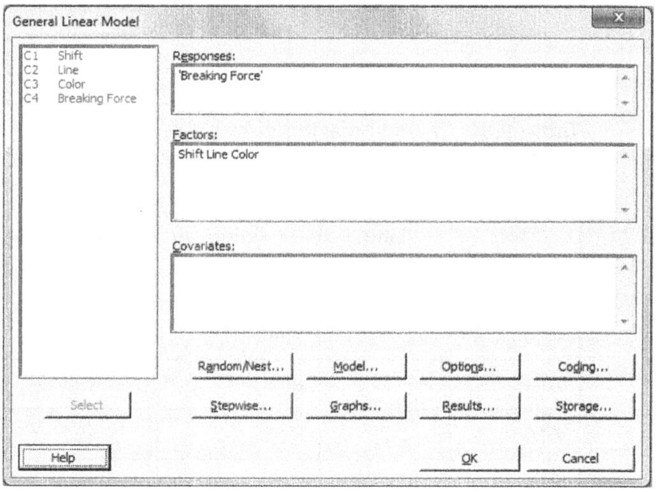

Figure 9.7 Fit General Linear Model dialog box.

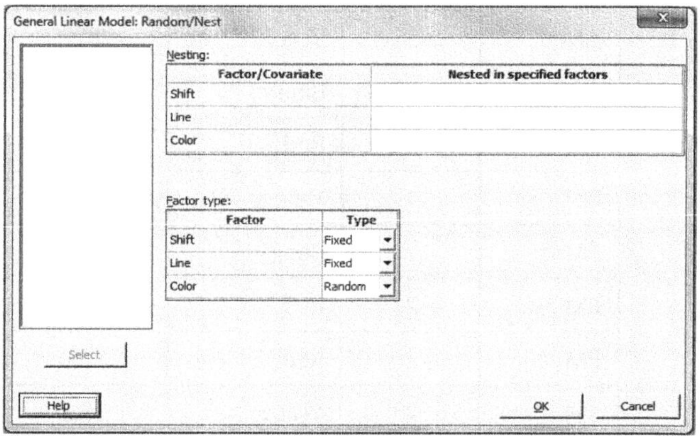

Figure 9.8 Random/Nest subdialog box in Fit General Linear Model.

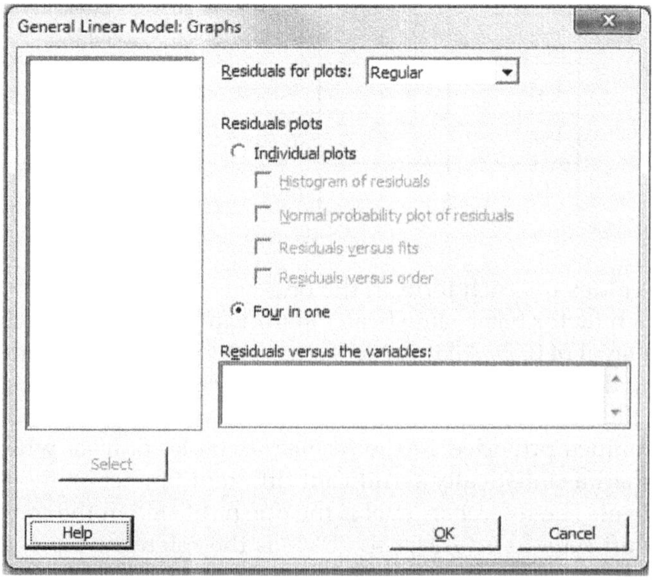

Figure 9.9 Graphs subdialog box in Fit General Linear Model.

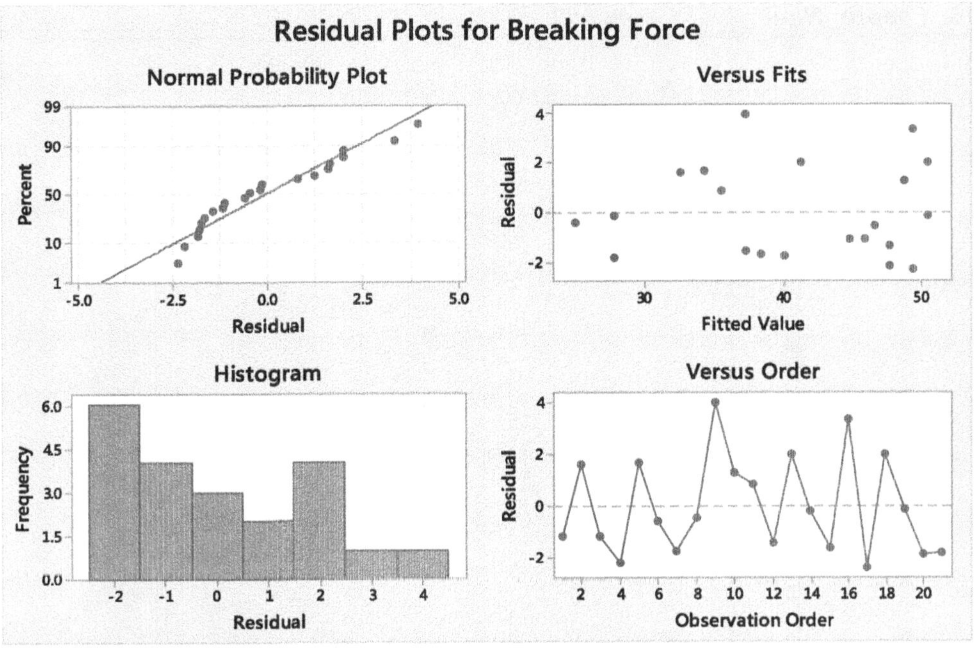

Figure 9.10 Residuals plot for crayon breaking force.

Session

General Linear Model: Breaking Force versus Shift; Line; Color

Method

Factor coding (-1; 0; +1)

Factor Information

Factor	Type	Levels	Values
Shift	Fixed	2	AM; PM
Line	Fixed	3	1; 2; 3
Color	Random	3	Blue; Red; Yellow

Analysis of Variance

Source	DF	Adj SS	Adj MS	F-Value	P-Value
Shift	1	42.42	42.417	9.01	0.009
Line	2	349.49	174.744	37.12	0.000
Color	2	749.71	374.855	79.62	0.000
Error	15	70.62	4.708		
Lack-of-Fit	12	50.51	4.209	0.63	0.757
Pure Error	3	20.11	6.703		
Total	20	1405.20			

Model Summary

S	R-sq	R-sq(adj)	R-sq(pred)
2.16976	94.97%	93.30%	90.13%

Coefficients

Term	Coef	SE Coef	T-Value	P-Value	VIF
Constant	40.610	0.481	84.44	0.000	
Shift					
AM	-1.444	0.481	-3.00	0.009	1.01
Line					
1	1.956	0.702	2.79	0.014	1.36
2	3.712	0.659	5.63	0.000	1.38
Color					
Blue	3.710	0.672	5.52	0.000	*
Red	4.825	0.672	7.18	0.000	*

Regression Equation

Breaking Force = 40.610 - 1.444 Shift_AM + 1.444 Shift_PM + 1.956 Line_1 + 3.712 Line_2
- 5.669 Line_3 + 3.710 Color_Blue + 4.825 Color_Red - 8.535 Color_Yellow

Equation treats random terms as though they are fixed.

Figure 9.11 Session window output for crayon breaking force model.

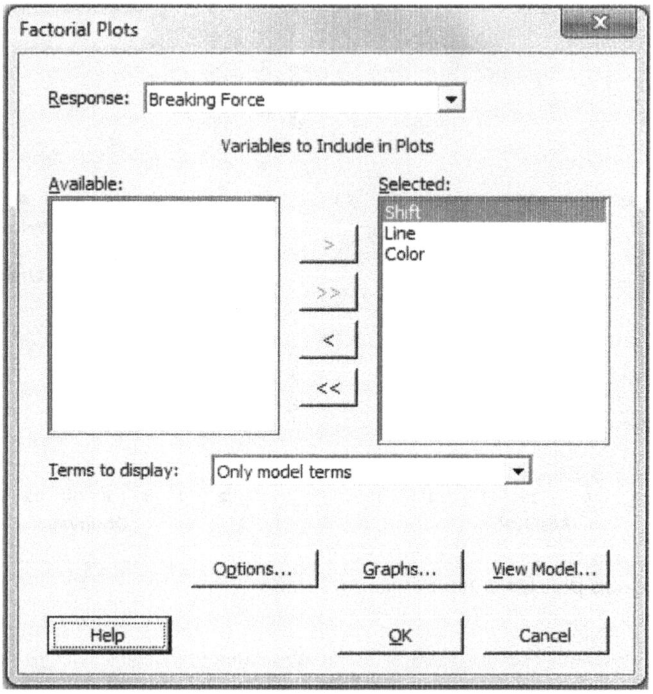

Figure 9.12 Factorial plots dialog box.

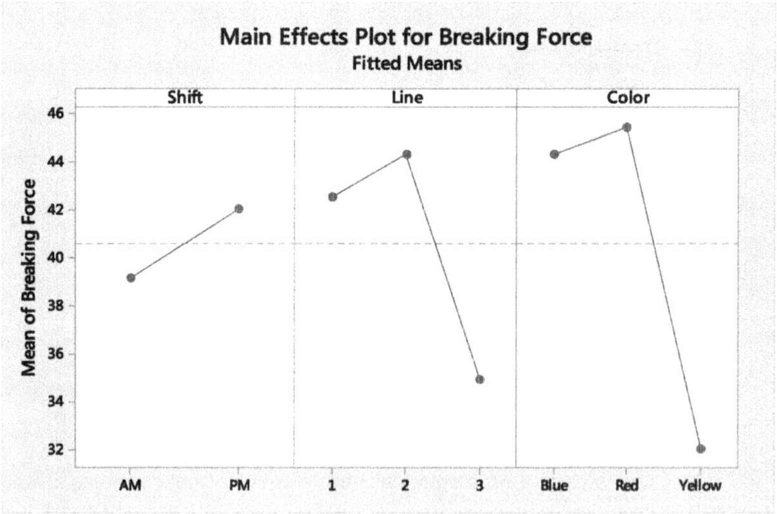

Figure 9.13 Main effects plot for crayon breaking force model.

regression equation, the impact of each level of each factor on breaking strength can be seen.

In Minitab: Go to **Stat > ANOVA > General Linear Model > Factorial Plots** and complete the dialog as shown in Figure 9.12 and click **OK**.

The resulting main effects plot is shown in Figure 9.13.

The plot in Figure 9.13 illustrates that crayons produced during the PM shift require a higher breaking force than those produced during the AM shift. Additionally, lines 1 and 2 produce higher breaking strengths than line 3. Finally, it can be seen that the color of the crayon has the greatest impact on the breaking strengths of the crayons.

QUESTIONS AND ANSWERS

Q: Do the data need to be normally distributed to perform an ANOVA or regression analysis?

A: The response data do not need to be normally distributed, but an assumption of the classical test is that within each group the data are normally distributed. This can be determined by checking the residuals for normality. Welch's test is robust to non-normality.

Q: What are residuals?

A: Residuals are the differences between observed values and the expected values from a model such as ANOVA or regression. For a good model, the residual should be small and meet certain criteria as outlined in Chapter 9 for ANOVA and Chapter 10 for regression.

Q: How critical is the requirement for common variance when using the Tukey comparison in ANOVA?

A: When performing ANOVA, it is recommended to use the Welch's test whenever there is evidence or suspicion that variances are unequal. Otherwise, the traditional test may give unreliable results, including p-values and multiple comparisons.

Chapter 10
Regression

SIMPLE LINEAR REGRESSION

Chapter 3 demonstrated the concept of correlation between two continuous variables. Often one of those variables is dependent on the other and it is desirable to quantify that association. The relationship between a single continuous independent variable (x) and a continuous dependent variable (y) can be analyzed using simple linear regression (Stephens 2004). Simple linear regression can provide both a formula for prediction and measures of the strength of the relationship.

Simple linear regression produces the formula $\hat{Y} = b_0 + b_1 x$ by finding the values of b_0 and b_1 that provide the best fit. For any given value of these coefficients, the difference between the y-value of each data point and the y-value of the fitted line is known as a *residual*. The best fit line minimizes the sum of the squared residuals for the entire data set. Figure 10.1 shows a best fit line with a dotted line representing one of the residuals.

The formulas for the coefficients are:

$$b_1 = \frac{\Sigma(x_i - \bar{x})(y_i - \bar{y})}{\Sigma(x_i - \bar{x})^2}$$

$$b_0 = \bar{y} - b_1\bar{x}$$

A one-way ANOVA table can be formed for regression using the following formulas for the sums of squares:

$$SS_{regression} = \Sigma(\hat{y}_i - \bar{y})^2$$

$$SS_{Error} = \Sigma(y_i - \hat{y}_i)^2$$

$$SS_{total} = \Sigma(y_i - \bar{y})^2$$

In the formulas, y_i is the *i*th y value, $\hat{y}_i$ is the predicted y value for the *i*th data point, and $\bar{y}$ is the mean y value. The one-way ANOVA table is then constructed as shown in Table 10.1.

The F statistic is used to perform a hypothesis test where the null hypothesis is that there is no linear relationship between x and y, and the alternative hypothesis is that there is a linear relationship. The p-value for the test is $P(F_{1,n-2} > F_{Regression})$. Although the test can indicate that there is a statistically significant relationship, it does not indicate the strength of that relationship or quality of prediction.

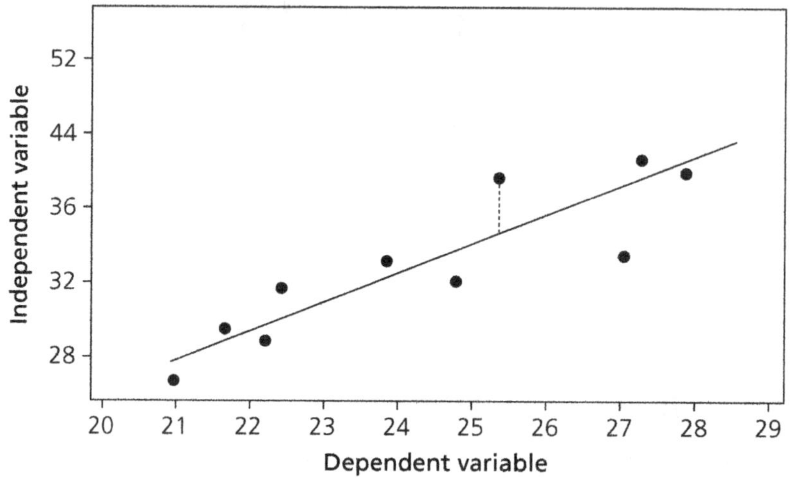

Figure 10.1 Best fit line.

Table 10.1 A one-way ANOVA table for regression.

Source	df	SS	MS	F
Regression	1	$\sum(\hat{y}_i - \bar{y})^2$	$SS_{Regression}/1$	$\frac{MS_{Regression}}{MS_{Error}}$
Error	n − 2	$\sum(y_i - \hat{y}_i)^2$	$SS_{Error}/n-2$	
Total	n − 1	$\sum(y_i - \bar{y})^2$		

One common metric for determining the strength of a relationship is R^2. This value indicates the percentage of the total variation in the y (response) variable that is explained by the x (predictor) variable. It is calculated as:

$$R^2 = \frac{SS_{Regression}}{SS_{Total}} * 100\%$$

In Chapter 3, the concept of correlation was introduced. A measure of linear correlation is the *Pearson correlation coefficient*,

$$r = \frac{\Sigma(x_i - \bar{x})(y_i - \bar{y})}{(n-1)s_x s_y},$$

where s_x is the standard deviation of the x variable and s_y is the standard deviation of the y variable. The statistic can take on values from –1 to 1, with values close to 0 indicating a weak or no relationship and values closer to –1 or 1 indicating a stronger linear relationship. Negative values occur when the y variable decreases as x variable increases, and positive values occur when the y variable increases as the x variable increases. The statistic R^2 is equal to the Pearson correlation coefficient squared, times 100%. R^2 is called the *coefficient of determination*, and it tells us how much of a change in the response variable y is explained by changes in the independent variable x (Nelson, Coffin, and Copeland 2003).

Recall that the "residual" for each data point is equal to the actual y-value minus the predicted y-value from the fitted line. There are several assumptions about the residuals. The first is that the residuals are normally distributed. The quality of the fitted line is somewhat robust to this assumption but if the residuals are skewed, then the fitted line may no longer represent the expected value of the y-variable at a given value of the x-variable. Further, extreme outliers may indicate a data point that is not representative of the true relationship because of data collection errors or environmental or other factors present when the data were collected. There should be no relationship between the residuals and the fitted values. One example of a relationship between the two would be the variance of the residuals increasing (or decreasing) as the fitted values increase, indicating that prediction is not as accurate for some fitted values as others. Another example would be a quadratic relationship, indicating that a regression model that includes a quadratic term may be necessary to fit the data well. There should also be no relationship between the residuals and the order in which the data were collected. A relationship would indicate that the conditions in which the data were collected were not stable over time, and other factors are influencing the model. Each of these assumptions about the residuals can be checked graphically.

Example

An engineer is studying the possible relationship between fuel consumption in liters used per 100 kilometers (l/km) and emissions in grams per kilometer (g/km) driven. The data are shown in Table 10.2.

Table 10.2 Regression data.

Fuel consumption (l/km)	Emissions (g/km)
5.40	172
5.57	177
3.41	109
4.08	115
4.83	154
4.41	139
5.50	154
5.33	149
3.08	99

Table 10.3 Table for calculating components of b_1.

X (Fuel consumption)	*Y* (Emissions)	$(x_i - \bar{x})(y_i - \bar{y})$	$(x_i - \bar{x})^2$
5.40	172	24.16	0.60
5.57	177	34.19	0.90
3.41	109	38.69	1.47
4.08	115	14.07	0.30
4.83	154	2.71	0.04
4.41	139	0.40	0.05
5.50	154	11.49	0.77
5.33	149	5.73	0.50
3.08	99	64.65	2.38
		Σ = 196.09	Σ = 7.01

The engineer defines fuel consumption as the x variable and emissions as the y variable. The mean of each is calculated as $\bar{x} = 4.623$ and $\bar{y} = 140.89$. The engineer then creates a table such as the one shown in Table 10.3 to assist in performing the calculations.

The coefficients of the regression equation are calculated as:

$$b_1 = \frac{\Sigma(x_i - \bar{x})(y_i - \bar{y})}{\Sigma(x_i - \bar{x})^2} = \frac{196.09}{7.01} = 27.97$$

$$b_0 = \bar{y} - b_1\bar{x} = 140.89 - 27.97 * 4.623 = 11.27$$

Therefore, the equation can be written:

$$\text{Emissions} = 11.57 + 27.97 * \text{Fuel Consumption}$$

To calculate sum of squares, a new table is created containing the predicted Y values as shown in Table 10.4.

The various sums of squares, correlation coefficient, best fit line, and coefficient of determination are then calculated as follows:

$$SS_{regression} = \Sigma(\hat{y}_i - \bar{y})^2 = (162.61 - 140.89)^2 + (167.36 - 140.89)^2 + \cdots + (97.72 - 140.89)^2$$
$$= 5480$$

$$SS_{Error} = \Sigma(y_i - \hat{y}_i)^2 = (172 - 162.61)^2 + (177 - 167.36)^2 + \cdots + (99 - 97.72)^2 = 637.4$$

$$SS_{total} = \Sigma(y_i - \bar{y})^2 = (172 - 140.89)^2 + (177 - 140.89)^2 + \cdots + (99 - 140.89)^2 = 6127$$

To calculate the correlation coefficient r, first the standard deviation of the X variable and Y variable are determined to be $s_x = 0.936$ and $s_y = 27.67$. Then the correlation coefficient is calculated:

Table 10.4 Table for calculations.

X (Fuel consumption)	*Y* (Emissions)	$\hat{y}_i = 11.57 + 27.97 * x_i$
5.40	172	162.61
5.57	177	167.36
3.41	109	106.95
4.08	115	125.69
4.83	154	146.67
4.41	139	134.92
5.50	154	165.40
5.33	149	160.65
3.08	99	97.72

$$r = \frac{\Sigma(x_i - \bar{x})(y_i - \bar{y})}{(n-1)s_x s_y} = \frac{(5.40 - 4.623)(172 - 140.89) + \cdots + (3.08 - 4.623)(99 - 140.89)}{(9-1) * 0.936 * 27.67}$$
$$= 0.946$$

The coefficient of determination is calculated:

$$R^2 = \frac{SS_{Regression}}{SS_{Total}} * 100\% = \frac{5480}{6127} * 100\% = 89.4\%$$

The regression equation Y = 11.5 + 27.99 X indicates that the expected exhaust for a given level of fuel consumption is equal to 27.99 times the fuel consumption plus 11.5.

Here is an example for a fuel consumption of 4 l/km:

$$27.97(4) + 11.57 = 123.45 \text{ g/km}$$

The R^2 tells us that 89.4% of the variability in emissions is determined by the level of fuel consumption.

In Minitab: Go to **Stat > Regression > Regression > Fit Regression Model** and enter the columns containing the response (y) and the continuous predictor (x) as shown in Figure 10.2. Then click on **Graphs** and select **Four in one** to view a histogram of residuals, a probability plot of residuals, residuals versus fits, and residuals versus order, as shown in Figure 10.3. This step is not necessary but can be helpful in assessing the data. Click **OK**.

The results are shown in Figure 10.4. The normal probability plot and histogram in Figure 10.5 indicate that the residuals may not be normally distributed. The residuals versus fits plot indicates no relationship between the two, and the residuals versus order plot also shows no relationship.

The results differ slightly from the hand calculation due to rounding.

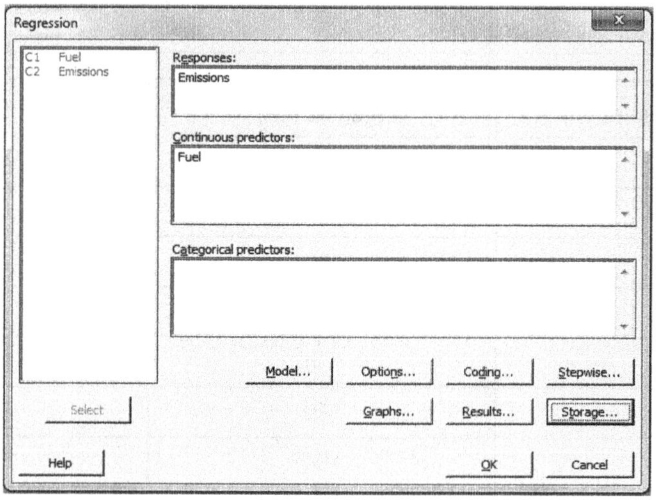

Figure 10.2 Dialog box for regression.

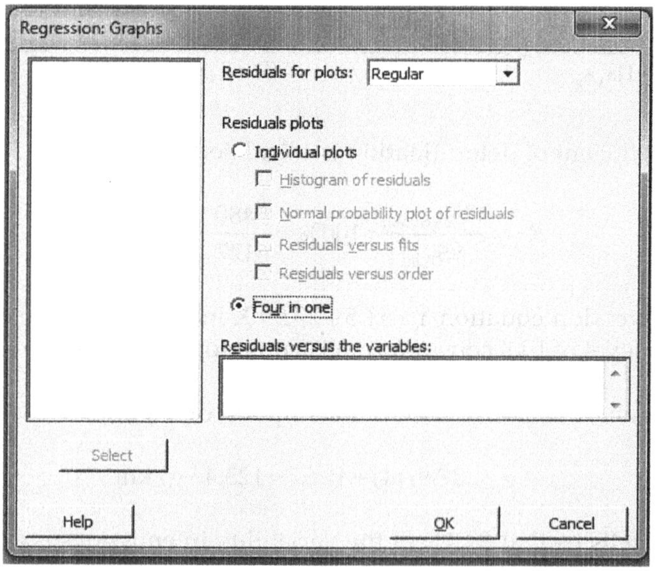

Figure 10.3 Graphs subdialog box for regression.

MULTIPLE REGRESSION

There often is more than one predictor that is of interest when creating a regression model. In such a case a single model can be formed that incorporates all predictors of interest and this is known as *multiple regression* (Minitab, "Multiple Regression"). For example, if you wanted to estimate the viscosity of lubricating oil during use in a piece of equipment, you might use temperature and the number of hours the oil has been in use as factors. Multiple regression allows for hypothesis tests to

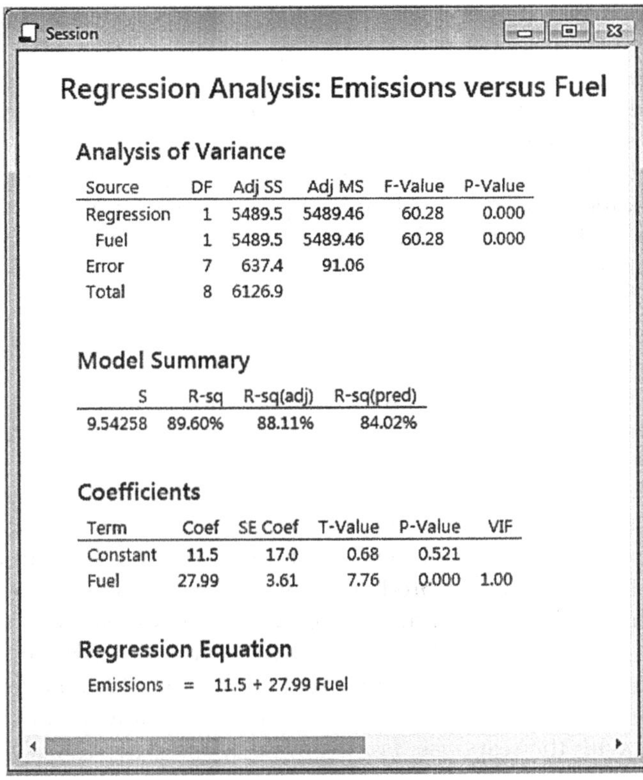

Session

Regression Analysis: Emissions versus Fuel

Analysis of Variance

Source	DF	Adj SS	Adj MS	F-Value	P-Value
Regression	1	5489.5	5489.46	60.28	0.000
Fuel	1	5489.5	5489.46	60.28	0.000
Error	7	637.4	91.06		
Total	8	6126.9			

Model Summary

S	R-sq	R-sq(adj)	R-sq(pred)
9.54258	89.60%	88.11%	84.02%

Coefficients

Term	Coef	SE Coef	T-Value	P-Value	VIF
Constant	11.5	17.0	0.68	0.521	
Fuel	27.99	3.61	7.76	0.000	1.00

Regression Equation

Emissions = 11.5 + 27.99 Fuel

Figure 10.4 Session window output for regression.

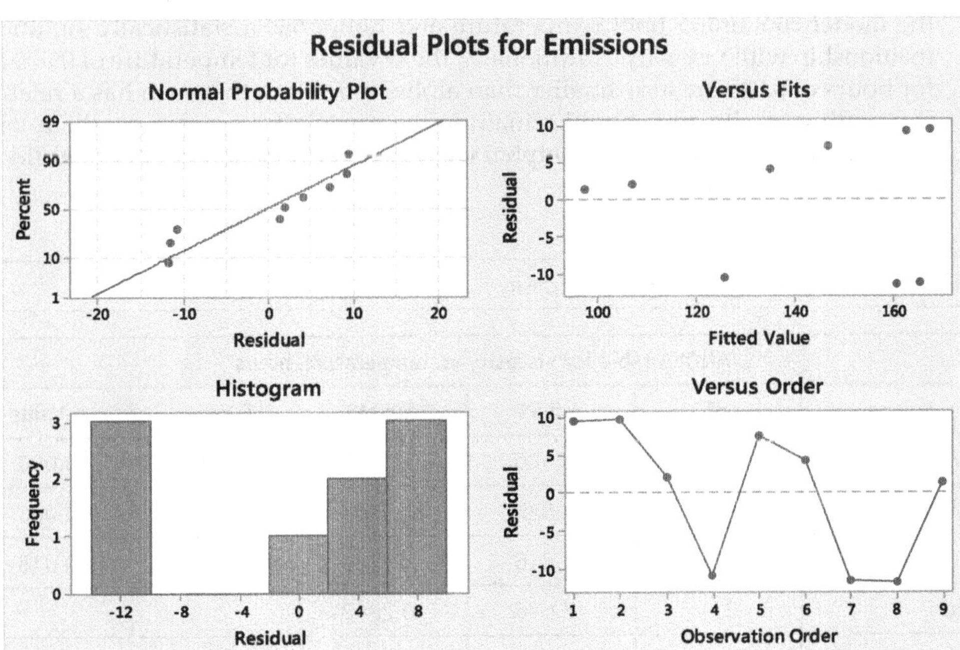

Figure 10.5 Residual plots for regression.

be performed on whether each predictor influences the response and provides the equation that describes the relationship and other diagnostics used with simple regression such as R-squared.

Calculating the estimated coefficients for a regression model with multiple predictors requires matrix algebra and should be done by software in all but the simplest of cases. As with simple regression, each predictor has a coefficient, and the combination of an intercept and all predictors multiplied by their coefficient forms a model. For example, in the case of the lubricating oil viscosity, we might find that the model to predict viscosity is estimated as:

$$viscosity = 6.506 - 03431 * temperature + 0.00859 * hours + error$$

Before including each predictor in a final model, however, an ANOVA table should be used to determine if the relationship is statistically significant. Terms that are not significant should be removed one at a time, and a new model and ANOVA table should be calculated each time. Only after all terms in the model have a statistically significant relationship, indicated by a p-value less than the desired alpha, should the model be used for prediction. Insignificant terms in the model affect the accuracy of prediction and the precision of the estimated error.

As with simple regression, an ANOVA table is created to show the sum of squares, mean square, F statistic, and p-value for each term in the model. Each predictor has its own row in the table, and there is an additional row for the overall regression indicating whether all terms combined show a statistically significant relationship with the response. For example, a one-way ANOVA table for the viscosity example is shown in Table 10.5.

From the one-way ANOVA table, it can be seen that the p-value for Regression of 0.000 is smaller than the desired alpha of 0.05, and we could conclude that the model containing both temperature and hours has a statistically significant relationship with viscosity. Additionally, the p-values for temperature of 0.000 and for hours of 0.018 are also smaller than alpha, indicating that each has a relationship with viscosity and should remain in the model. If, for example, the p-value for hours had been larger than alpha, we would create a new model that did not include hours as a predictor.

Table 10.5 One-way ANOVA table for viscosity.

ANOVA table for viscosity vs. temperature, hours					
Source	**df**	**Adj SS**	**Adj MS**	**F-Value**	**p-Value**
Regression	2	15.586	7.7930	27.41	0.000
Temperature	1	12.522	12.5218	44.04	0.000
Hours	1	1.816	1.8160	6.39	0.018
Error	27	7.676	0.2843		
Total	29	23.262			

The *error term* in the model represents the unexplained variation in the response after the effect of the predictors has been accounted for. An assumption of regression is that the error is normally distributed. It has a mean on 0, meaning that the term does not affect the predicted value of the response for given values of the predictors. However, it does have variance equal to the MS(Error) term from the one-way ANOVA table and, equivalently, a standard deviation equal to the square root of that number.

Multiple regression provides a value of R^2 with the same interpretation as in simple regression, as the percentage of the variation in the response explained by the predictors. However, R^2 has a potential flaw with multiple predictors. Its value must increase each time an additional predictor is included in a model regardless of whether that predictor was statistically significant or had a meaningful effect on the response. To counter this, an adjusted version of the statistic was created that provides a penalty for an increasing number of predictors. R^2(adjusted) is calculated as:

$$R^2(adjusted) = 1 - \left(\frac{MS(Error)}{MS(Total)} \right)$$

A model that provides an R^2(adjusted) that is much smaller than R^2 likely has at least one term that should be removed because it is statistically not significant or does not have a meaningful effect on the response.

Additionally, there is one other form of R^2 that is often used for assessing the predictive ability of the model. Predicted R^2 is calculated by systematically removing each data point from the data set and then calculating a model based on the remaining data. A prediction of the response of the missing data point based on the predictor values for that data point is then made. Next, squared difference between the prediction and the actual value is calculated. The sum of those squared differences is a statistic called PRESS, and the formula for predicted R^2 is:

$$Predicted\ R^2 = 1 - \frac{PRESS}{SS(Total)}$$

A model that results in a predicted R^2 that is much lower than R^2 likely has too many predictors relative to the size of the data set. This situation is called *overfitting* and results in poor prediction. A good model will provide values for all three of the R^2 statistics that are very similar.

Because regression models are almost always based on observed processes and not a designed experiment, the predictors are not controlled during the data collection. One result is that the predictors may be correlated with one another. For example, if a logging company wanted to estimate the board feet of lumber that could be obtained from a tree, it might measure the circumference near ground level, the height of the trunk, and the circumference of the largest branch. These three measurements could be used as predictors in a model with a response of board feet. However, trees with a larger circumference near the ground are likely to be taller and have bigger branches, so all three predictors are correlated with one another. The condition of having correlated predictors in known as *multicollinearity* and is very important to understand.

When multicollinearity is present, it is mathematically impossible to estimate the coefficients for each predictor independently of one another. Therefore the value of each coefficient may be illogical and not helpful. In the tree example, it is obvious that as the value for each predictor increases, the number of board feet expected from the tree also increases, and therefore each should have a positive coefficient. However, because of multicollinearity, a model created from data collected for this example might include one or two negative coefficients and the magnitude of all the coefficients might be illogical. Additionally, when all three are put into a single model, one or more of them may come up as not statistically significant, although alone each predictor is very likely to have a statistically significant relationship with the response.

Fortunately, multicollinearity typically does not affect the predictive ability of the model. While the p-values and coefficients may be very inaccurate, the model is likely to still provide good prediction of the response for a given set of predictor values. In the presence of multicollinearity, it may be tempting to just keep a single factor and do simple regression, but the multiple regression model often will still remain the more useful model in making prediction. However, it will not be useful in understanding the nature of the relationship between the predictor and response.

To determine the presence and severity of multicollinearity in multiple regression, there is a statistic known as Variance Inflation Factor (VIF). For each predictor x_i, a regression model is created in which x_i is the response and all other predictors are used in a model to predict x_i. The VIF is then calculated as:

$$VIF(x_i) = \frac{1}{1 - R^2(x_i)}$$

If x_i is highly correlated with other predictors, then the model will have a high R^2 and the VIF will then be high. If it is not correlated, then R^2 will be very low and the VIF will be close to 1, with values between 2 and 5 indicating moderate multicollinearity. In cases where multicollinearity is moderate or severe, predictions made by the model may be accurate but p-values should not be relied on, because an indication of which predictors have a relationship with the response and coefficients should not be interpreted for direction or magnitude.

Multiple regression makes the same assumptions about the residuals as simple regression and the same plots are used to assess whether these assumptions have been reasonably met.

Example

A filtration process to remove particles from a slurry uses variable pressure to keep the flow rate constant. An engineer wants to study the effect of pressure, slurry temperature, and the amount of time since the last filter cleaning on the amount of residual matter in the output. Over a period of a week the engineer collects data at random times on the response and each predictor and wants to create a regression model. The data are shown in Table 10.6.

In Minitab: Go to **Stat > Regression > Regression > Fit Regression Model** and complete the dialog as shown in Figure 10.6 and click **OK**.

The session window output is shown in Figure 10.7.

Observe the p-values for each factor in Figure 10.7. The p-value for a temperature of 0.947 is much higher than the alpha level of 0.05 and there is no significant

Table 10.6 Data for filtration process.

Pressure	Temperature	Time since cleaning	Residual matter
1.32	71.1	2.41	21.7
1.48	72.6	2.74	21.2
1.54	67.0	2.70	21.7
1.49	69.5	2.61	21.8
1.49	77.0	2.61	21.7
1.46	69.5	2.59	21.5
1.45	72.2	2.52	21.7
1.54	69.7	2.81	21.4
1.55	71.9	2.82	21.4
1.48	75.2	2.77	21.5
1.37	71.6	2.41	21.8
1.58	73.6	2.85	21.5
1.54	73.2	2.84	21.4
1.46	70.4	2.39	22.3
1.48	70.0	2.78	21.5
1.42	69.4	2.42	21.7
1.56	68.8	2.73	21.6
1.54	76.4	2.84	21.4
1.50	72.8	2.44	21.9
1.32	72.8	2.37	21.7

evidence of a relationship with residual matter. To determine whether this is due to multicollinearity, the VIF for temperature can be checked and is found to be 1.05. Because this is very close to 1, there is no meaningful correlation found between temperature and the other predictors and we rule out multicollinearity as hiding a relationship between temperature and residual matter. Temperature should be removed from the model, and a new model calculated with the remaining factors.

In Minitab: Go to **Stat > Regression > Regression > Fit Regression Model** and complete the dialog as shown in Figure 10.8 and click **OK**.

Click on **Graphs** and complete the dialog as shown in Figure 10.9 and click **OK** in each dialog.

The graphical output is shown in Figure 10.10.

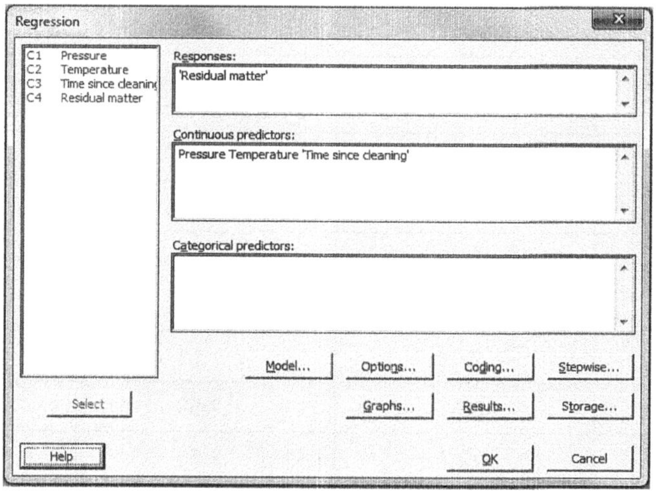

Figure 10.6 Dialog box for fit regression model.

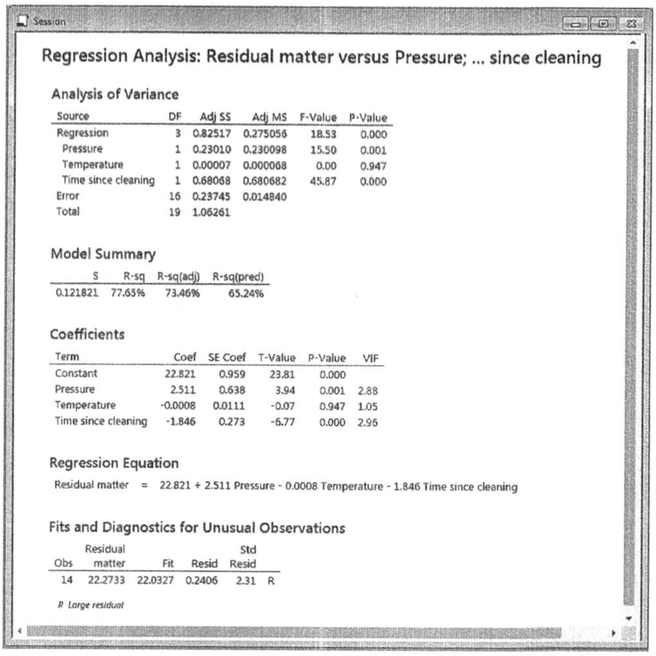

Session

Regression Analysis: Residual matter versus Pressure; ... since cleaning

Analysis of Variance

Source	DF	Adj SS	Adj MS	F-Value	P-Value
Regression	3	0.82517	0.275056	18.53	0.000
Pressure	1	0.23010	0.230098	15.50	0.001
Temperature	1	0.00007	0.000068	0.00	0.947
Time since cleaning	1	0.68068	0.680682	45.87	0.000
Error	16	0.23745	0.014840		
Total	19	1.06261			

Model Summary

S	R-sq	R-sq(adj)	R-sq(pred)
0.121821	77.65%	73.46%	65.24%

Coefficients

Term	Coef	SE Coef	T-Value	P-Value	VIF
Constant	22.821	0.959	23.81	0.000	
Pressure	2.511	0.638	3.94	0.001	2.88
Temperature	-0.0008	0.0111	-0.07	0.947	1.05
Time since cleaning	-1.846	0.273	-6.77	0.000	2.96

Regression Equation

Residual matter = 22.821 + 2.511 Pressure - 0.0008 Temperature - 1.846 Time since cleaning

Fits and Diagnostics for Unusual Observations

Obs	Residual matter	Fit	Resid	Std Resid	
14	22.2733	22.0327	0.2406	2.31	R

R Large residual

Figure 10.7 Session window output for multiple regression.

The graphical output is a four-in-one plot containing four graphs that help assess whether the assumptions about the residuals are met. In Figure 10.10, the normal probability plot and histogram both suggest the residuals are approximately normally distributed and that the assumption is met. The versus fits plot shows no obvious pattern and suggests the assumption of independence between residuals and fits is likely met. Finally, the versus order plot shows no obvious trend or pattern over time, suggesting that the residuals were stable throughout the data collection.

The session window output is shown in Figure 10.11.

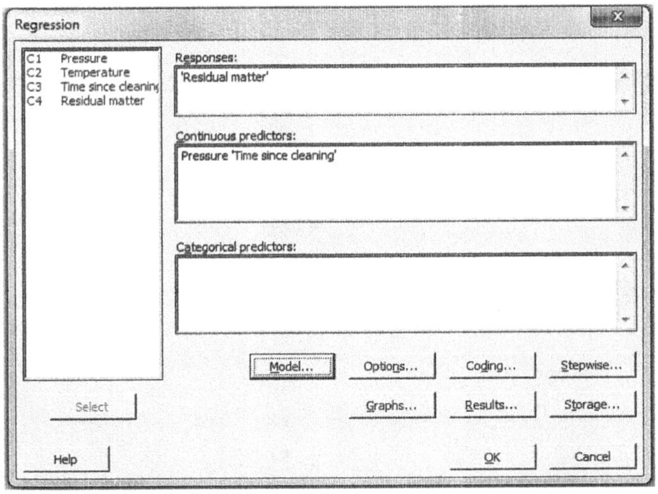

Figure 10.8 Dialog box for regression.

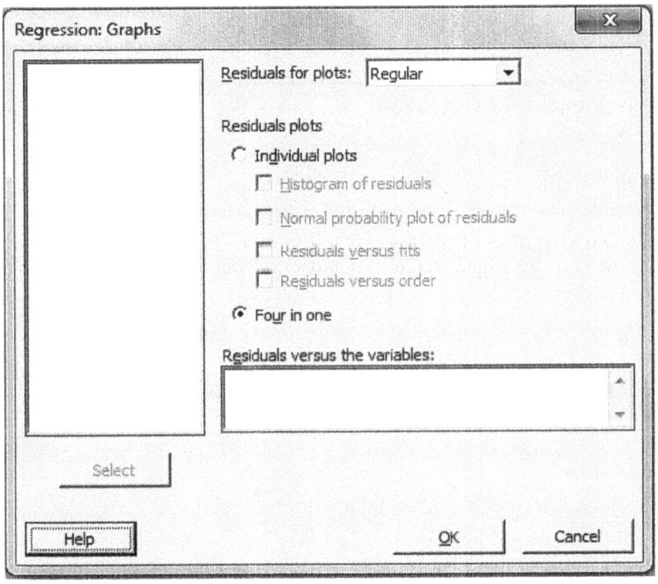

Figure 10.9 Graphs subdialog box for regression.

The p-values for the predictors are both statistically significant at 0.001 and 0.000; therefore, there is strong evidence of a relationship between both pressure and the time since cleaning and the response variable residual matter. The R^2 value of 77.65% and adjusted R^2 value of 75.02% are close to one another and reasonably high, which suggests that the two predictors in the model explain most of the variation in residual matter observed. The predicted R^2 value of 67.14% is a little lower than R^2, suggesting that overfitting might be an issue, although not a serious one.

Each predictor shows a VIF of 2.81, indicating that moderate multicollinearity is present and the two predictors are correlated with one another. Because of this, the direction and magnitude of the coefficients for each term should not be

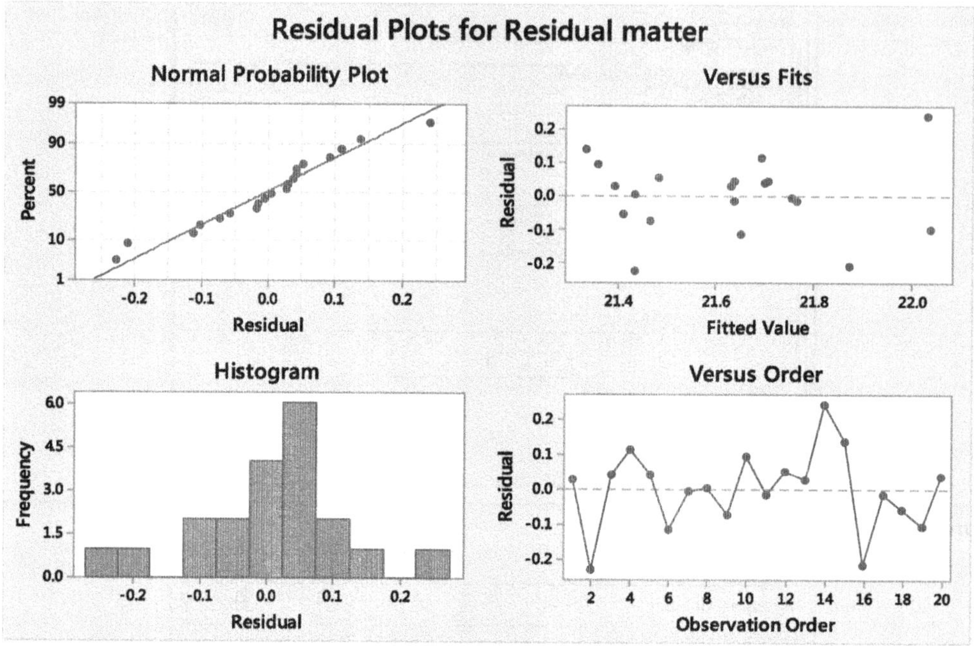

Figure 10.10 Four-in-one plot for regression.

Session

Regression Analysis: Residual matter versus Pressure; ... since cleaning

Analysis of Variance

Source	DF	Adj SS	Adj MS	F-Value	P-Value
Regression	2	0.8251	0.41255	29.53	0.000
Pressure	1	0.2366	0.23663	16.94	0.001
Time since cleaning	1	0.7190	0.71903	51.46	0.000
Error	17	0.2375	0.01397		
Total	19	1.0626			

Model Summary

S	R-sq	R-sq(adj)	R-sq(pred)
0.118201	77.65%	75.02%	67.14%

Coefficients

Term	Coef	SE Coef	T-Value	P-Value	VIF
Constant	22.768	0.542	41.99	0.000	
Pressure	2.517	0.612	4.12	0.001	2.81
Time since cleaning	-1.850	0.258	-7.17	0.000	2.81

Regression Equation

Residual matter = 22.768 + 2.517 Pressure - 1.850 Time since cleaning

Fits and Diagnostics for Unusual Observations

Obs	Residual matter	Fit	Resid	Std Resid	
2	21.2065	21.4346	-0.2280	-2.03	R
14	22.2733	22.0326	0.2407	2.38	R

R Large residual

Figure 10.11 Session window output for multiple regression.

interpreted in terms of the predictor's relationship with the response. However, the model may be used to make reasonable prediction for residual matter for given values of pressure and time since cleaning. The model provided is:

$$Residual\ matter = 22.768 + 2.517 * Pressure - 1.850 * Time\ since\ cleaning + error$$

The error term S has a standard deviation of 0.118, as seen in the output. As with all models, the equation should only be used for predictor values in the range of the data used to create the model. For example, the data set contains values of pressure from 1.32 to 1.58. The regression equation can be used to predict residual matter for a pressure value of 1.39 but should not be used for a value of 1.21 because that is well outside of the range observed.

BINARY LOGISTIC REGRESSION

When performing regression, the response of interest may be binary instead of continuous. In other words, the response can only be one of two values, such as yes or no or defective or not defective. In this case binary logistic regression can be performed to determine which factors likely affect the odds of each value and to estimate the probability of each value given a set of predictor values.

One of the two values is chosen as the reference value, and all probabilities are determined relative to that when performing binary logistic regression. One might choose defective as being the reference value if a model is created to determine the probability of a part being defective or not based on some set of predictors. Then the result will be a predictive equation for the probability of being defective. The probability of being not defective is simply one minus the probability of being defective. It is important to understand that although the data collected on the response are binary, the model will always predict a probability between 0 and 1. Thus binary logistic regression would never predict an event to happen but might predict that it is highly likely to happen.

Recall that when performing regression, the model that is estimated corresponds to the response being a linear equation containing the series of predictors. However, when the response is binary the equation can no longer be linear because we are modeling a probability that is bounded at zero and one. The most common model is known as the *logit function*. If p is the probability that the reference value occurs, then the logit model is:

$$\ln\left(\frac{p}{1-p}\right) = f(x) + error$$

The result is a curve that initially is very close to either zero or one and then follows a continually steeper slope until reaching 0.5, then becoming less steep until following very close to the other bound. An example of this curve for a single predictor, which results in higher odds of the reference event as that predictor increases, is shown in Figure 10.12.

In the figure you can see that even smaller x values would result in probabilities so small than the chance of the reference event happening is nearly 0, and that very large values of x would result in the reference event almost certainly occurring.

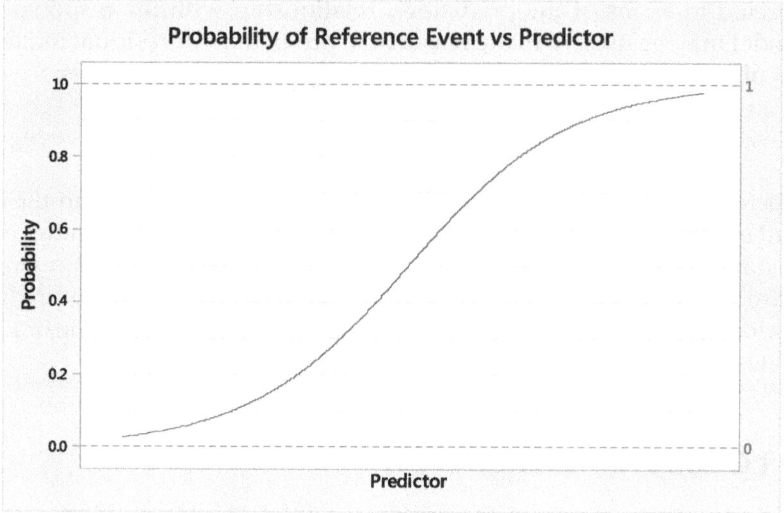

Figure 10.12 Logit model from binary logistic regression.

As with linear regression, each predictor has a coefficient and a p-value associated with that term that has the same interpretation. The ANOVA table is often referred to as a *deviance table*. Although binary logistic regression can also produce an R-Sq value, this value does not have the same interpretation and should not be used to determine model quality. Good models may have very low R-Sq values depending on the data collected and predicted probabilities.

There are three goodness-of-fit tests that can be used to determine model quality, and in each the null hypothesis of the test is that the data are well-fit by the model and the alternative hypothesis is that the data are not well-fit by the model. The Pearson and deviance tests are based on the magnitude of the residuals and how well they match what would be expected by the binomial distribution. The Hosmer-Lemeshow test is performed by dividing the data into groups based on the estimated probabilities where each group has approximately the same number of data points. Then a Chi-square test is performed on the number of observed events in each group compared to the number expected.

It is not uncommon for one or two tests to show statistical significance while the others do not. If all tests agree, then the resulting decision is straightforward. When the tests do not all agree, the p-values may be compared for multiple models and the model with the highest p-values selected. The Pearson and deviance tests often result in very similar p-values, so these will most often agree. A single outlier will affect these tests much more than the Hosmer-Lemeshow test, so if these tests are showing statistical significance, then a residuals plot can be used to determine if there are unusual data points.

Example

Outgoing product shipments are regularly being delivered late to the customer and you wish to study whether delays are related to the mode of transportation or distance. You randomly select 10 shipments each by truck and by rail and then

Table 10.7 Delayed shipment data.

Mode	Distance	Delayed
Truck	62	No
Truck	399	Yes
Truck	398	Yes
Truck	246	No
Truck	362	Yes
Truck	201	No
Truck	378	No
Truck	186	No
Truck	288	No
Truck	329	Yes
Rail	282	Yes
Rail	139	No
Rail	148	No
Rail	303	Yes
Rail	209	No
Rail	214	No
Rail	250	No
Rail	38	No
Rail	315	No
Rail	110	No

collect data on how far each shipment needed to travel and whether it was late. The data are shown in Table 10.7.

In Minitab: Go to **Stat > Regression > Binary Logistic Regression > Fit Binary Logistic Model** and complete the dialog as shown in Figure 10.13 and click **OK**.

The deviance and coefficient tables from the session window output are shown in Figure 10.14.

The p-values indicate that we have strong evidence of a relationship between the distance and the likelihood of being delayed, but no statistically significant relationship was found with the mode of transportation. The VIFs for each term are somewhat close to 1, indicating that multicollinearity is not severe. If the VIFs were high, approaching 5 or more, then it would not be possible to determine which term was really affecting the odds of being late.

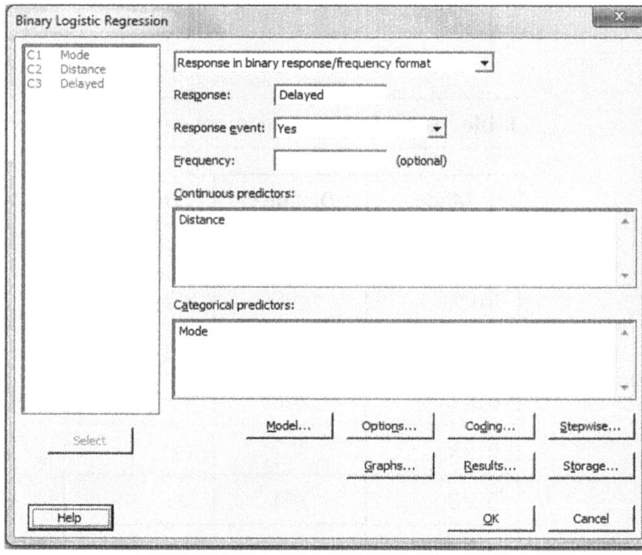

Figure 10.13 Fit Binary Logistic Model dialog box in Minitab.

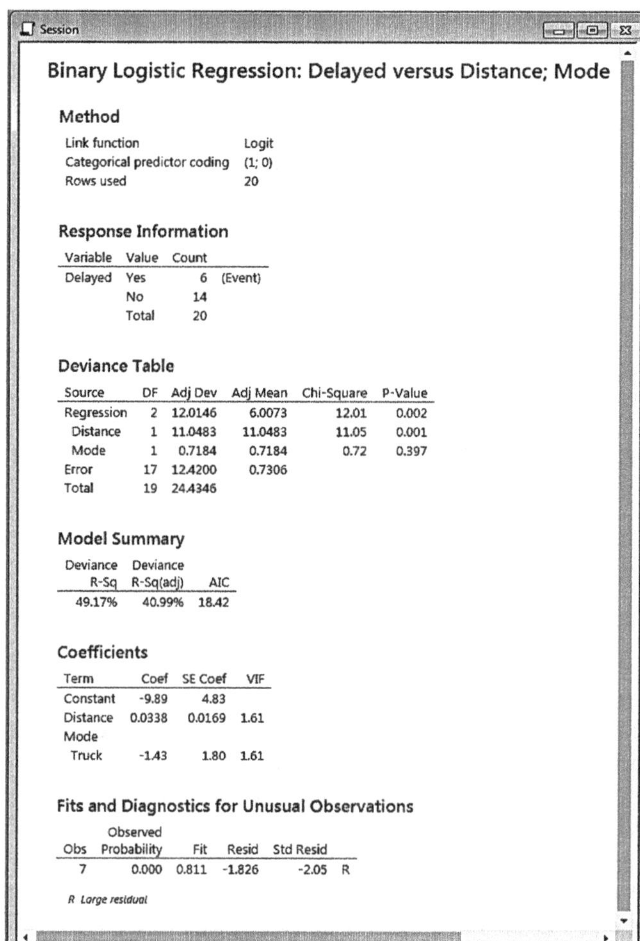

Session

Binary Logistic Regression: Delayed versus Distance; Mode

Method

Link function	Logit
Categorical predictor coding	(1; 0)
Rows used	20

Response Information

Variable	Value	Count	
Delayed	Yes	6	(Event)
	No	14	
	Total	20	

Deviance Table

Source	DF	Adj Dev	Adj Mean	Chi-Square	P-Value
Regression	2	12.0146	6.0073	12.01	0.002
Distance	1	11.0483	11.0483	11.05	0.001
Mode	1	0.7184	0.7184	0.72	0.397
Error	17	12.4200	0.7306		
Total	19	24.4346			

Model Summary

Deviance R-Sq	Deviance R-Sq(adj)	AIC
49.17%	40.99%	18.42

Coefficients

Term	Coef	SE Coef	VIF
Constant	-9.89	4.83	
Distance	0.0338	0.0169	1.61
Mode			
Truck	-1.43	1.80	1.61

Fits and Diagnostics for Unusual Observations

Obs	Observed Probability	Fit	Resid	Std Resid	
7	0.000	0.811	-1.826	-2.05	R

R Large residual

Figure 10.14 Deviance and coefficients tables.

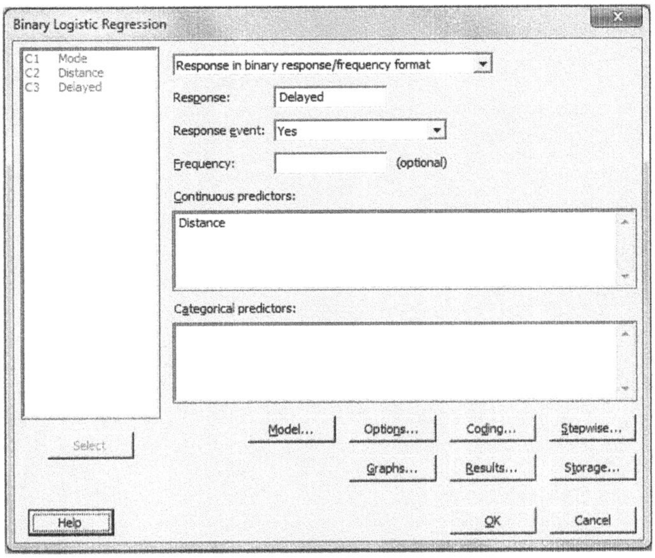

Figure 10.15 Fit Binary Logistic Model dialog box.

To create a model containing only the distance in Minitab, go to **Stat > Regression > Binary Logistic Regression > Fit Binary Logistic Model** and complete the dialog as shown in Figure 10.15 and click **OK.**

The deviance and coefficients tables are shown in Figure 10.16.

Again, the p-value for the distance term is highly significant and indicates a strong relationship. The positive coefficient indicates that as distance increases, the likelihood of the delivery being late goes up as well. The model can be written as either of the following equations:

$$\ln\left(\frac{p}{1-p}\right) = -8.53 + 0.0270 * distance$$

$$p(delayed) = \frac{e^{-8.53+0.0270*distance}}{1+e^{-8.53+0.0270*distance}}$$

The results of the goodness-of-fit tests from the session window are shown in Figure 10.17.

All goodness-of-fit tests indicate that the data are well-fit by the model. As a result of the analysis, it can be determined that there is a statistical relationship between distance and the likelihood of being late but there was not evidence of the mode of transportation having a relationship.

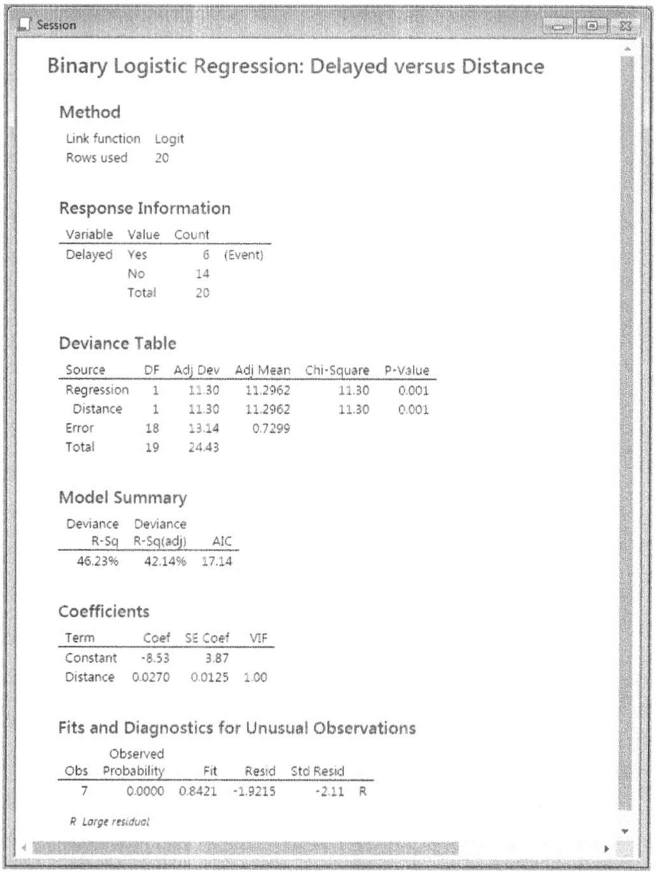

Session

Binary Logistic Regression: Delayed versus Distance

Method

Link function	Logit
Rows used	20

Response Information

Variable	Value	Count	
Delayed	Yes	6	(Event)
	No	14	
	Total	20	

Deviance Table

Source	DF	Adj Dev	Adj Mean	Chi-Square	P-Value
Regression	1	11.30	11.2962	11.30	0.001
Distance	1	11.30	11.2962	11.30	0.001
Error	18	13.14	0.7299		
Total	19	24.43			

Model Summary

Deviance R-Sq	Deviance R-Sq(adj)	AIC
46.23%	42.14%	17.14

Coefficients

Term	Coef	SE Coef	VIF
Constant	-8.53	3.87	
Distance	0.0270	0.0125	1.00

Fits and Diagnostics for Unusual Observations

Obs	Observed Probability	Fit	Resid	Std Resid	
7	0.0000	0.8421	-1.9215	-2.11	R

R Large residual

Figure 10.16 Deviance and coefficients tables.

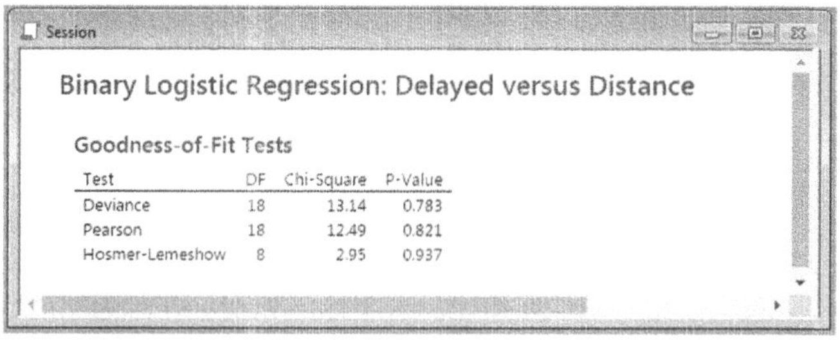

Session

Binary Logistic Regression: Delayed versus Distance

Goodness-of-Fit Tests

Test	DF	Chi-Square	P-Value
Deviance	18	13.14	0.783
Pearson	18	12.49	0.821
Hosmer-Lemeshow	8	2.95	0.937

Figure 10.17 Goodness-of-fit test results.

QUESTIONS AND ANSWERS

Q: How many values do I need to determine if a correlation exists?

A: Detecting a moderate correlation is highly likely with about 40 or more data points. Weaker correlation obviously requires more, whereas strong correlation might be easily detected with just 10 points.

Q: How many values do I need to estimate the influence of one factor on another?

A: Determining a specific value is difficult because the answer depends on how closely the coefficient on that factor is desired to be estimated, as well as the strength of the relationship. However, the guidelines for determining whether correlation exists would also tend to produce reasonably precise estimates.

Q: What is the difference between adjusted R square and R square?

A: The adjusted statistic includes a penalty for the number of terms in the model. R-squared always increases as additional terms are added, even when they have no influence on the response. Therefore, if the adjusted statistic is much smaller than the R-squared statistic, it can indicate whether such variables are in the model.

Q: What does R square predicted tell me?

A: R-squared predicted is a measure of how well the model would predict data points not used to create it. It is a good indicator that the model is not overfit, and provides some indication that the model can be trusted for prediction. If R square predicted is well below R-squared, typically removing insignificant terms will improve it. Otherwise the factors in the model may not be truly related to the response.

Q: When should I do regression instead of a General Linear Model?

A: Regression is typically used when all factors are continuous, although categorical terms can be included. General Linear Model is preferred when most or all factors are categorical, or when factors are random effects or nested in other factors.

Q: Can I simply measure the same part multiple times if I don't have enough values for a regression analysis?

A: No! Like most analyses, regression assumes each data point is independent of the other data points. Multiple measurements of the same part are not independent of one another, as each has the same expected X and Y value. Therefore they do not represent any additional information beyond what the first measurement provided.

Chapter 11
Measurement System Analysis

Very often data are collected using a process that introduces variation in the data. For example, suppose technicians measure a certain part using calipers: Not only might different technicians measure differently, but a single technician may not get the same value even if measuring the exact same part multiple times. Measurement system variation causes quality levels to be understated and lowers power for hypothesis tests and models like regression. Before using data, you must establish that the data do not contain a large amount of measurement variation.

A measurement system is statistically evaluated using measurement system analysis (MSA) (Chrysler, Ford, and General Motors 2010). This allows us to assess precision, accuracy, stability, bias, and linearity (see Table 11.1).

An example of a measurement problem could be a micrometer that measures a diameter as 14.8 millimeters (mm) when the diameter is actually 14.7 mm or when more precision is needed than the measuring device can provide. The difference between precision and accuracy is shown in Figure 11.1. The difference between bias and the true value is shown in Figure 11.2, and Figure 11.3 depicts a linearity problem over the length of the values being assessed.

The same part measured at different times should give the same result. A micrometer has a stability problem if it measured a part as 7.1 mm yesterday, but the same micrometer measures the same part as 7.2 mm today. Figure 11.4 shows a part that is checked three times a day for a week and the results are consistent during any given day, but there is a stability problem across the days.

Table 11.1 Aspects assessed during an MSA.

Precision	How closely together observations are taken under the exact same conditions
Accuracy	The difference between observed measurements and the true value
Stability	The ability to perform measurements on the same part consistently over time
Bias	The difference between the average measurement and the true value
Linearity	The relationship between bias and the true value over the range of values for the item being measured

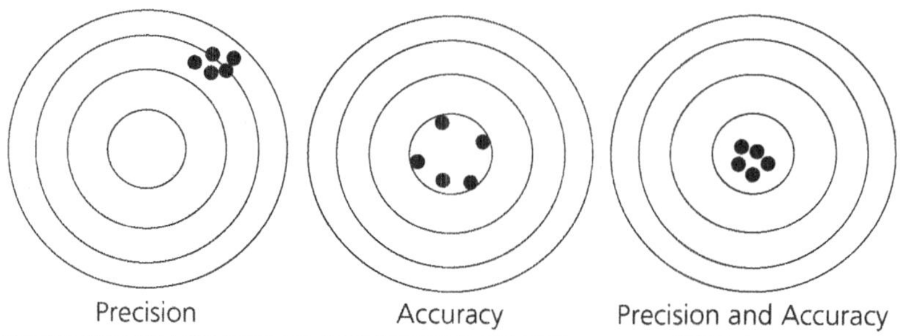

Figure 11.1 Precision and accuracy.

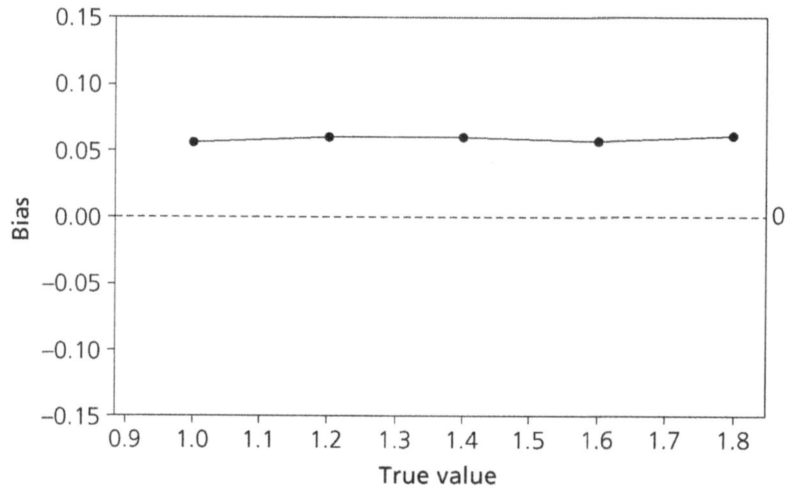

Figure 11.2 Bias compared to the true value.

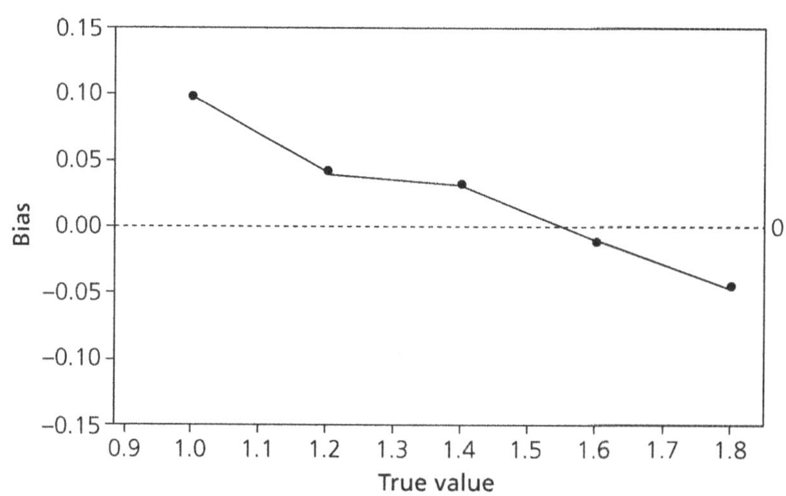

Figure 11.3 A linearity problem.

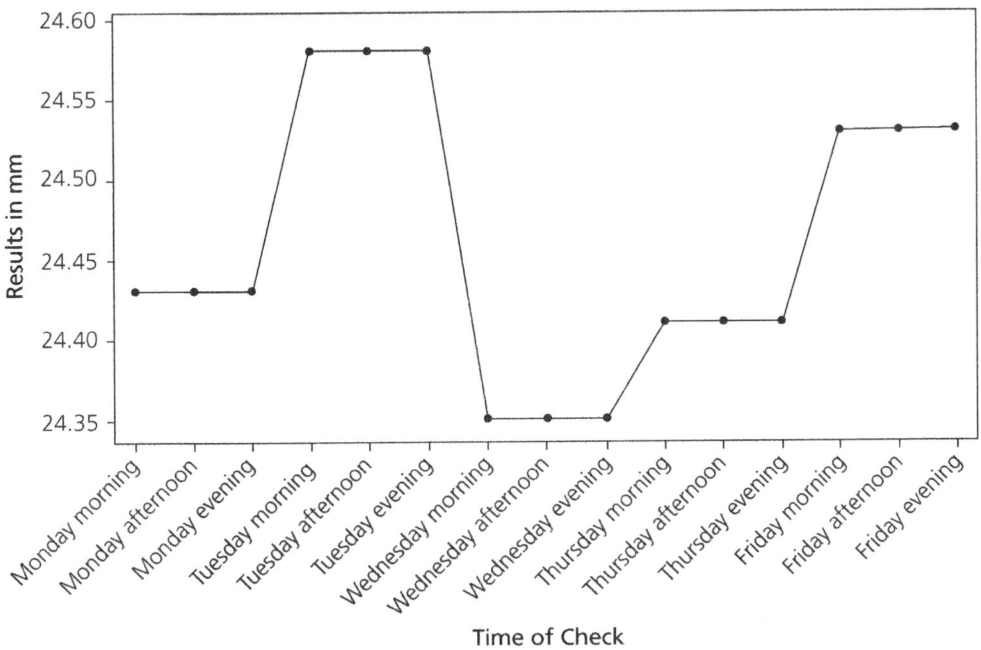

Figure 11.4 Stability.

Two forms of measurement system variation that are frequently of interest are repeatability and reproducibility (R&R). *Repeatability* corresponds to the ability of operators to measure the same part consistently under the same conditions. *Reproducibility* is the ability of multiple operators to measure the same parts consistently (Griffith 2003).

TYPE 1 GAGE STUDY

A type 1 gage study is performed to evaluate the bias and repeatability of a measuring system. Additionally, stability can be assessed using this type of study. The same part with a known reference value is measured multiple times by the same person using the same gage. This allows for a relatively quick evaluation of two aspects of the measurement system before doing a lengthier study. An inadequate measurement system with regards to bias or repeatability needs to be replaced or improved before a full gage R&R study is performed. A type 1 gage R&R study should be performed using 25 or more measurements.

Bias is equal to the mean of the measurements minus the known reference value and represents the average amount that measurements deviate from the true value. It is critical that the part being measured has a known standard value, typically called a gold standard. A 1-sample t test can be performed to determine whether there is statistical evidence that the mean differs from this reference value.

The index Cg compares the tolerance to the measurement variation due to repeatability, and Cgk compares the tolerance to measurement variation and bias. Both should be equal to or greater than 1.33 or the measurement system is not capable of measuring adequately. Cg is equal to

$$Cg = \frac{\left(\frac{K}{100}\right)(Tolerance)}{Study\ Variation},$$

where the tolerance is equal to the upper specification limit minus the lower specification limit; K is equal to a percent of the tolerance, generally 20%; and study variation is the standard deviation times 6. Cgk is equal to

$$\frac{\left(\frac{K}{200}\right)(Tolerance) - |\bar{x} - x_m|}{(Study\ Variation)/2},$$

where $\bar{x}$ is the mean of the measurements and x_m is the reference value.

Two other measurements compare the percentage of the tolerance covered by six standard deviations of measurement error. These are:

$$\%Var(Repeatability) = \frac{Study\ Variation}{Tolerance} * 100\%$$

$$\%Var(Repeatability\ and\ Bias) = \frac{\frac{K * Study\ Variation}{2}}{\frac{K * Tolerance}{200} - |\bar{x} - x_m|} * 100\%$$

The measurements gathered in the study can also be plotted in the order in which they were collected to assess stability. Any patterns or trends in the data indicate that the measurement system does not perform consistently over time.

Example

A quality engineer measures a gage block 25 times. The gage block has a known value of 9.504 and the quality engineer would like to use the measuring device for a part that has specifications of +/– 0.2. The mean of the measurements is 9.4963 and the standard deviation of the measurements is 0.00923. The Cg is

$$\frac{\left(\frac{20}{100}\right)(0.4)}{(6)(0.00923)} = 1.44$$

and Cgk is equal to

$$\frac{\left(\frac{20}{200}\right)(0.4) - |9.4963 - 9.504|}{(6)(0.00923)/2} = 1.17.$$

The Cg value of 1.44 is greater than 1.33 and indicates that the repeatability in the measurement system is sufficiently small. However, the Cgk value of 1.17 is smaller than 1.33 and indicates that the combination of bias and repeatability is too large and should be corrected. This could be done by either correcting the bias or reducing the repeatability.

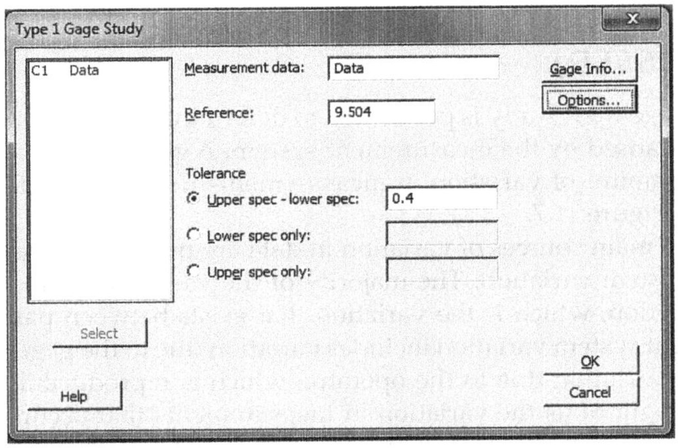

Figure 11.5 Dialog box for a type 1 gage study.

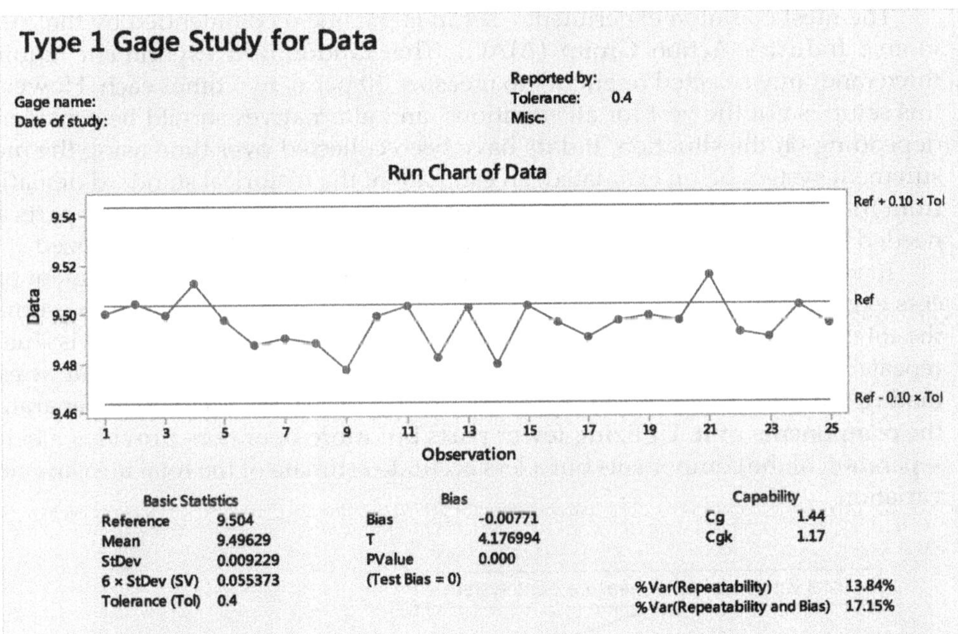

Figure 11.6 Graphical output for a type 1 gage study.

In Minitab: Go to **Stat > Quality Tools > Gage Study > Type 1 Gage Study** and enter the column containing the measurement values, the reference value, and the tolerance range, as shown in Figure 11.5, and click **OK**.

Minitab provides a run chart of the results as shown in Figure 11.6. The chart does not exhibit any patterns or trends that would indicate a lack of stability. The graph also shows a bias of –0.00771, a Cg of 1.44, and a Cgk of 1.17.

GAGE REPEATABILITY AND REPRODUCIBILITY STUDY (CROSSED)

A crossed gage R&R study is performed to determine the amount of process variation that is caused by the measurement system. A gage R&R study (crossed) estimates the amount of variation in measurements due to each of the components depicted in Figure 11.7.

The two main sources of variation in data are part-to-part variation and measurement system variation. The majority of the variation should consist of part-to-part variation, which is the variation that exists between parts naturally. The measurement system variation includes variation due to the gage, which is repeatability, and variation due to the operator, which is reproducibility. Repeatability is often thought of as the variation in measurements that occurs when the same operator measures the same part multiple times. Reproducibility consists of operator biases and operator*part variation, which occurs when some or all operators exhibit bias that varies for different parts.

The most common experimental setup is the one recommended by the Automotive Industry Action Group (AIAG). This randomized experiment requires three randomly selected operators to measure 10 parts, two times each. However, this setup is not the best for all situations, and alternatives should be considered depending on the situation. If data have been collected over time using the measurement system being evaluated, an estimate of the historical standard deviation from that data can be used in the analysis. If that is the case, then fewer parts are needed in the experiment and, if possible, more operators should be utilized.

If no historical estimate of standard deviation from the same measurement process exists, then it should be determined whether it is more valuable to estimate the total measurement system variation or to estimate what percentage is due to repeatability versus reproducibility. The standard experiment is efficient at estimating the total measurement system variation but is not accurate at separating the components of it. Utilizing fewer parts but more operators provides a better separation of the components but a less accurate estimate of the total measurement variation.

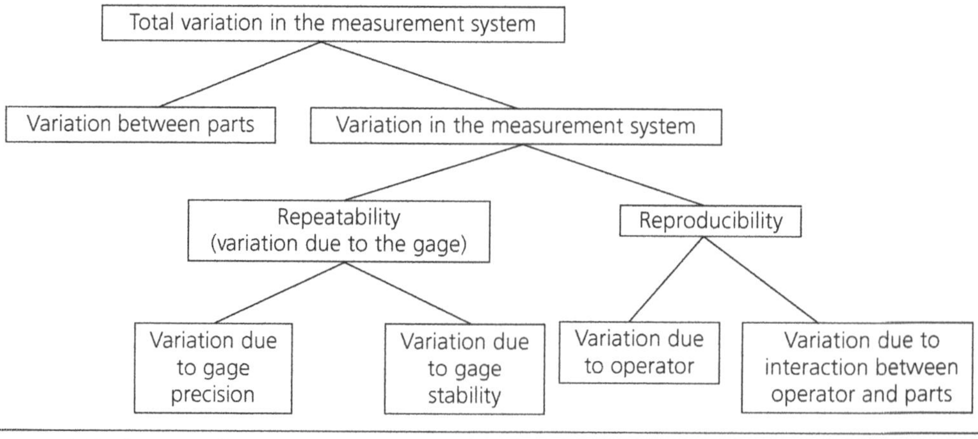

Figure 11.7 Sources of measurement variation.

When no historical estimate exists, it is critical that the parts selected for the experiment are chosen at random over time to provide the best chance of the parts representing the distribution of parts created over time. Part selection schemes that attempt to choose parts at specific measurement values result in estimates that are completely inaccurate, so these schemes should always be avoided in favor of the random selection of parts.

The data resulting from the experiment can be evaluated using the one-way ANOVA method or Xbar and R method. The one-way ANOVA method uses a one-way ANOVA analysis and indicates which factors are statistically significant. The Xbar and R method uses the mean and range of values for each operator and each part. The one-way ANOVA method is recommended as it provides operator*part interactions and more accurately assesses the components of variation.

When analyzing data in a gage R&R experiment, the operator*part interaction is first checked for statistical significance. A one-way ANOVA table is generated that includes terms for Operator, Part, and Operator*Part (see Table 11.2) and a second one-way ANOVA table (see Table 11.3) is produced without the Operator*Part term if it is found to not be statistically significant.

In the tables, p is the number of parts, o is the number of operators, and r is the number of replicates. The one-way ANOVA table is used to calculate statistics on the adequacy of the measurement system. First, variance components for each term are calculated. If the interaction term is in the model, they are calculated as:

$$\text{VarComp(Operators)}: \hat{\sigma}_O^2 = \frac{MS_O - MS_{O*P}}{pr}$$

$$\text{VarComp(Operator} * \text{Part)}: \hat{\sigma}_{O*P}^2 = \frac{MS_{O*P} - MS_{repeatability}}{r}$$

$$\text{VarComp(Parts)}: \hat{\sigma}_P^2 = \frac{MS_P - MS_{O*P}}{or}$$

Table 11.2 One-way ANOVA table for an MSA that includes Operator*Part.

ANOVA table for MSA				
Source	**df**	**SS**	**MS**	**F**
Part	$p-1$	SS (parts)	$\frac{MS(parts)}{p-1}$	$\frac{MS(part)}{MS(part*operator)}$
Operator	$o-1$	SS (operator)	$\frac{MS(operator)}{o-1}$	$\frac{MS(operator)}{MS(part*operator)}$
Part x Operator	$(p-1)(o-1)$	SS (operator parts)	$\frac{MS(op.par.)}{(p-1)(0-1)}$	$\frac{MS(part*operator}{MS(repeatability)}$
Repeatability	$op(r-1)$		$\frac{MS(repeatability)}{op(r-1)}$	
Total	$n-1$	SS (Total)		

Table 11.3 One-way ANOVA table for an MSA that does not include the interaction.

ANOVA table for MSA				
Source	**df**	**SS**	**MS**	**F**
Part	$p-1$	SS (parts)	$\frac{MS(parts)}{p-1}$	$\frac{MS(parts)}{MS(repeatability)}$
Operator	$o-1$	SS (operator)	$\frac{MS(operator)}{o-1}$	$\frac{MS(operator)}{MS(repeatability)}$
Repeatability	$op(r-1)$		$\frac{MS(repeatability)}{op(r-1)}$	
Total	$n-1$	SS (Total)		

If the interaction term is not in the model, these variance components are calculated as:

$$\text{VarComp(Operators)}: \hat{\sigma}_O^2 = \frac{\text{MS}_O - \text{MS}_{repeatability}}{\text{pr}}$$

$$\text{VarComp(Parts)}: \hat{\sigma}_P^2 = \frac{\text{MS}_P - \text{MS}_{repeatability}}{\text{or}}$$

The remaining components of variation are calculated as:

$$\text{VarComp(Repeatability)} = \text{MS}_{repeatability}$$

$$\text{VarComp(Reproducibility)} = \text{VarComp(Operator)} + \text{VarComp(Operator * Part)}$$

$$\text{VarComp(Gage R\&R)} = \text{VarComp(Repeatability)} + \text{VarComp(Reproducibility)}$$

$$\text{VarComp(Total)} = \text{VarComp(Gage R\&R)} + \text{VarComp(Parts)}$$

The overall variance of the data is equal to VarComp(Total). One common measure of measurement system variation is %Contribution, which is calculated as:

$$\%Contribution = \frac{VarComp(Gage\ R\&R)}{VarComp(Total)} * 100\%$$

%Contribution is interpreted as the percentage of variation in the data that is due to measurement system variation. Guidelines from AIAG provide an example of what is considered acceptable, with less than 1% considered excellent, between 1% and 9% considered marginal, and greater than 9% consisted poor (Chrysler, Ford, and General Motors 2010) as shown in Table 11.4.

Although these guidelines provide an idea of the quality of a measurement system, the situation and the actual value should be taken into consideration. Although %Contributions of 1.5% and 8.5% both fall in the marginal range, clearly

Table 11.4 %Contribution assessment based on AIAG guidelines.

%Contribution	Assessment
<1%	Excellent
>1% and <9%	Marginal
>9%	Poor

Table 11.5 %StudyVariation assessment based on AIAG guidelines.

%Study Variation	Assessment
<10%	Excellent
>10% and <30%	Marginal
>30%	Poor

one is a much higher quality system. Further, depending on the application more or less stringent criteria may want to be considered. For example, in a very high quality process the effect of a poor measurement system may be very minimal and values up to 20% may be acceptable.

Another metric often used is %StudyVariation, which is calculated similarly to %Contribution but using standard deviations instead of variances, as follows:

$$\%\,StudyVariation = \frac{\sqrt{VarComp(Gage\ R\&R)}}{\sqrt{VarComp(Total)}} * 100\%$$

The AIAG guidelines for %StudyVariation are mathematically equivalent to %Contribution: Less than 10% is acceptable, 10–30% is marginal, and greater than 30% is poor (Chrysler, Ford, and General Motors 2010) (see Table 11.5).

%StudyVariation can be thought of as the ratio of the standard deviation of measurement error to the standard deviation of the data. However, unlike %Contribution, the %StudyVariation for gage R&R and part-to-part do not add to total. Because of this, many practitioners prefer to use %Contribution because the interpretation is simpler.

Another common metric is %Tolerance, which compares the measurement system variation to the range of the specification limits. The formula is:

$$\%\,Tolerance = \frac{6 * \sqrt{VarComp(Gage\ R\&R)}}{Tolerance} * 100\%$$

Small values of %Tolerance indicate that the measurement system variation is very small relative to the tolerance and is likely not having a meaningful effect on estimates of product quality. Higher values indicate that measurement system variation is likely causing quality levels to be understated, acceptable parts to be rejected, and defective parts to be accepted.

Discrimination, also known as resolution, is the measuring device's ability to read changes in a reference value. Resolution should be 1/10th of a parts specification; therefore, a part with an upper specification limit (USL) of 22.8 and lower specification limit (LSL) of 22.2 would have a range equal to $22.8-22.2=0.6$. Therefore, the resolution would be $0.6\times0.1=0.06$.

Another metric of the quality of a measurement system is the *number of distinct categories* (NDC). The NDC is thought of as the number of separate categories that measurement data can be divided into. Measuring devices should have at least five distinct categories. A measuring device with only two distinct categories is only useful for OK/Not-OK checks. A measuring device with three to four distinct categories can only give rough estimates and should not be used for statistical process control. The automotive standard for MSA recommends five or more distinct categories (Chrysler, Ford, and General Motors 2010). The number of distinct categories is calculated as:

$$1.41\,\frac{\text{Standard deviation of parts}}{\text{Standard deviation of gage}}$$

It is important to ensure measuring devices have an updated calibration. A measuring device may lose the ability to make accurate or precise measurements over time. For example, the measuring device may lose the zero position or could be inaccurate due to damage or wear. Using an uncalibrated measuring device risks getting inaccurate results.

Example

An MSA is performed using three operators and 10 randomly-selected parts. All parts are then checked three times by each operator in random order. Table 11.6 shows the results for all 90 measurements. The data is then evaluated using a one-way ANOVA such as the one shown in Figure 11.6.

Table 11.7 shows the ANOVA table that would be used to complete the required calculations with interactions; Table 11.8 shows the ANOVA table without interactions.

Because the interaction term was removed from the model, the variance components are calculated as:

$$\text{VarComp(Operators)}: \hat{\sigma}_O^2 = \frac{MS_O - MS_{repeatability}}{pr} = \frac{0.17596 - .00444}{30} = 0.0057$$

$$\text{VarComp(Parts)}: \hat{\sigma}_P^2 = \frac{MS_P - MS_{repeatability}}{or} = \frac{1.09089 - 0.00444}{9} = 0.1207$$

$$\text{VarComp(Repeatability)} = MS_{repeatability} = 0.00444$$

$$\text{VarComp(Reproducibility)} = \text{VarComp(Operator)} = 0.0057$$

$$\text{VarComp(Gage R\&R)} = \text{VarComp (Repeatability)} + \text{VarComp (Reproducibility)} = 0.01014$$

$$\text{VarComp(Total)} = \text{VarComp (Gage R\&R)} + \text{VarComp (Parts)} = 0.13084$$

Table 11.6 MSA data.

Operator	Trial	Part number									
		1	2	3	4	5	6	7	8	9	10
A	1	1.18	0.89	1.53	1.24	0.81	1.09	1.28	0.98	1.83	0.63
	2	1.22	0.85	1.47	1.25	0.77	1.04	1.33	1.01	1.74	0.66
	3	1.29	0.89	1.50	1.29	0.80	1.01	1.30	1.02	1.75	0.64
B	1	1.11	0.92	1.48	1.08	0.89	1.01	1.24	0.87	1.68	0.52
	2	1.16	0.67	1.39	1.42	0.68	1.15	1.26	1.11	1.79	0.54
	3	1.10	0.85	1.53	1.15	0.65	1.10	1.36	0.97	1.81	0.58
C	1	1.09	0.62	1.37	1.13	0.59	0.98	1.09	0.93	1.67	0.58
	2	1.04	0.70	1.44	1.15	0.72	0.86	1.08	0.89	1.56	0.49
	3	1.03	0.76	1.30	1.12	0.60	0.92	1.15	0.92	1.70	0.36

Table 11.7 One-way ANOVA with interactions for MSA example.

ANOVA table for MSA					
Source	df	SS	MS	F	P
Part	9	9.818	1.0909	492.291	0.000
Operator	2	0.352	0.1760	79.406	0.000
Part x Operator	18	0.0399	0.0022	0.434	0.974
Repeatability	60	0.3065	0.00511		
Total	89	10.516			

Table 11.8 One-way ANOVA for MSA example.

ANOVA table for MSA					
Source	df	SS	MS	F	P
Part	9	9.8180	1.09089	245.614	0.000
Operator	2	0.3519	0.17596	39.617	0.000
Repeatability	78	0.3464	0.00444		
Total	89	10.5163			

%Contribution is calculated as:

$$\%\,Contribution = \frac{VarComp(Gage\ R\&R)}{VarComp(Total)} * 100\% = \frac{0.01014}{0.13084} * 100\% = 7.75\%$$

This falls within the marginal range but very close to the poor range. Finally, the number of distinct categories is calculated as:

$$\#\text{distinct categories} = 1.41\frac{Standard\ deviation\ of\ parts}{Standard\ deviation\ of\ gage} = 1.41 * \frac{\sqrt{.1206}}{\sqrt{.0108}} = 4.7$$

The NDC is less than 5, so the measurement system should be improved.

In Minitab: Go to **Stat > Quality Tools > Gage Study > Create Gage R&R Study Worksheet** and enter the number of parts, number of operators, and number of replicates. The name of the operators can also be entered; in this example, they will be identified as A, B, and C (see Figure 11.8). Minitab then creates a worksheet with a randomized run order such as the one shown in Figure 11.9. The study then needs to be performed and the results must be entered into column C4. The column with the measurement data can be renamed. In this example, it will be renamed "measurement." Then click **OK**.

Once the measurement data is entered go to **Stat > Quality Tools > Gage Study > Gage Study (Crossed)** and enter the part numbers, operators, and measurement data as shown in Figure 11.10.

In Minitab, the results under gage R&R %Contribution are shown in Figure 11.11. The total gage R&R, consisting of repeatability and reproducibility, and the part-to-part add up to a total of 100%.

The output in Figure 11.11 shows the %Contribution for total gage R&R to be 7.76%, which is on the high end of the marginal range of values. If data quality is especially important for the application, it would be worth considering improving the measurement system.

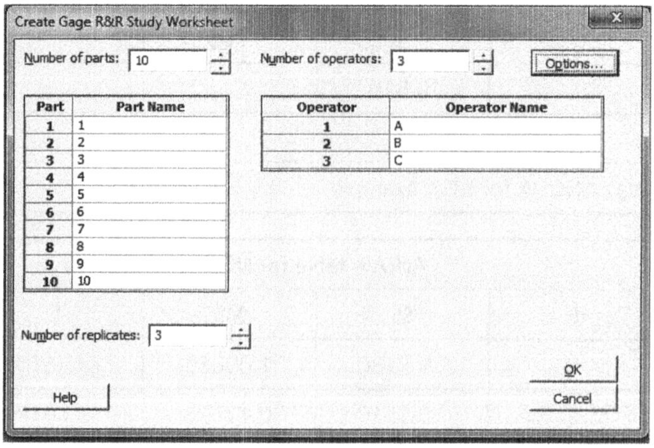

Figure 11.8 Dialog box for creating a gage study worksheet.

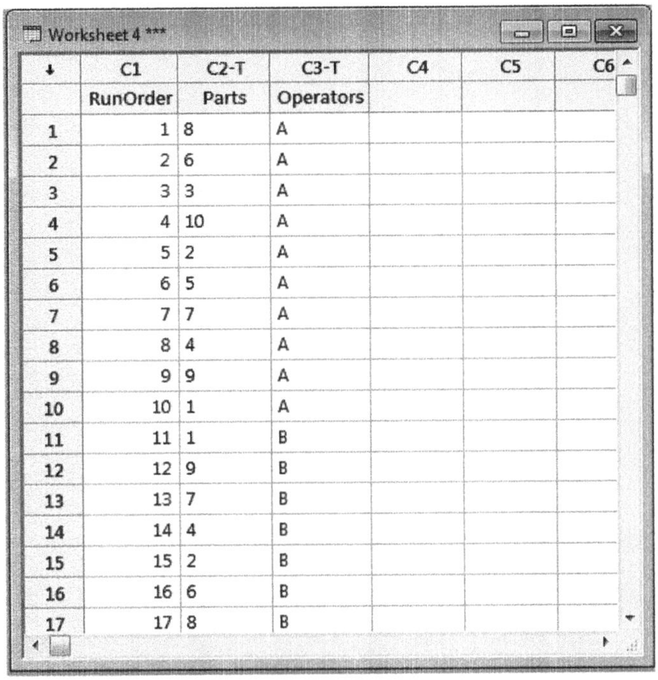

Worksheet 4 ***

↓	C1	C2-T	C3-T	C4	C5	C6
	RunOrder	Parts	Operators			
1	1	8	A			
2	2	6	A			
3	3	3	A			
4	4	10	A			
5	5	2	A			
6	6	5	A			
7	7	7	A			
8	8	4	A			
9	9	9	A			
10	10	1	A			
11	11	1	B			
12	12	9	B			
13	13	7	B			
14	14	4	B			
15	15	2	B			
16	16	6	B			
17	17	8	B			

Figure 11.9 Minitab gage R&R study worksheet.

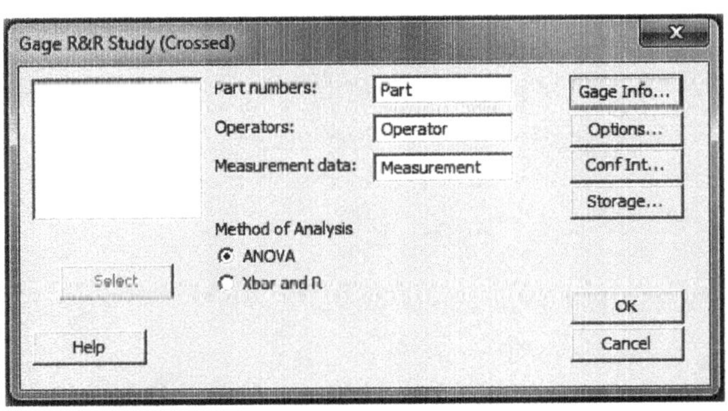

Figure 11.10 Dialog box for analyzing a crossed gage study.

The number of distinct categories is four, indicating that the system should be improved to better distinguish parts.

In Figure 11.12, the components of variation graphically depict the sources of variation. The R chart indicates whether or not parts are measured consistently; a value above the upper control limit (UCL) indicates an operator did not measure consistently, as is seen in operator B's measurements of part 4. Unlike an Xbar chart used in statistical process control, the Xbar chart on this graph indicates the ability to distinguish parts, and ideally most points will fall outside of the control limits. This indicates that the variation between parts is much higher than the variation within measurements of each part taken by each operator. Measurement by part shows all values arranged by part number to illustrate differences in parts.

Session

Gage R&R Study - ANOVA Method

Two-Way ANOVA Table With Interaction

Source	DF	SS	MS	F	P
Part	9	9.8180	1.09089	492.291	0.000
Operator	2	0.3519	0.17596	79.406	0.000
Part * Operator	18	0.0399	0.00222	0.434	0.974
Repeatability	60	0.3065	0.00511		
Total	89	10.5163			

α to remove interaction term = 0.05

Two-Way ANOVA Table Without Interaction

Source	DF	SS	MS	F	P
Part	9	9.8180	1.09089	245.614	0.000
Operator	2	0.3519	0.17596	39.617	0.000
Repeatability	78	0.3464	0.00444		
Total	89	10.5163			

Gage R&R

Variance Components

Source	VarComp	%Contribution (of VarComp)
Total Gage R&R	0.010159	7.76
Repeatability	0.004441	3.39
Reproducibility	0.005717	4.37
Operator	0.005717	4.37
Part-To-Part	0.120716	92.24
Total Variation	0.130875	100.00

Gage Evaluation

Source	StdDev (SD)	Study Var (6 × SD)	%Study Var (%SV)
Total Gage R&R	0.100791	0.60474	27.86
Repeatability	0.066644	0.39987	18.42
Reproducibility	0.075613	0.45368	20.90
Operator	0.075613	0.45368	20.90
Part-To-Part	0.347442	2.08465	96.04
Total Variation	0.361767	2.17060	100.00

Number of Distinct Categories = 4

Gage R&R for Measurement

Figure 11.11 Session window for a crossed gage R&R study.

Measurement by operator shows a boxplot of each operator's values, and parts by operator interaction shows the interaction between parts and operators. If the interaction is significant, then the lines connecting each operator's measurements will not be parallel.

LINEARITY AND BIAS

Performing a gage R&R study is important in determining how consistently measurements are taken, but it does not determine whether those measurements are

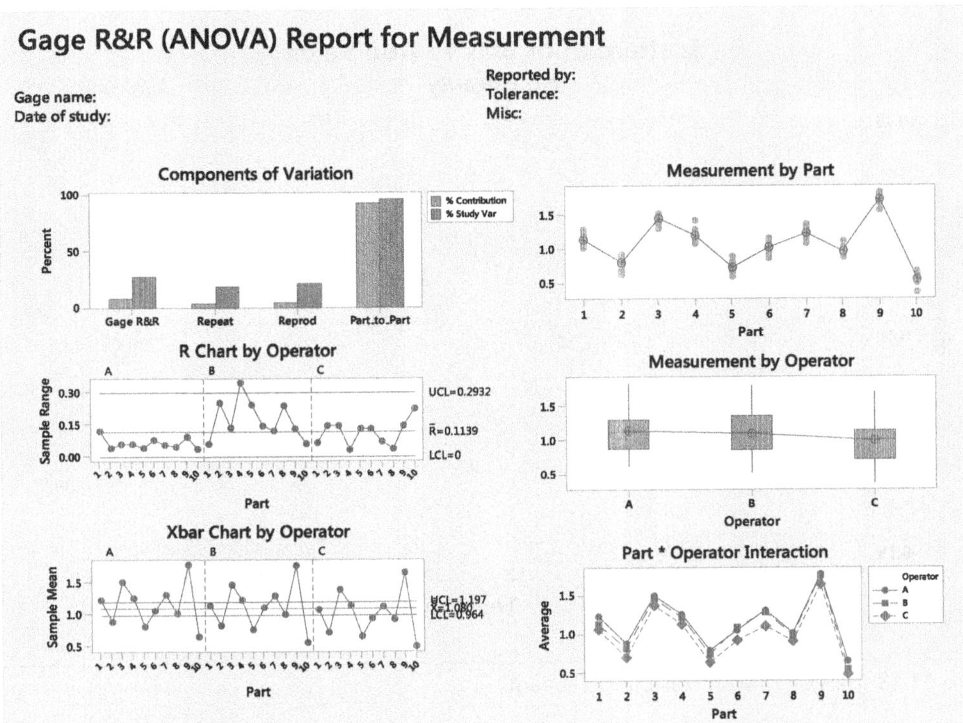

Figure 11.12 Minitab gage R&R report.

accurate across the range of values you are likely to experience. A linearity and bias study should be performed to determine whether measurements accurately reflect the true values.

Recall that bias is the difference between the average measurement, and the true value and linearity is the relationship between the bias and the true value. For example, Figure 11.13 illustrates a measurement system that exhibits a bias of about .06 across the entire range of true values of interest.

On the other hand, Figure 11.14 shows a measurement system in which the bias is changing linearly across the true values of interest.

Gage R&R should be used to ensure that operators are able to measure consistently before performing a linearity and bias experiment. Then true standards must be identified, as establishing bias requires samples that have known values. A true standard often must be purchased from a standards organization and should never be selected by measuring parts with the same measurement system that is being evaluated. In some cases a measurement system may be available that is certain to be unbiased and extremely precise and this may be used to select standards.

To perform the experiment, standards should be selected across the range of values in which you wish to ensure that the measurement system is accurate. The order of measurements should be randomized, and a single operator should perform all measurements. At least five measurements should be taken for each standard part.

Bias is calculated by taking each measurement and subtracting the true value.

$$Bias = measurement - true\ value$$

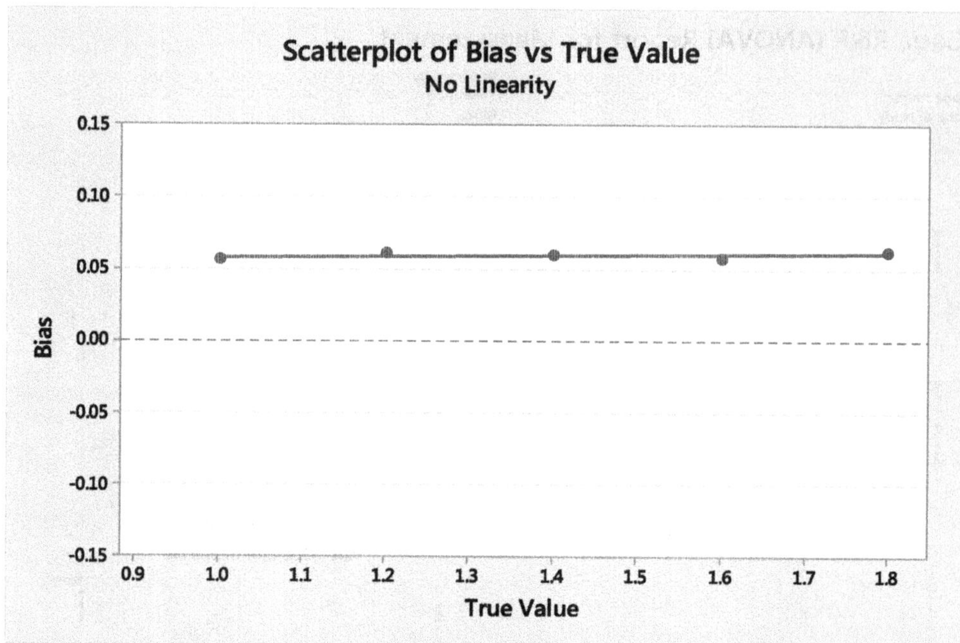

Figure 11.13 A system with bias but no linearity.

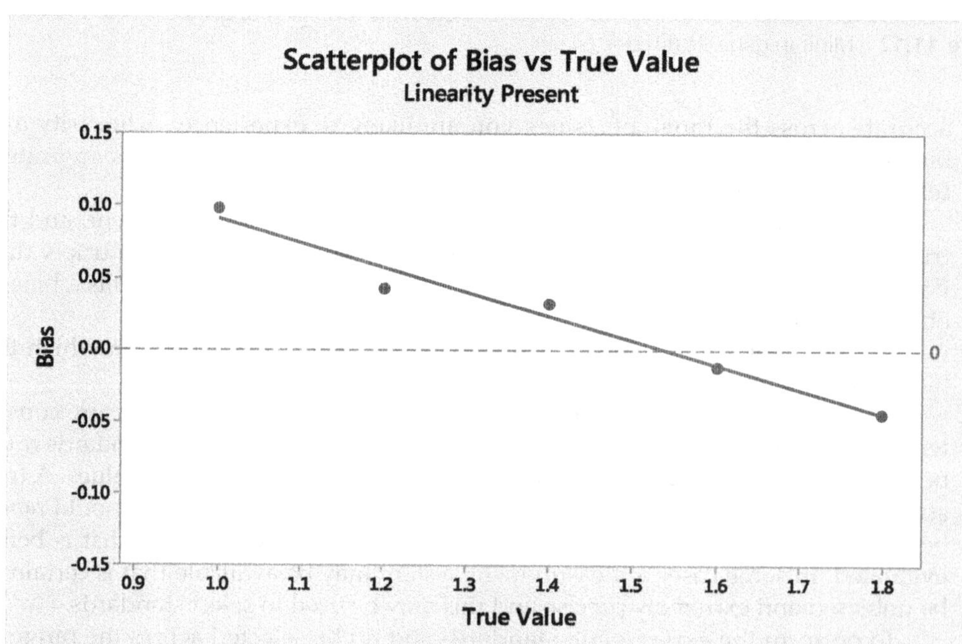

Figure 11.14 A system with bias and linearity present.

Then the bias for each part is found by averaging the bias for all measurements of that part.

$$Bias_{part\ i} = \frac{\sum Biases\ for\ part\ i}{n_i}$$

The number of measurements taken of that part is identified as n_i. Finally, the overall bias is found by summing the biases of all measurements of all parts and dividing by the total number of measurements.

$$Average\ bias = \frac{\sum Biases}{N}$$

From the already performed gage R&R, the process variation may be calculated using the overall standard deviation.

$$Process\ variation = 6 * s_{overall}$$

Then %Bias is found by taking the absolute value of the average bias and dividing by the process variation.

$$\%\ Bias = \frac{|Average\ bias|}{Process\ variation} * 100\%$$

Regression is performed with the biases of each measurement as the response variable and the true values of each measurement as the predictor to determine whether linearity is present. There is evidence that linearity is present and the amount of bias is dependent on the true value if the p-value for regression is statistically significant. Linearity is defined as the absolute value of the slope times the process variation.

$$Linearity = |\ Slope\ | * Process\ variation$$

The % Linearity is found by dividing linearity by process variation or by taking the absolute value of the slope and multiplying by 100%, as follows:

$$\%\ Linearity = \frac{Linearity}{Process\ variation} * 100\%$$

or

$$\%\ Linearity = |\ Slope\ | * 100\%$$

Example

Blocks of wax are produced with a stated weight of 16 ounces. The process has a historical overall standard deviation of 0.05 ounces and gage R&R has already been performed to ensure that operators measure consistently. Standard blocks

Table 11.9 Linearity and bias study data.

Block 15.8	Block 15.9	Block 16	Block 16.1	Block 16.2
15.84	15.93	16.02	16.11	16.20
15.84	15.92	16.04	16.13	16.21
15.83	15.93	16.02	16.11	16.20
15.86	15.93	16.03	16.10	16.21
15.86	15.93	16.00	16.11	16.20

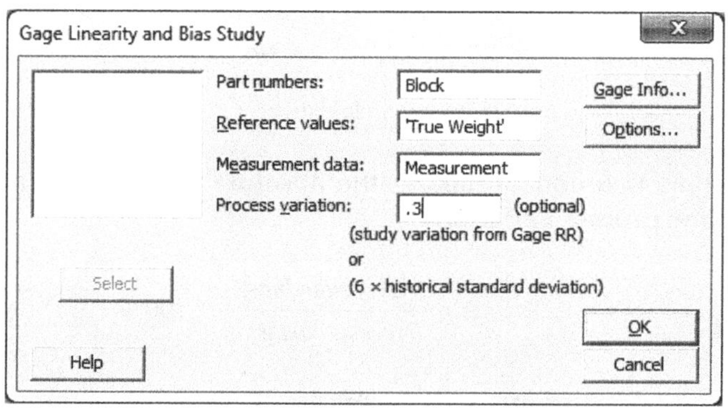

Figure 11.15 Gage linearity and bias study dialog box in Minitab.

with weights of 15.8, 15.9, 16, 16.1, and 16.2 are obtained and an operator has taken five measurements of each in random order. The data are shown in Table 11.9.

The data are entered in Minitab in columns representing the part number, the measurement, and the true value. Go to **Stat > Quality Tools > Gage Study > Gage Linearity and Bias Study** and complete the dialog as shown in Figure 11.15.

The output is shown in Figure 11.16.

The p-value for the slope in gage linearity is rounded to 0.000, indicating there is strong statistical significance that linearity is present. Both the slope and a visual check of the regression fit indicate that as parts get larger, the bias decreases. It can be seen that parts at the lower end of the range of interest have a positive bias and are being measured as heavier than they truly are, while at the highest reference value of interest there is almost no bias exhibited.

Although the average bias has a p-value of 0.000, we know from the linearity portion of the output in Figure 11.16 that the bias is dependent on the true value. Therefore the average bias across all reference values is not as informative as the biases of each reference value individually. Those all exhibit statistically significant p-values except for that of the 16.2 ounce reference block, which we cannot conclude is biased.

Although the measurement system was shown to be consistent using gage R&R, actions should be taken to correct the linearity and bias problem.

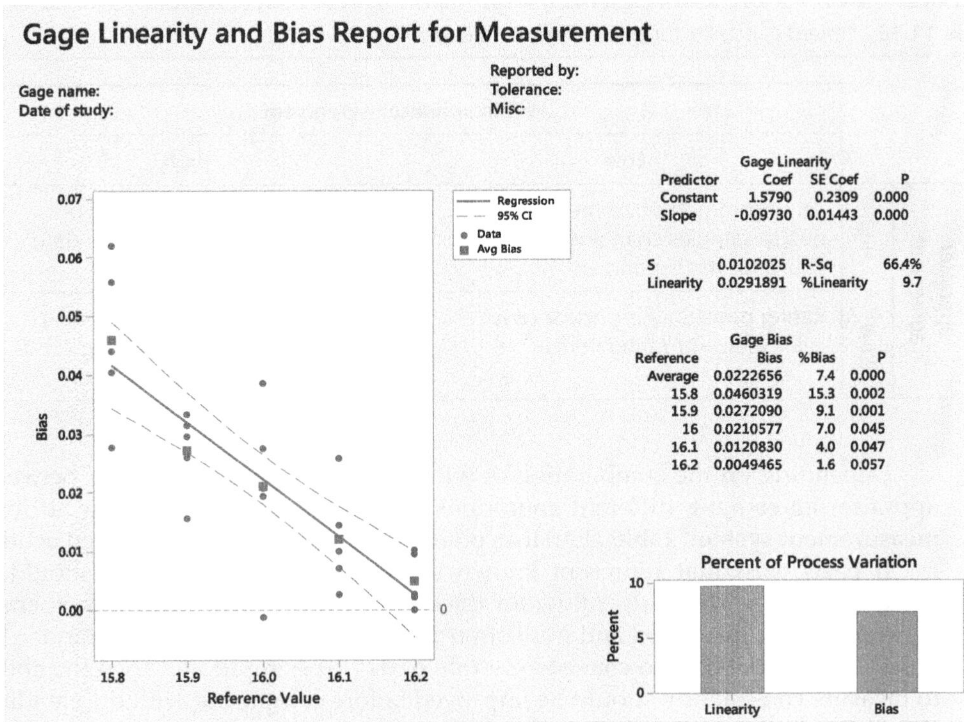

Figure 11.16 Gage linearity and bias study output for wax blocks.

ATTRIBUTE AGREEMENT ANALYSIS

Some measurements result in a categorical assessment of an item rather than a continuous measurement. For example, baked bread may be categorized into any of 10 different shades of brown, or patients arriving at an emergency room may be categorized into five categories of urgency. Attribute agreement analysis is a study of how consistently appraisers rate items on a categorical scale as well as whether they can correctly rate known standards if they exist.

If only a few appraisers exist, all should be included in the study to identify any that may perform significantly better or worse than the others. To assess the consistency of the appraisers, up to 50 different parts should be used in the study. Each part should be replicated as well so that each operator has rated each part twice. A study performed in this manner can assess the percentage of parts in which individual appraisers agreed with themselves across both replicates and the percentage of parts in which all appraisers agreed with themselves across both replicates and with each other.

The study is typically done by randomizing the order of the appraisers within each replicate and then randomizing the order of the parts within each appraiser's replicate. For example, one appraiser would rate all parts in a random order, followed by another appraiser until all appraisers have measured each part once. Then the order of the appraisers is randomized again and the order of parts within each appraiser is randomized again.

Table 11.10 Potential causes for poor attribute measurement system.

<table>
<tr><th colspan="2"></th><th colspan="2">Within appraiser agreement</th></tr>
<tr><th colspan="2"></th><th>Low</th><th>High</th></tr>
<tr><th rowspan="2">Between appraiser agreement</th><th>High</th><td>An uncommon situation, potentially due to samples changing during the course of the experiment</td><td>Measurement system is good but should be checked versus standard if possible</td></tr>
<tr><th>Low</th><td>Rating procedure is unclear or is insufficient for rating items</td><td>Appraisers are interpreting testing procedure differently or rating conditions differ between appraisers</td></tr>
</table>

Depending on the combination of within-appraiser agreement and between-appraiser agreement, different conditions may be the cause of a poor attribute measurement system. Table 11.10 lists potential causes and recommended actions.

If parts exist that represent known standards, then those parts should be used. This will additionally allow for the study to assess how well each operator agreed with that standard and how often all operators agree with the standard. If appraisers are not able to consistently rate parts across replicates, then the ability to measure consistently should be improved before comparing with the standard. If individual appraisers are able to measure consistently across replicates but do not match the standard a high percentage of time, then the appraisers may not be following the standard procedure. If all appraisers consistently measure consistently with each other but do not match the standard, then the standards likely are not consistent with the measuring procedure as understood by the appraisers.

The simplest and often most informative measure of agreement—whether within appraiser, between appraiser, each appraiser versus standard, or all appraisers versus standard—is simply the percentage of samples in which the ratings matched a corresponding confidence interval. The percentage matched is calculated as:

$$\%\ matched = \frac{n_{matched}}{n_{total}} * 100\%$$

The formula for the 95% confidence interval on the proportion matched can be found in Chapter 6. When the proportion matched is 0, the lower limit on the confidence interval is 0 and only an upper limit is calculated; similarly, when the proportion is 1, the upper limit is 1 and only a lower limit is calculated. In both cases the values of F used in the formula corresponds to α instead of $\alpha/2$.

When the ratings used are nominal, meaning they have no natural ordering, another common statistic for assessing the measurement system is known as the *kappa statistic*. When no known standard is available, it is calculated as

$$kappa = \frac{P_o - P_e}{1 - P_e},$$

where P_o = the proportion matched, and P_e = the expected proportion by chance = Σp_j^2. Then

$$p_j = \frac{1}{Nn} \Sigma_{i=1}^{N} x_{ij},$$

where N is the number of parts, n is the number of appraisers, and x_{ij} is the number of raters who assigned the ith part to the jth rating.

Kappa values closer to 1 indicate a better ability to consistently rate items. Generally, values above 0.7 are considered marginal and those above 0.9 are considered good. However, the value of the kappa statistic changes depending on the number of possible ratings, the number of parts, and the distribution of ratings, so those guidelines should be considered loose.

Additionally, a p-value can be created based on the kappa statistic. This p-value corresponds to a hypothesis test with the null hypothesis being that appraisers' ratings are due purely to chance and an alternative hypothesis being that they are not due to random chance. It is important to consider that even a poor measurement system can likely outperform random chance, so significant p-values should not be taken as an indication of a good measurement system. Also, p-values may be generated for every combination of appraiser and rating as well as across appraisers and versus the standard, so a complete attribute agreement analysis results in many p-values.

When ratings are ordinal instead of nominal, meaning there is a natural ordering to them, a statistic known as *Kendall's coefficient* is a common statistic for evaluating the measurement system. Kendall's coefficient of concordance takes the ordering into account: Inconsistent ratings that were very close to one another are penalized less than those that were rated further from one another. For example, if the appearance of apples is rated poor, okay, good, or excellent, an appraiser who rates a single apple poor once and excellent once performs worse than one who rates that apple good once and excellent once. When comparing to a standard, Kendall's correlation coefficient does the same, but compares how closely ratings matched the standard instead of the other ratings.

The Kendall's coefficient ranges from 0 to 1, with values closer to 1 indicating a better measurement system. As with kappa, a p-value can be calculated for Kendall's—but again it corresponds to a test of whether ratings are random or not. As with kappa, even a poor measurement system may show statistical significance, so the p-value alone is not a good indication that the measurement system performs well.

When evaluating an attribute measurement system, a very important consideration is the implications of incorrectly categorizing an item. If the implications are moderate to severe, appraisers need to correctly rate items nearly 100% of the time. Determining that appraisers are achieving this high level of performance will require a large number of parts when performing the study. Only this will result in a confidence interval for the percentage matched that is narrow enough to provide confidence of a high-quality measurement system.

Another important consideration is that often items that are measured with attribute measurements are unique and easy for an appraiser to recognize. In this case it may be necessary to separate the replicates of the experiment over time to try to prevent appraisers from recognizing the item and recalling what their previous rating of it was.

Example

Whiskey is aged in barrels that may have flaws, resulting in differing qualities of product that can be detected by observing the color of the whiskey that is poured from each barrel. The color of the whiskey is appraised on a scale of 1–5. To ensure that the three appraisers are assessing the color consistently and accurately, an attribute agreement analysis is performed on 10 samples that include known standards across the entire scale. The data are found in Table 11.11.

In Minitab: Go to **Stat > Quality Tools > Attribute Agreement Analysis**, complete the dialog as shown in Figure 11.17, and click **OK**.

The within-appraisers portion of the session window output is shown in Figure 11.18.

The assessment agreement table shows that all appraisers agreed with themselves on both replicates of the 10 parts at least seven times. However, the confidence intervals on the true percentage of time each appraiser agrees with himself are very wide with such a small number of parts.

The severity of an incorrect rating should be considered when determining whether the system needs to be improved. In the case of the whiskey, missing a defective barrel by misclassifying the color could be a very expensive mistake as that barrel will be mixed with many others in a large batch, potentially ruining it.

Because the categories of color follow a natural order, it is important to consider Kendall's coefficient of concordance instead of the kappa statistics. Each appraiser has a significant p-value and a coefficient close to 1, an indication that although individual appraisers may not always agree with themselves across replicates, they do not tend to miss by much. In other words, an individual appraiser misclassifying a sample that should be labeled a 2 is much more likely to label it a 1 or a 3 than a 5.

The each appraiser versus standard portion of the session window output is shown in Figure 11.19.

The assessment agreement table indicates that the appraiser named Kam not only agreed with himself across replicates of all 10 parts, but also agreed with the known standard. However, Bruce and Michael only met this criteria 6 and 7 times, respectively. In the case of Kam and Michael, all within-appraiser agreements also agreed with the standard. However, Bruce agreed with himself 8 out of 10 times but only 6 out of 10 times with the standard. The number of samples was too small to say with any certainty, but this may be an indication that Bruce needs additional training. Consistent with the within-appraiser agreement, all Kendall's coefficients are high and the corresponding p-values are highly significant. The between-appraisers portion of the output is shown in in Figure 11.20.

On only 3 of the 10 samples did the appraisers not only agree with themselves but also with each other. This is an indication that the current measurement system is not capable of producing consistent measurements while using multiple appraisers. Kendall's coefficient is very close to 1 with a very low p-value, demonstrating that while appraisers may not be able to rate very consistently they generally are able to come close to the true value. The all appraisers versus standard portion of the session window output is shown in Figure 11.21.

As with the between-appraisers assessment, on 3 of 10 samples all appraisers not only agreed with themselves but also with the standard. Therefore, there were no instances in which all appraisers agreed but failed to match the standard.

Table 11.11 Attribute agreement analysis data.

Run	Sample	Appraiser	Color	Standard	Run	Sample	Appraiser	Color	Standard
1	3	Bruce	4	4	31	7	Bruce	5	5
2	10	Bruce	2	2	32	6	Bruce	2	3
3	4	Bruce	1	1	33	1	Bruce	4	5
4	5	Bruce	3	3	34	3	Bruce	4	4
5	8	Bruce	2	3	35	2	Bruce	1	1
6	7	Bruce	5	5	36	4	Bruce	1	1
7	9	Bruce	3	4	37	9	Bruce	4	4
8	6	Bruce	2	3	38	10	Bruce	2	2
9	2	Bruce	1	1	39	5	Bruce	3	3
10	1	Bruce	4	5	40	8	Bruce	3	3
11	9	Michael	4	4	41	6	Michael	3	3
12	3	Michael	4	4	42	10	Michael	2	2
13	6	Michael	3	3	43	5	Michael	4	3
14	4	Michael	1	1	44	1	Michael	5	5
15	1	Michael	5	5	45	4	Michael	1	1
16	2	Michael	1	1	46	3	Michael	5	4
17	8	Michael	3	3	47	8	Michael	3	3
18	10	Michael	3	2	48	7	Michael	5	5
19	5	Michael	3	3	49	9	Michael	4	4
20	7	Michael	5	5	50	2	Michael	1	1
21	5	Kam	3	3	51	4	Kam	1	1
22	1	Kam	5	5	52	7	Kam	5	5
23	10	Kam	2	2	53	9	Kam	4	4
24	2	Kam	1	1	54	5	Kam	3	3
25	9	Kam	4	4	55	6	Kam	3	3
26	7	Kam	5	5	56	3	Kam	4	4
27	6	Kam	3	3	57	10	Kam	2	2
28	8	Kam	3	3	58	2	Kam	1	1
29	3	Kam	4	4	59	1	Kam	5	5
30	4	Kam	1	1	60	8	Kam	3	3

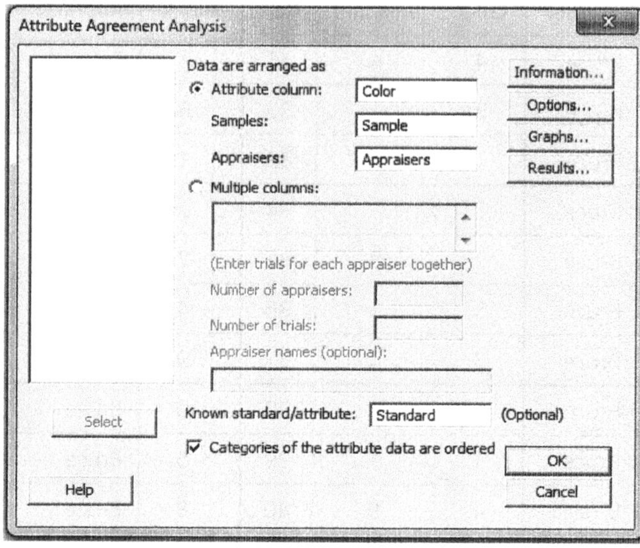

Figure 11.17 Minitab dialog box for attribute agreement analysis.

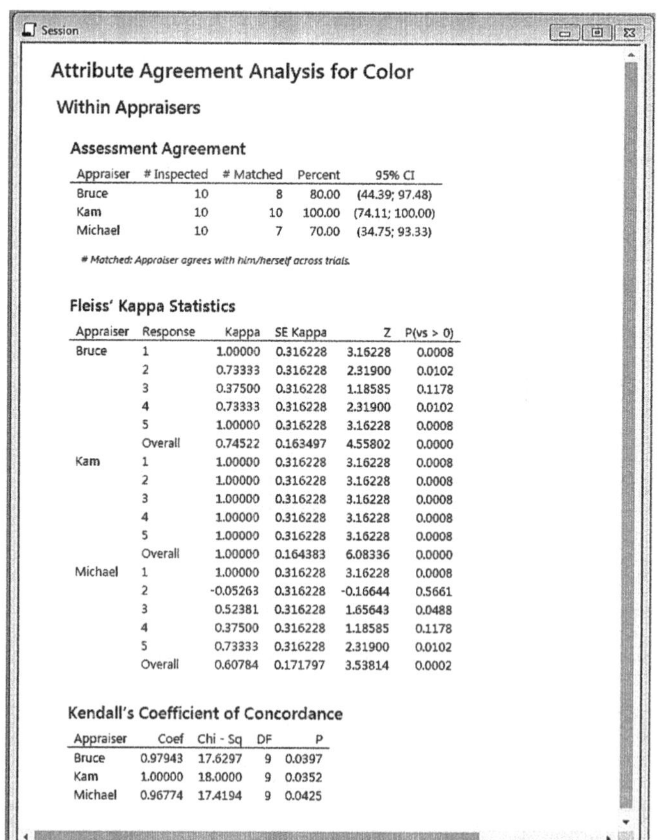

Session

Attribute Agreement Analysis for Color

Within Appraisers

Assessment Agreement

Appraiser	# Inspected	# Matched	Percent	95% CI
Bruce	10	8	80.00	(44.39; 97.48)
Kam	10	10	100.00	(74.11; 100.00)
Michael	10	7	70.00	(34.75; 93.33)

 # Matched: Appraiser agrees with him/herself across trials.

Fleiss' Kappa Statistics

Appraiser	Response	Kappa	SE Kappa	Z	P(vs > 0)
Bruce	1	1.00000	0.316228	3.16228	0.0008
	2	0.73333	0.316228	2.31900	0.0102
	3	0.37500	0.316228	1.18585	0.1178
	4	0.73333	0.316228	2.31900	0.0102
	5	1.00000	0.316228	3.16228	0.0008
	Overall	0.74522	0.163497	4.55802	0.0000
Kam	1	1.00000	0.316228	3.16228	0.0008
	2	1.00000	0.316228	3.16228	0.0008
	3	1.00000	0.316228	3.16228	0.0008
	4	1.00000	0.316228	3.16228	0.0008
	5	1.00000	0.316228	3.16228	0.0008
	Overall	1.00000	0.164383	6.08336	0.0000
Michael	1	1.00000	0.316228	3.16228	0.0008
	2	-0.05263	0.316228	-0.16644	0.5661
	3	0.52381	0.316228	1.65643	0.0488
	4	0.37500	0.316228	1.18585	0.1178
	5	0.73333	0.316228	2.31900	0.0102
	Overall	0.60784	0.171797	3.53814	0.0002

Kendall's Coefficient of Concordance

Appraiser	Coef	Chi - Sq	DF	P
Bruce	0.97943	17.6297	9	0.0397
Kam	1.00000	18.0000	9	0.0352
Michael	0.96774	17.4194	9	0.0425

Figure 11.18 Within-appraisers portion of output.

Attribute Agreement Analysis for Color

Each Appraiser vs Standard

Assessment Agreement

Appraiser	# Inspected	# Matched	Percent	95% CI
Bruce	10	6	60.00	(26.24; 87.84)
Kam	10	10	100.00	(74.11; 100.00)
Michael	10	7	70.00	(34.75; 93.33)

 # Matched: Appraiser's assessment across trials agrees with the known standard.

Fleiss' Kappa Statistics

Appraiser	Response	Kappa	SE Kappa	Z	P(vs > 0)
Bruce	1	1.00000	0.223607	4.47214	0.0000
	2	0.49142	0.223607	2.19770	0.0140
	3	0.46667	0.223607	2.08700	0.0184
	4	0.55417	0.223607	2.47831	0.0066
	5	0.60784	0.223607	2.71836	0.0033
	Overall	0.62185	0.113414	5.48296	0.0000
Kam	1	1.00000	0.223607	4.47214	0.0000
	2	1.00000	0.223607	4.47214	0.0000
	3	1.00000	0.223607	4.47214	0.0000
	4	1.00000	0.223607	4.47214	0.0000
	5	1.00000	0.223607	4.47214	0.0000
	Overall	1.00000	0.116236	8.60317	0.0000
Michael	1	1.00000	0.223607	4.47214	0.0000
	2	0.47368	0.223607	2.11838	0.0171
	3	0.75678	0.223607	3.38441	0.0004
	4	0.68750	0.223607	3.07459	0.0011
	5	0.86667	0.223607	3.87585	0.0001
	Overall	0.80639	0.119095	6.77096	0.0000

Kendall's Correlation Coefficient

Appraiser	Coef	SE Coef	Z	P
Bruce	0.91026	0.175682	5.11802	0.0000
Kam	1.00000	0.175682	5.62885	0.0000
Michael	0.94192	0.175682	5.29827	0.0000

Figure 11.19 Each appraiser vs. standard output.

Attribute Agreement Analysis for Color

Between Appraisers

Assessment Agreement

# Inspected	# Matched	Percent	95% CI
10	3	30.00	(6.67; 65.25)

 # Matched: All appraisers' assessments agree with each other.

Fleiss' Kappa Statistics

Response	Kappa	SE Kappa	Z	P(vs > 0)
1	1.00000	0.0816497	12.2474	0.0000
2	0.48077	0.0816497	5.8882	0.0000
3	0.52273	0.0816497	6.4021	0.0000
4	0.54828	0.0816497	6.7150	0.0000
5	0.71058	0.0816497	8.7027	0.0000
Overall	0.65425	0.0415671	15.7396	0.0000

Kendall's Coefficient of Concordance

Coef	Chi - Sq	DF	P
0.967799	52.2611	9	0.0000

Figure 11.20 Between-appraisers output.

Session

Attribute Agreement Analysis for Color

All Appraisers vs Standard

Assessment Agreement

# Inspected	# Matched	Percent	95% CI
10	3	30.00	(6.67; 65.25)

 # Matched: All appraisers' assessments agree with the known standard.

Fleiss' Kappa Statistics

Response	Kappa	SE Kappa	Z	P(vs > 0)
1	1.00000	0.129099	7.7460	0.0000
2	0.65504	0.129099	5.0739	0.0000
3	0.74115	0.129099	5.7409	0.0000
4	0.74722	0.129099	5.7880	0.0000
5	0.82484	0.129099	6.3892	0.0000
Overall	0.80941	0.067129	12.0575	0.0000

Kendall's Correlation Coefficient

Coef	SE Coef	Z	P
0.950726	0.101430	9.33670	0.0000

Figure 11.21 All appraisers vs. standard output.

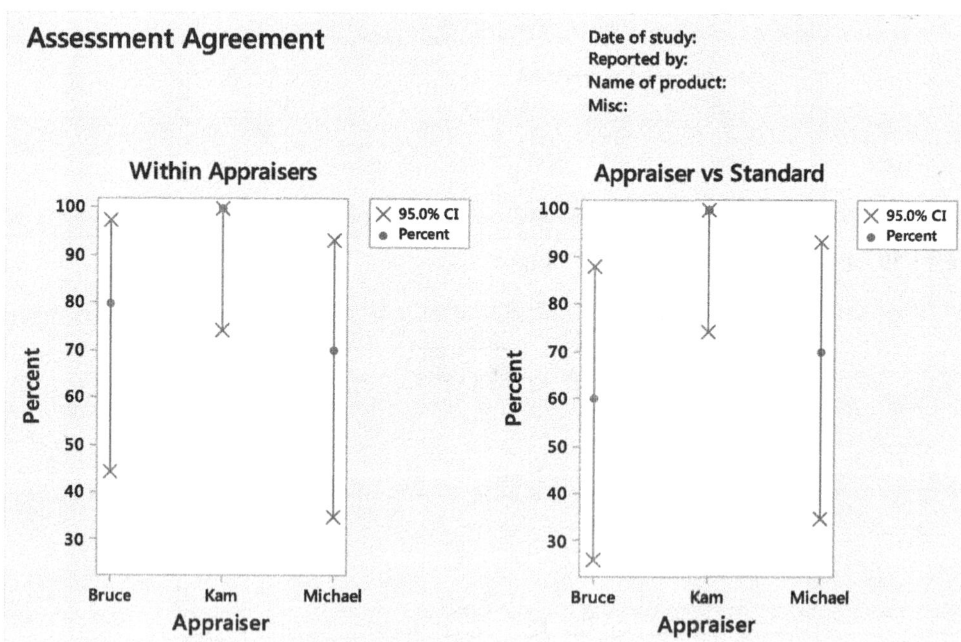

Figure 11.22 Assessment agreement graph.

Kendall's coefficient is again high with a very low p-value. A graph of two of the agreement measures and the corresponding confidence intervals is found in Figure 11.22.

The graph clearly illustrates that while there were differences between appraisers, those differences fall well within the very wide confidence intervals (CI), and it is not clear that any appraisers are truly better or worse than others. A study using a larger number of samples would provide more narrow intervals and better clarity.

Based on the expensive consequences of misclassifying the color of the whiskey and the results of the study, the measurement system should be improved to gain more consistency among appraisers. A larger study that used more samples would then provide better clarity on any remaining issues in classifying the color.

QUESTIONS AND ANSWERS

Q: How many evaluators should I use for a crossed gage R&R study?

A: Only evaluators who are or will be performing the evaluations in practice should be included. If feasible, all the evaluators should be included. In some cases, however, there are too many for this to be a reasonable study. In those cases at least three evaluators should be used.

Q: How should I select parts for a crossed gage R&R study?

A: If no historical estimate of standard deviation using the same measurement system is available, then parts must be selected at random to get as accurate an estimate of part-to-part standard deviation as possible. If a historical estimate is available and will be used, random selection is still a good option but not as critical.

Q: What do I do if I need to evaluate a measuring device but I don't have a reference standard?

A: Without a standard, a gage R&R study can still be performed to assess the variation in data due to the measurement system. However, that study will not indicate whether the system is accurate—only how precise. If no standard exists, then it may be feasible to use a measurement system that has been validated against standards.

Q: I don't have a standard but need to do a gage R&R study. How do I benchmark off a measurement system that has been validated against standards?

A: The parts used in the gage R&R study should be measured by a system that has been validated as very precise and accurate against a standard. Then those measurements can be used as reference values.

Chapter 12
Process Capability

The capability of a process is its ability to produce items that are within specifications and not defective. Generally a sample is used to determine the parameters of the underlying distribution, and that distribution is used to calculate the percentage of parts expected to be outside of specifications or other metrics. A *capability study* compares the voice of the customer to the voice of the process (Kahn 2013), with the specification limits representing the voice of the customer and the spread of the data representing the voice of the process.

Capability is often thought of in two dimensions: short-term capability and long-term capability. *Short-term or within-subgroup* capability corresponds to the inherent variation in the characteristics of interest. *Long-term or overall* capability refers to the variation that is inherent as well as variation between subgroups and due to the drifting of the process over time.

The first step in calculating capability is to determine if the process has natural subgroups or can be sampled in subgroups. Natural subgroups occur when items are produced in batches or molds or from lots of raw material. Items produced within a single one of these would be considered to be a "subgroup" of all data and should be indicated as such. Otherwise, items can be sampled in subgroups for more continuous processes by collecting multiple items simultaneously to form a subgroup and then collecting subgroups of this manner over time. Generally, it is desirable to have at least 5 items per subgroup. If data cannot be collected in subgroups, then the subgroup size is considered to be 1. In all cases the data should be entered in the order in which the items were produced.

The second step in calculating capability is to determine the time frame in which data can be collected. Because overall capability is expected to contain long-term process drifts, it is important that the data be collected over a long enough period of time to capture these drifts (often weeks or even months). Otherwise, if data cannot be collected over a long enough period of time to include this variation, then the analysis can only report short-term or within-subgroup capability.

There are three criteria that must be met to correctly perform a capability study (see Table 12.1). The process must be in statistical control, which can be evaluated by means of a control chart. A process that is not in a state of statistical control is not predictable, and therefore any resulting capability values apply only to the data set that was used to perform the calculation and it gives no indication as to what to expect in terms of future performance. Worse than being simply wrong, the resulting capability value may give false confidence in a process and improvement opportunities could be missed. The data also must be normally distributed, or if performing capability for non-normal data, the data must fit the

Table 12.1 Requirements for a capability study.

Requirement	Evaluation method	Consequences of failing to meet the requirement	Actions to take
Process must be in a state of control	Evaluate using an I chart of individual values or an xbar chart of subgroups	An incorrect conclusion because future samples may not be reflected by the evaluated sample	Study the process to identify and remove the source or sources of special-cause variation
Data must be normally distributed	Evaluate with a probability plot	An incorrect conclusion because the metrics are based on the properties of the distribution	Either study the process to determine why it is not normally distributed if normality is expected and improve it, or assess using a capability study for non-normally distributed data
Sufficient data must be available (100 or more values)	Check the number of values available	Imprecise estimates due to too few samples to precisely estimate distribution parameters	Collect more values or include confidence intervals to understand uncertainty of estimates

chosen distribution well. If the data deviate from the assumed distribution by any meaningful amount, then the resulting statistics likely no longer reflect the actual quality to be expected from the process. Sufficient data must be available to minimize sampling error; ideally, there should be 100 or more values from randomly selected samples and these samples must represent the population. Not only is it difficult to assess with much certainty the distribution and stability with too little data, but estimates of the parameters of that distribution, and therefore capability estimates, can be imprecise. For smaller samples, Minitab can calculate confidence intervals for capability statistics so that their precision is known. In Minitab, to add a confidence interval when in the **Capability Analysis** dialog box, select **Options** and then place a checkmark next to **Include confidence intervals**.

There are multiple formulas available for calculating the standard deviation when performing a capability study for normally distributed data (Breyfogle 2014). The formulas are shown in Table 12.2.

NORMAL CAPABILITY ANALYSIS

The short-term capability of a process is measured using the statistics Cp and CpK. Cp compares the tolerance of the process to the variation within subgroups and represents the capability of the process if it were centered and sources of between-subgroup and long-term variation were removed. Cpk compares the distance from the mean to the closest specification limit with the within-subgroup variation and

Table 12.2 Standard deviation formulas for capability studies.

Type of variation	Subgroup size	Formula	Explanation of standard deviation calculation
Within subgroup or short-term	>1	$\frac{S_{pooled}}{c_4(N-k+1)}$ With $S_{pooled} = \sqrt{\frac{\sum s_i^2}{k}}$	Uses the pooled standard deviation of subgroups when subgroups are used
Within subgroup or short-term	1	$\frac{\frac{\sum R_i}{n-1}}{d_2(2)}$	Uses the moving range method when subgroups are not used
Overall or long term	n/a	$\sqrt{\frac{\sum (x_i - \bar{x})^2}{n-1}}$	Uses the sample standard deviation for long term data

is reduced as the mean of the process moves away from the midpoint of the specification limits. The formulas for normally distributed data are:

$$Cp = \frac{USL - LSL}{6 * s_{short\text{-}term}}$$

$$Cpk = \text{the minimum value of} \left(\frac{USL - \bar{x}}{3 * s_{short\text{-}term}}, \frac{\bar{x} - LSL}{3 * s_{short\text{-}term}} \right)$$

The within-subgroup or short-term standard deviation $s_{short\text{-}term}$ is estimated using the pooled standard deviation when subgroup sizes are larger than 1, and it is estimated using the moving range when subgroup size equals 1. The formulas are:

$$s_{short\text{-}term} = \frac{\sqrt{\frac{\Sigma s_i^2}{k}}}{c_4(N-k+1)} \text{ when subgroup size} > 1,$$

where s_i^2 = the standard deviation of subgroup i, k = the number of subgroups, N = the total sample size, and c_4 = an unbiasing constant.

$$s_{short\text{-}term} = \frac{\frac{\Sigma R_i}{N-1}}{d_2(2)} \text{ when subgroup size} = 1,$$

where R_i = the range of each pair of consecutive points, and d_2 = an unbiasing constant.

A Cp of 1 corresponds to a distribution where the specifications are 6 short-term standard deviations apart. For a process that is centered, Cpk is equal to

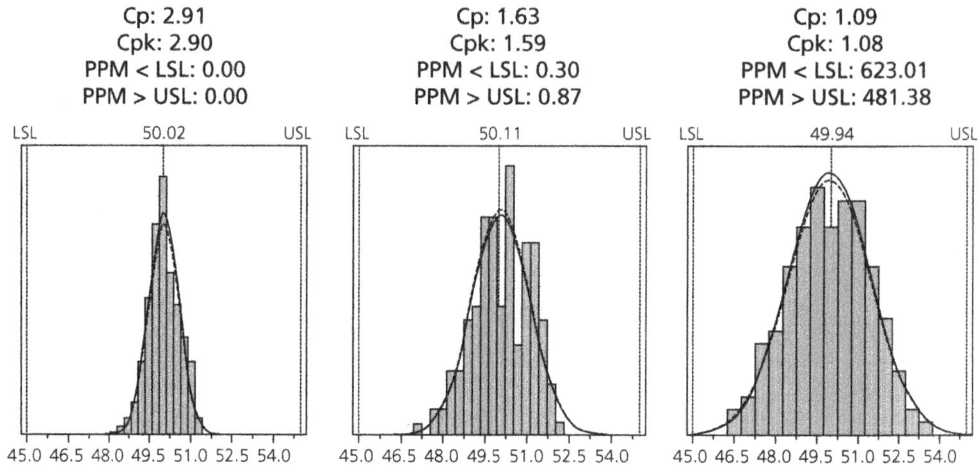

Figure 12.1 Centered process examples.

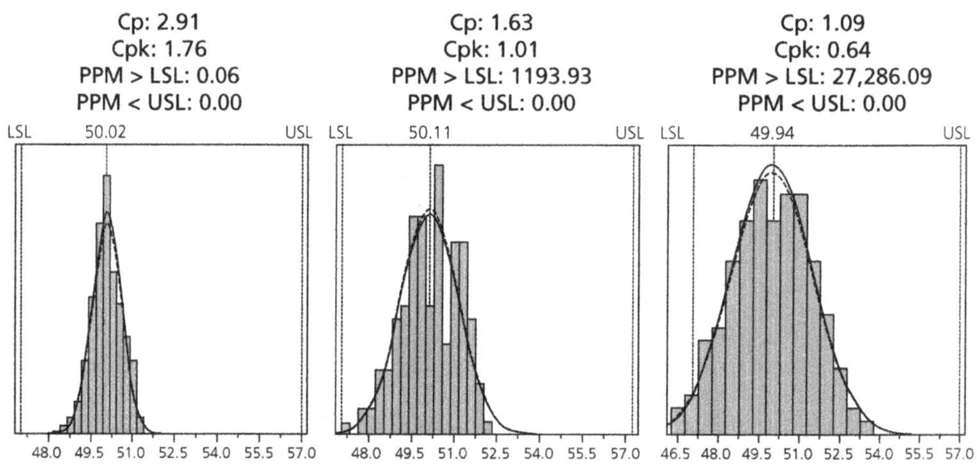

Figure 12.2 Off-center process examples.

Cp. It is generally desirable to have Cpk values of at least 1.33. Figure 12.1 shows examples of Cp and Cpk for centered processes with different variances.

Figure 12.2 shows examples of Cp and Cpk with off-centered processes and the same variances as in Figure 12.1.

It can be seen that for a particular variance, Cp does not change for the off-center processes but Cpk is greatly reduced.

The overall capability or performance of a process is measured using the statistics Pp and Ppk. Both Pp and Ppk are used for long-term data and tell us what the process is actually producing over time. Pp compares the overall variation of the process to the specification limits (Barsalou 2015b). Like Cpk, Ppk takes into consideration the position of the process mean relative to the specification limits and is reduced as the process moves away from center. For normally distributed data the formulas are:

$$Pp = \frac{USL - LSL}{6 * s_{overall}}$$

$$Ppk = minimum\left(\frac{USL - \bar{x}}{3 * s_{overall}}, \frac{\bar{x} - LSL}{3 * s_{overall}}\right)$$

The overall standard deviation is equal to the sample standard deviation. As with Cpk, values of 1.33 or higher for Ppk indicate good quality.

For processes that have a target value, another statistic for assessing overall process performance is Cpm. It assesses variation of the data relative to the target (Preis 2009) and assumes the target is the midpoint between the upper specification limit (USL) and lower specification limit (LSL). The formula for target value T is:

$$Cpm = \frac{minimum(USL - T, T - LSL)}{3\sqrt{\frac{\Sigma\Sigma(x_{ij} - T)^2}{(\Sigma n_i) - 1}}} = \frac{(USL - LSL)/2}{3\sqrt{s^2 + (\bar{x} - T)^2}}$$

As with the previous indices, values of 1.33 or higher are desirable for Cpm. For example, the Cpm for a process with USL = 24, LSL = 20 and a target value of 22 with an actual process mean of 22.353 and a standard deviation of 0.781 is:

$$Cpm = \frac{minimum(USL - T, T - LSL)}{3\sqrt{\frac{\Sigma\Sigma(x_{ij} - T)^2}{(\Sigma n_i) - 1}}} = \frac{(USL - LSL)/2}{3\sqrt{s^2 + (\bar{x} - T)^2}}$$

$$= \frac{(USL - LSL)/2}{3\sqrt{0.781^2 + (22.353 - 22.0)^2}} = 0.778$$

The resulting Cpm is only 0.778; the process is clearly off center or has too much variation.

Example

Table 12.3 shows 10 subgroups with the measurements from 10 parts in each. The lower specification limit is 20.0 and the upper specification limit is 24.0. The target value for the process is 22.0.

The individual plots of subgroup values are shown in Figure 12.3. Notice subgroup 7 has more variation than the other subgroups.

The mean of the data is found to be 22.109. The short-term standard deviation must then be calculated as

$$s_{short\text{-}term} = \frac{\sqrt{\frac{\Sigma s_i^2}{k}}}{c_4(N - k + 1)}$$

$$= \frac{\sqrt{\frac{.575^2 + .310^2 + .484^2 + .428^2 + .436^2 + .410^2 + .925^2 + .394^2 + .426^2 + .570^2}{10}}}{c_4(91)}$$

$$= \frac{0.521}{0.9972} = 0.526,$$

Table 12.3 Capability study data.

	Subgroup 1	Subgroup 2	Subgroup 3	Subgroup 4	Subgroup 5	Subgroup 6	Subgroup 7	Subgroup 8	Subgroup 9	Subgroup 10
	21.92	21.77	21.30	21.16	21.53	22.80	23.30	21.61	21.94	22.32
	22.27	21.94	21.87	22.52	22.75	21.87	24.05	22.86	21.99	21.94
	22.02	21.81	22.22	21.57	22.01	21.84	23.03	21.88	21.79	21.92
	23.50	22.08	22.47	21.70	22.21	22.22	22.03	21.72	22.98	21.68
	22.14	21.79	21.53	22.01	22.11	22.12	20.94	21.85	21.43	22.46
	23.02	22.17	22.07	21.98	22.05	22.22	22.59	22.18	22.06	20.79
	22.21	21.39	21.92	22.53	21.28	22.81	22.44	21.85	21.99	22.19
	21.98	21.43	22.94	21.64	21.73	22.65	21.36	22.12	21.83	22.70
	21.65	21.90	22.44	22.05	22.39	21.97	21.78	22.54	22.53	22.27
	22.81	22.40	21.77	21.65	21.63	22.89	22.35	22.35	21.95	22.73
Mean	22.35	21.87	22.05	21.88	21.97	22.34	22.39	22.10	22.05	22.10
StDev	0.575	0.310	0.484	0.428	0.436	0.410	0.925	0.394	0.425	0.570

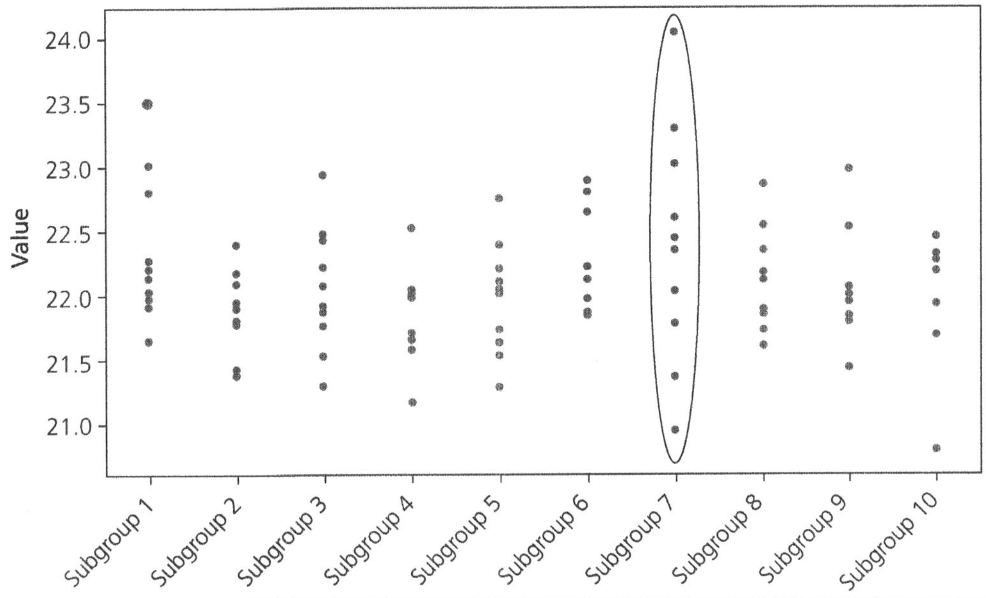

Figure 12.3 Individual value plot of each subgroup.

where $c_4(91) = 0.9972$, which is read from Table D.1 in Appendix D. Note: For values over 30, you should use the value c4(30) = 0.9914; however, Minitab uses the exact value for values over 30.Then Cp and Cpk are calculated as:

$$\text{Cp} = \frac{USL - LSL}{6 * s_{short\text{-}term}} = \frac{24 - 20}{6 * 0.522} = 1.28$$

$$\text{Cpk} = \frac{minimum(\bar{X} - \text{LSL, USL} - \bar{X})}{3s_{short\text{-}term}} = \frac{24 - 22.109}{3 * 0.522} = 1.27$$

The Cp is 1.28 and the Cpk is 1.21.

Pp and Ppk can be calculated using the overall mean of 22.103 and overall standard deviation of 0.529, as follows:

$$\text{Pp} = \frac{USL - LSL}{6s_{overall}} = \frac{24 - 20}{6(0.529)} = 1.26$$

$$\text{Ppk} = \frac{minimum(\bar{X} - LSL, USL - \bar{X})}{3s_{overall}} = \frac{24 - 22.103}{3(0.529)} = 1.20$$

The Pp is 1.26 and the Ppk is 1.20.

In Minitab: Go to **Stat > Quality Tools > Capability Analysis > Normal** and enter the column containing the data as well as the subgroup size. Use a subgroup size of 1 if there were no subgroups in the data collection. Then enter the lower and upper specification limits (see Figure 12.4). If a Cpm or confidence intervals are desired, select **Options** and then enter a target value to include Cpm or check **Include confidence intervals** as shown in Figure 12.5. Then click **OK**.

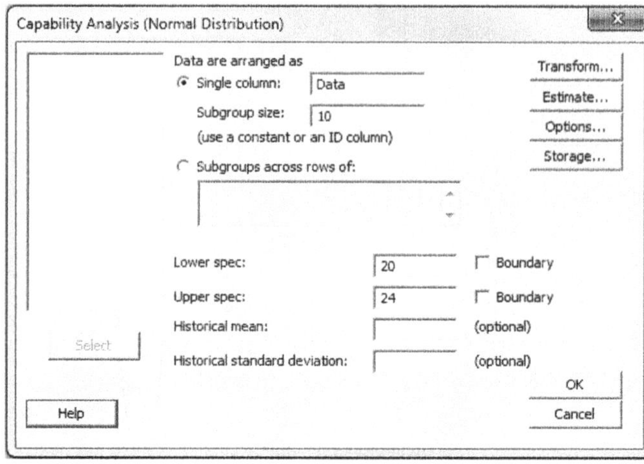

Figure 12.4 Dialog box for a capability study.

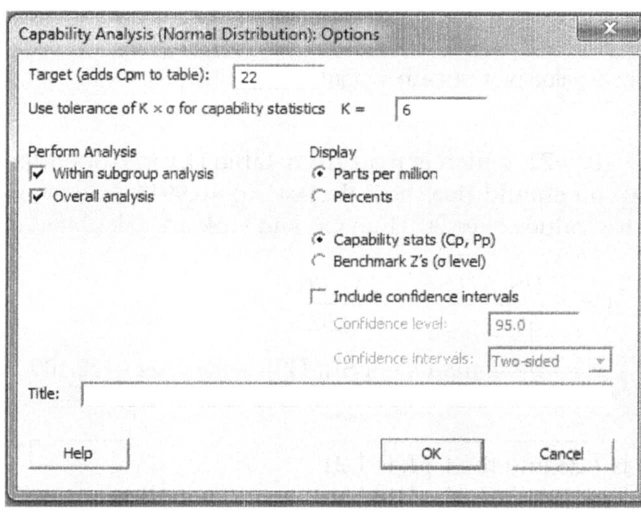

Figure 12.5 Options for a capability study.

The resulting graph shown in Figure 12.6 lists Cp, Cpk, Pp, Ppk, and Cpm. It also includes the expected overall performance in parts per million (PPM) out of specification based on the normal distribution fit to the data. The expected within performance is the number of units expected to be out of specification if there were no difference between subgroups or long-term process drift. The observed performance is the number of units out of specification in PPM based on the actual values observed rather than the distribution.

MINITAB CAPABILITY SIXPACK

Because it is important to assess the stability of a process and how well the data fit the distribution before doing capability analysis, Minitab offers a capability six-pack that automatically includes a histogram and probability plot to assess the

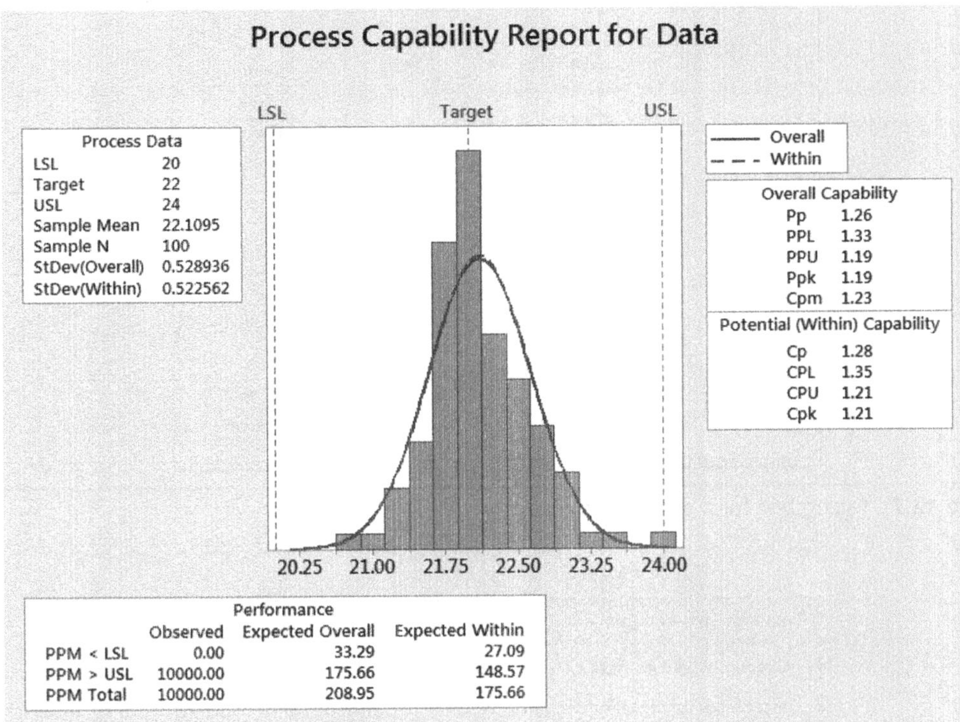

Figure 12.6 Graph for a capability study.

normality of the data. There is also an I-MR chart when subgroup size is 1, or an Xbar-R or Xbar-S chart when subgroups are used. This makes it possible to quickly test the requirements for doing capability analysis. Only the last 25 subgroups or observations are shown, although all data are used in the analysis. There is also a capability plot that shows a comparison of the overall and within performance to the specification limits.

In Minitab: Go to **Stat > Quality Tools > Capability Analysis > Capability Sixpack >Normal** and enter the column containing the data as well as the subgroup size. Use a subgroup size of 1 if there were no subgroups in the data collection. Then enter the lower and upper specification limits (see Figure 12.7). If a Cpm is desired, the target value can be entered by selecting **Options** (shown in Figure 12.8). Click **OK**.

The resulting graph is shown in Figure 12.9. The required assumptions are easily evaluated by using the capability sixpack. In this example, we can easily see that there is too much subgroup variation in subgroup 7. This indicates the presence of special cause variation; therefore, the process should be improved to eliminate the special cause variation and the study should then be repeated.

NON-NORMAL CAPABILITY ANALYSIS

Just as a normal capability analysis can be used to assess the quality of a product or process that produces normally distributed data, it can also do so for data following other distributions. First, however, it is important to consider whether the

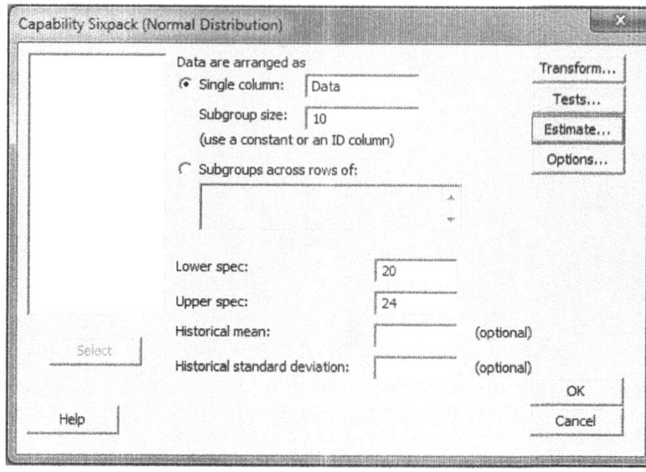

Figure 12.7 Dialog box for a capability sixpack.

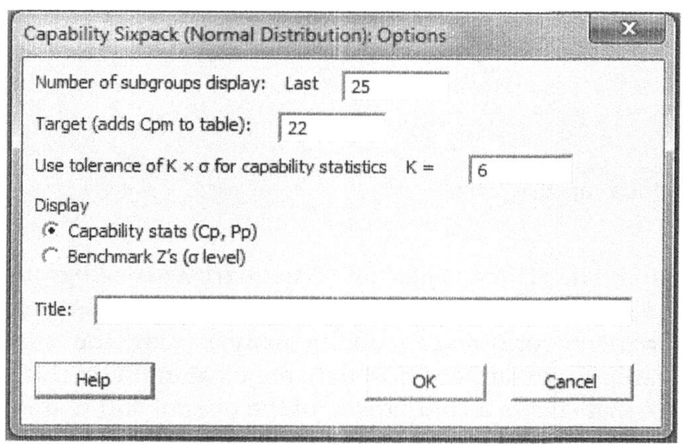

Figure 12.8 Options for a capability sixpack.

data collected are produced by a process in which a non-normal distribution is expected. A process that produces normally distributed data when in control but is not stable may generate data that appear to follow some non-normal distribution. For more information about the most common non-normal distributions and when they typically occur, review Chapter 3.

One difference between performing capability analysis on normal and non-normal data is that short-term statistics are not calculated for non-normal capability. If knowing the short-term capability of a non-normal process is desirable, the sample used should be collected within a single batch or in quick succession and the overall statistics should be treated as short-term.

Pp and Ppk are used to assess the capability. Once a distribution is fit to the data, the probabilities of that distribution falling above the upper specification and below the lower specification are calculated as p_2 and p_1, respectively. Pp is then calculated by first finding the Z-score associated with each:

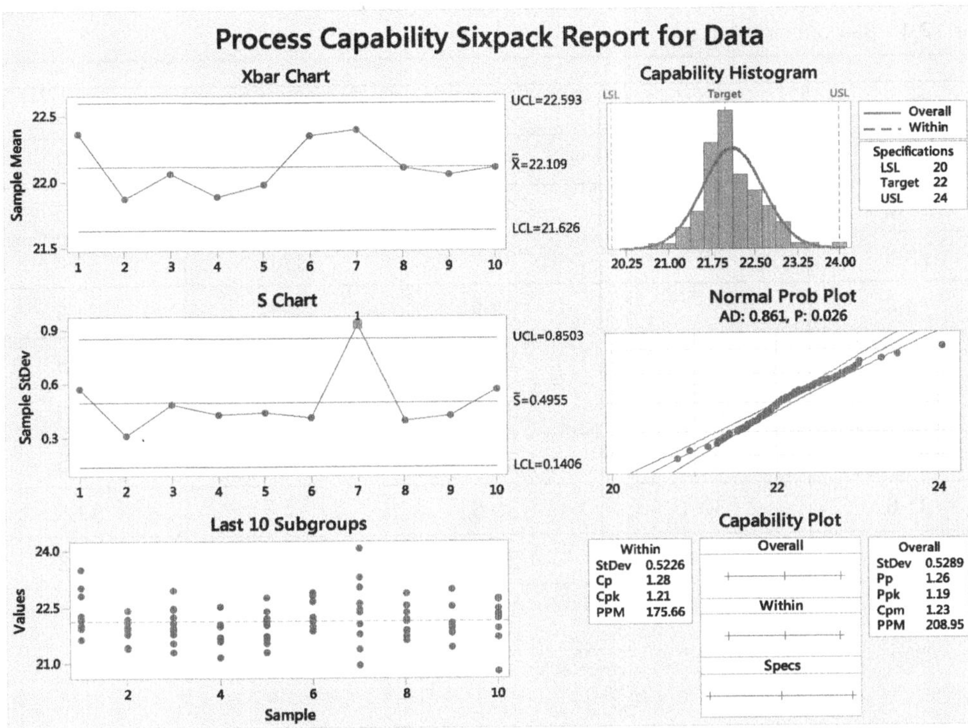

Figure 12.9 Graph for a capability sixpack.

$$P(Z < Z_{USL}) = p_2$$
$$P(Z < Z_{LSL}) = p_1$$
$$Pp = \frac{Z_{USL} - Z_{LSL}}{6}$$

Ppk is calculated as:

$$Ppk = minimum\left(\frac{-Z_{LSL}}{3}, \frac{Z_{USL}}{3}\right)$$

As with normal capability analysis, 1.33 is a commonly used minimum for a high-quality process.

Example

The force required to tear open each of 50 plastic candy bags is measured. The lower specification is 30, as lower values may tear during shipping. The upper specification is 50, as higher values may be too difficult to open. Historically, the opening force has been well-fit by a 3-parameter Weibull distribution. The data are found in Table 12.4.

In Minitab: Go to **Stat > Quality Tools > Capability Analysis > Non-normal**, complete the dialog as shown in Figure 12.10, and click **OK**.

The output is shown in Figure 12.11.

Table 12.4 Bag opening force.

32.0	35.4	30.9	31.3	28.9
37.8	34.0	50.9	33.9	31.2
31.6	31.9	38.6	44.3	31.2
28.5	29.1	39.7	33.0	34.4
34.4	33.5	32.9	31.5	31.9
38.7	34.0	28.4	29.2	34.3
32.5	28.4	34.0	30.3	34.0
49.7	33.2	38.5	37.6	33.5
31.6	30.1	30.6	34.3	33.8
29.3	27.0	34.6	42.7	35.8

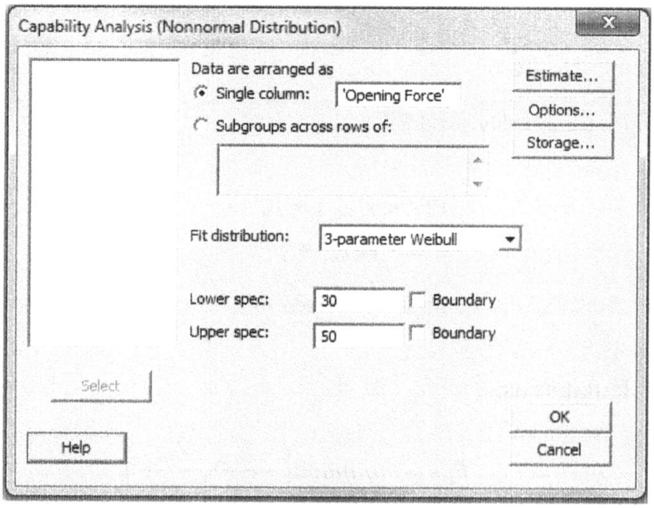

Figure 12.10 Minitab dialog box for non-normal capability analysis.

The output shows a Pp value of 0.56 and a Ppk value of 0.26. As with normal capability analysis, a Ppk much smaller than Pp indicates a process that is not centered—or, more accurately, one that is expected to produce more out-of-spec parts on one side than the other. Either value in this example is small and demonstrates that the process is producing poor quality.

The expected overall performance shows that 219,513 parts per million, or 21.9%, of the bags are expected to open with a force outside of the specifications. Most of that is below the lower specification, so the current process is likely to result in many bags tearing open during shipping and handling.

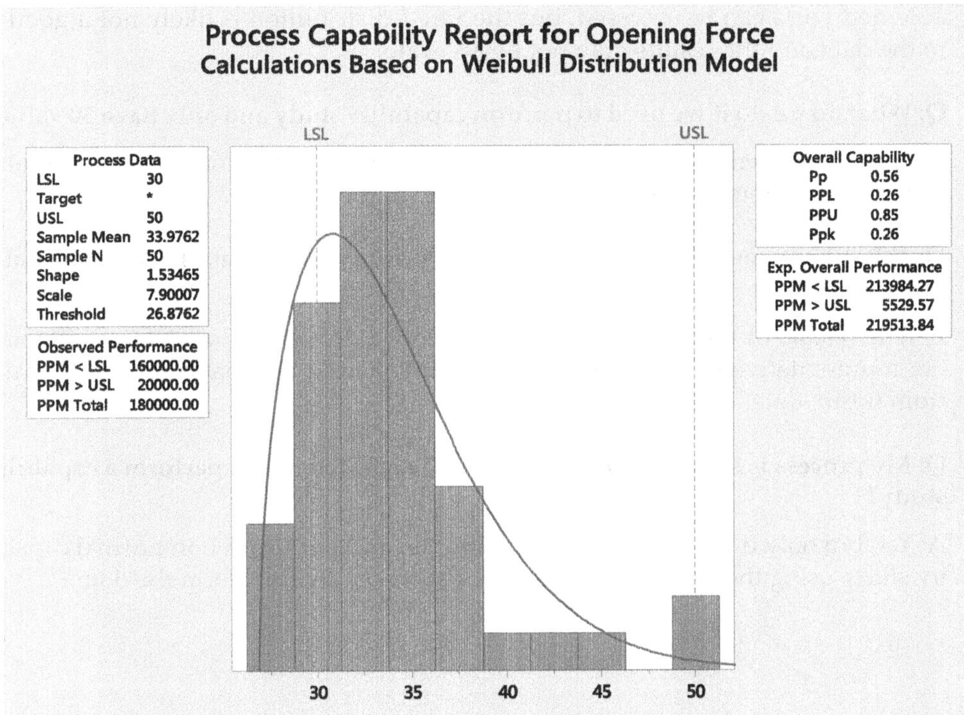

Figure 12.11 Output for non-normal capability analysis.

QUESTIONS AND ANSWERS

Q: In capability studies, what is within-subgroup variation?

A: This is a measure of how much variability occurs among data sharing a subgroup. Typically, when subgroups are selected rationally, the variation within those subgroups will be less than that of data collected over time, represented by the overall variation.

Q: In capability studies, what is overall variation?

A: Overall variation is the variation for the entire data set, and it is calculated the same way as a sample standard deviation. It does not take subgroups into account and incorporates trends over time if the data were collected over a period of time.

Q: What do I do if I want to perform a capability study on a process that is 100% inspected, resulting in the nonconforming parts being removed from the data before the analysis is performed?

A: To provide the best estimate of capability, the values of the scrapped parts are needed. If those values are unknown, the performance of the process can still be assessed if the number of scrapped and accepted parts are known. To do so, use the perfect defective and calculate DPMO (defects per million opportunities), PPM (parts per million), or a Z-score. Otherwise the capability of only the accepted parts can be assessed, but the fitted distribution is likely not a good fit to the data and the results may not be accurate.

Q: What do we do if we need to perform capability study and only have 30 values?

A: It is recommended that you report your results with a confidence interval to provide an understanding of the uncertainty of the statistics.

Q: What happens if I perform a capability study using data from an unstable process?

A: The results of the capability analysis may be invalid and are almost certainly not representative of the data that were not collected or what can be anticipated from future data.

Q: My process is stable but not normally distributed. Can I perform a capability study?

A: Yes, but only if you identify the distribution and perform a non-normal capability study using the correct distribution or successfully transform the data.

Chapter 13
Statistical Process Control

Statistical process control (SPC) is used to statistically monitor the performance of a process and to provide an early warning if a change occurs. This is preferable to 100% inspection to detect problems after they have occurred. Detecting a trend in the data also makes it possible to take actions before parts are out of specification, such as when tool wear is detected by a control chart.

You could say that statistical process control was created on May 16, 1924, when Dr. Walter A. Shewhart submitted a memo on SPC to his superiors at Western Electric's Hawthorne plant. Shewhart experimented with numbered chips in a kitchen bowl to create SPC and further described the principles in his book *Economic Control of Quality of Manufactured Product*, which was first published in 1931 (Shewhart 1980).

There is variation in manufactured products and the delivery of services and we expect observations to reflect that variation. However, SPC can differentiate between the normal variation in the process (common cause variation) and variation due to changes in the process (special cause variation) (Borror 2009). *Common cause variation* is inherit in the system and is always present. Possible causes include variation in raw materials, maintenance of machines, procedures, environmental changes, and employee training. *Special cause variation*, also known as assignable cause variation, is periodic and may be unpredictable. Possible causes include incorrect material, wrong machine setting, and tool wear.

SPC allows us to see whether a process is showing the variation expected from common causes and therefore should not be changed or is showing special causes that need to be identified and corrected. Tampering is responding to common cause variation as if it were special cause variation. For example, a point may appear high relative to most others and an operator may be tempted to adjust the process. If SPC did not indicate that the high value was due to special cause variation, then the process should not be modified. Tampering *increases* variation! Mark Durivage (2015) warns that improper use of a control chart could lead to spending money on unneeded equipment, taking unnecessary actions, failing to take actions when necessary, and blaming operators for issues that are inherent in the process and not in the operator's control.

The type of SPC chart used is determined by the type of data being analyzed and subgroup size. Variable (continuous) data are measured on a continuous scale, such as 8 mm, 14 mb, or 78 dB. Attribute (discrete) data represent a characteristic or a count. It could be the number of defects, such as one failure on a part or nine scratches on a piece of sheet metal. It can also be the number or proportion of defective units, such as the number of parts with one or more defects. One part with many defects still counts as one defective.

Continuous data may or may not be collected in subgroups. Subgroups are groups of observations taken under very similar conditions, such as from the same batch or mold or consecutively. Many SPC charts depend on a rational subgroup to estimate the variation inherent in the process. The observations in a rational subgroup should be independent of each other; this means that one observation should not influence the other.

SPC uses control charts to monitor the process and identify when special cause variation occurs. A control chart is created by plotting the data in time order. Most control charts use control limits that are placed +/− 3 within-subgroup standard deviations from the process mean. A point falling outside of these limits is an indication of special cause variation. A process in statistical control will have nearly all data points between the control limits. This will result in a very low false detection rate. For example, if a process produces stable data from the normal distribution, 99.73% of points will fall within the limits and only 0.27% will fall outside and indicate a special cause even though there was none. Although this provides some context for how infrequently a stable process will produce a false alarm, control charts do not require normality and maintain similar rates for many common distributions.

The lower control limit should be set to zero when the control limit is calculated as being a negative number when the data cannot actually have negative values (Committee E-11 on Quality and Statistics, 2002). In this situation, it is recommended that another test be included, such as eight points in a row below the center line.

It is often advised to "set" the limits at their current values and discontinue updating them as additional data is collected after a process has exhibited stability for a longer period of time. This allows for better detection of a process that has gradually drifted away from its initial state.

Control limits are independent of specification limits, which means a process can be in control but out of specification or out of control, but still in specification. They "are established by the inherent variation in a process" and "are *not* related to blueprint (specification) limits" (Griffith 2003, 474). To monitor how often measurements are falling within specifications, a capability analysis should be used. Figure 13.1 shows both a process that is in control and in specification and a process that is in control but out of specification.

Aside from a point falling outside of the control limits, other rules exist for detecting special cause variation. Common rules include the Western Electric rules (see Table 13.1) and Nelson rules (see Figure 13.2).

Many of these rules are very similar and likely to detect the same types of special cause variation, and they should not be all used simultaneously. Doing so greatly increases the false alarm rate while only moderately improving the power to detect changes. Using tests 1 and 2 from the Nelson rules is likely to be very powerful in detecting special cause variation without a large increase in the false detection rate (Minitab, "Variables Control Charts").

Figure 13.3 shows a flowchart for selecting the correct control chart for a given situation. Each of these charts will be discussed in more detail in this chapter.

Some types of SPC charts require the use of a table with SPC constants for setting the upper control limit (UCL) and lower control limit (LCL). The correct constant is identified in the control chart's formula and depends on the subgroup size (see Table 13.2).

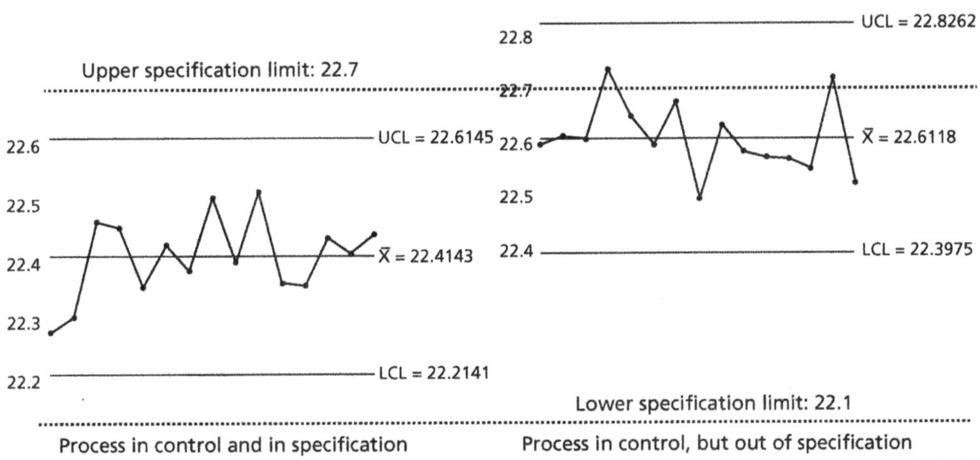

Figure 13.1 Control vs. specification.

Table 13.1 Western Electric rules.

Rule
Any point above + 3 sigma limit
2 Out of the last 3 points above + 2 sigma limit
4 Out of the last 5 points above the + 1 sigma limit
8 Consecutive points above the center line
8 Consecutive points below the center line
4 Out of the last 5 points below the −1 sigma limit
2 Out of the last 3 points below the −2 sigma limit
Any point below the −3 sigma limit
Trend rules: 6 in a row trending up or down
Trend rules: 14 in a row alternating up and down

Source: NIST/SEMATECH (April 2012, 6.3.2), available at http://www.itl.nist.gov/div898/handbook/.

VARIABLE DATA CONTROL CHARTS

I-MR Chart

An individuals-moving range (I-MR) chart is used when the data were not collected in subgroups and therefore the subgroup size is one. It consists of an individuals chart that plots the observed data points in time order and detects when the location of the process has likely changed, and a moving range chart that plots the absolute differences between consecutive points and detects when the spread of the data has likely changed. The formulas for the centerline and average moving range used for within-subgroup variation for the individuals chart are:

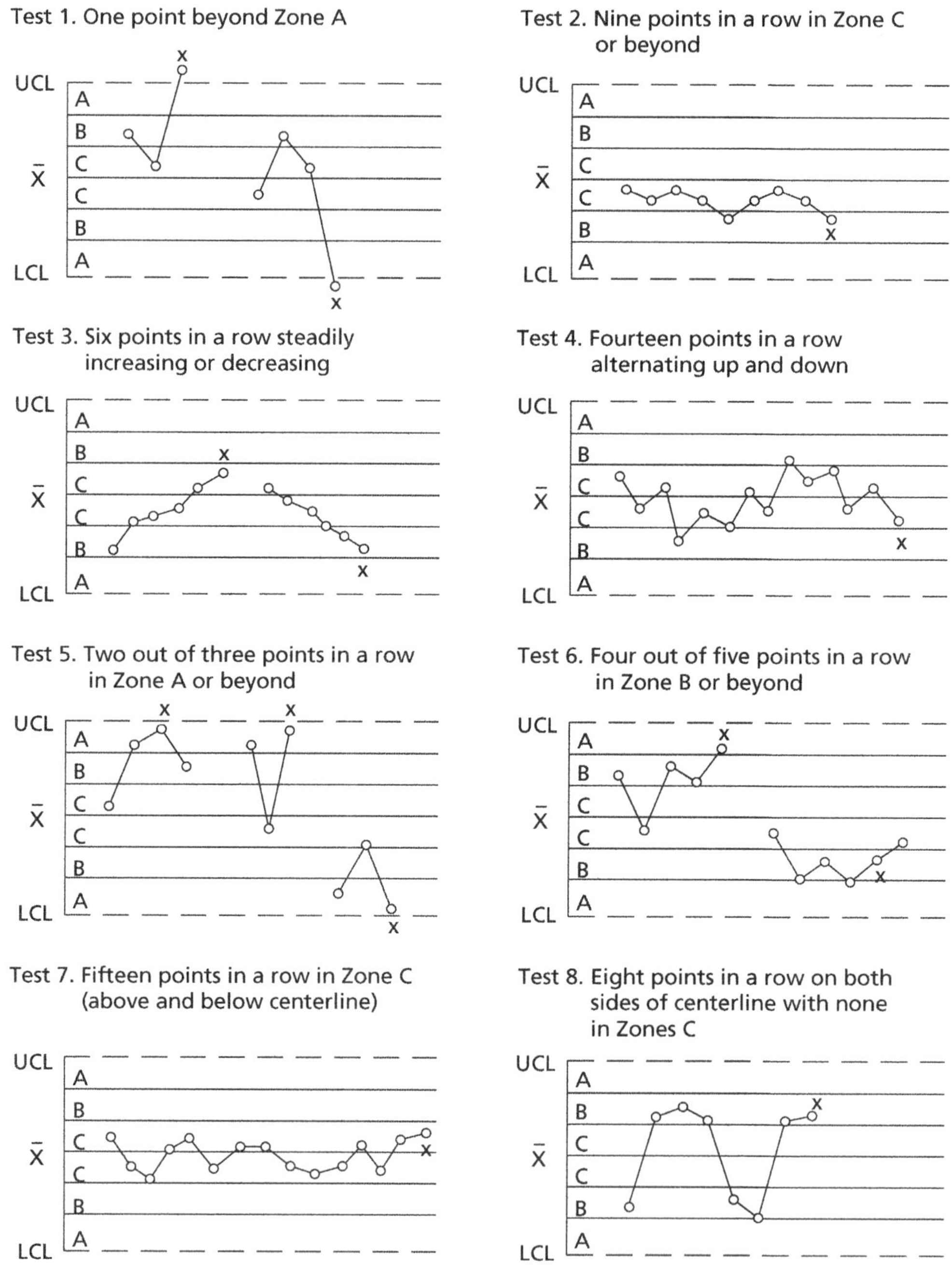

Figure 13.2 Nelson rules.

Source: Lloyd S. Nelson, "Technical Aids," Journal of Quality Technology 16, no 4 (October 1984): 238–239.

$$\text{Centerline} = \bar{X} = \frac{\sum X_i}{n}$$

$$\text{Standard Deviation} = \overline{MR} = \frac{\sum MR_i}{n-1}$$

MR_i is the absolute difference between each pair of consecutive points. The control limits are then calculated using the unbiasing constant $d_2(2) = 1.128$, which is always used to form I-MR charts. The formulas are:

$$\text{Lower control limit (LCL)} = \bar{X} - 3\frac{\overline{MR}}{d_2(2)} = \bar{X} - 2.66 * \overline{MR}$$

$$\text{Upper control limit (UCL)} = \bar{X} + 3\frac{\overline{MR}}{d_2(2)} = \bar{X} + 2.66 * \overline{MR}$$

The moving range chart plots the absolute values of the moving ranges instead of the individual data points and the centerline and control limits are:

$$\text{Center line} = \overline{MR}$$

$$\text{LCL} = 0$$

$$\text{UCL} = d_4(2) * \overline{MR} = 3.267 * \overline{MR}$$

Because the limits of the individuals chart are determined using the moving range, the MR chart should first be checked for stability before interpreting the individuals chart.

Example

Suppose we randomly select one part every hour for five hours of production and check the diameter in millimeters (mm). The results are 31.91, 32.29, 30.44, 29.74, and 32.90.

The moving ranges, data mean, and average moving range are shown in Table 13.3. The centerline and control limits for the individuals chart are:

$$\text{Centerline} = \bar{X} = 31.456$$

$$LCL = \bar{X} - 2.66 * \overline{MR} = 31.456 - 2.66 * 1.523 = 27.405$$

$$LCL = \bar{X} + 2.66 * \overline{MR} = 31.456 + 2.66 * 1.523 = 35.507$$

In some cases, a process naturally produces a distribution that is highly skewed and bounded at some minimum value. Often times in these cases the LCL of the individuals chart will be below that bound and the chart will be unable to detect a special cause by observing a point below that limit. When this type of data is observed, it is first important to consider whether that process would be expected to produce highly skewed and bounded data such as those described in Chapter 3 regarding the exponential distribution. If the process would not seem to naturally produce such data, it is possible that the data appear skewed because of an out-of-control process. If the process would likely produce such data, then subgrouping should be considered if possible or the use of other tests to detect special causes.

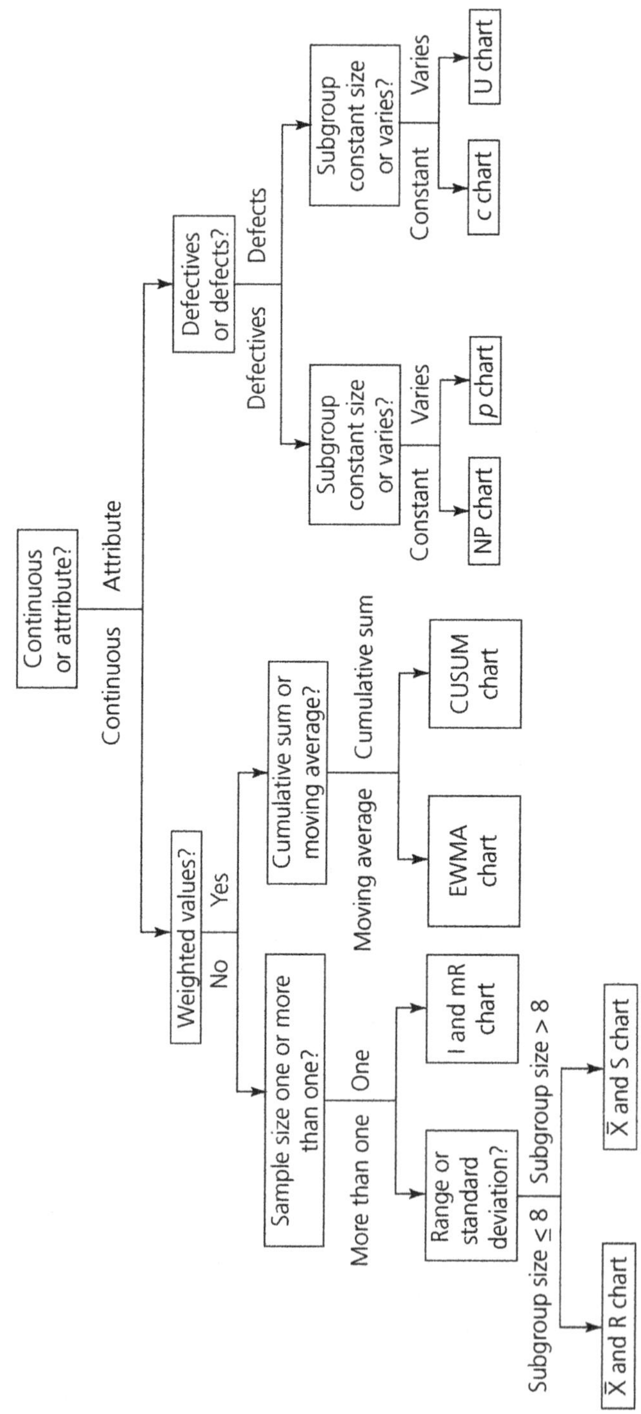

Figure 13.3 Flowchart for control chart selection.

Table 13.2 SPC constants.

Statistical process control unbiasing constants							
n	d_2	D_3	D_4	c_4	B_3	B_4	n
2	1.128	0	3.26729	0.7979	0	2.60632	2
3	1.693	0	2.57425	0.8862	0	2.27598	3
4	2.059	0	2.28188	0.9213	0	2.08775	4
5	2.326	0	2.11449	0.94	0	1.96363	5
6	2.534	0	2.00395	0.9515	0.0289	1.87417	6
7	2.704	0.0755917	1.92441	0.9594	0.1129	1.80583	7
8	2.847	0.1361433	1.86386	0.965	0.1786	1.75144	8
9	2.97	0.1840404	1.81596	0.9693	0.2318	1.70683	9
10	3.078	0.2230994	1.7769	0.9727	0.2759	1.66937	10
11	3.173	0.2556256	1.74437	0.9754	0.3134	1.63734	11
12	3.258	0.2831492	1.71685	0.9776	0.3456	1.60954	12
13	3.336	0.3071942	1.69281	0.9794	0.3737	1.58511	13
14	3.407	0.3281479	1.67185	0.981	0.3985	1.56343	14
15	3.472	0.3466014	1.6534	0.9823	0.4206	1.54401	15
16	3.532	0.3630521	1.63695	0.9835	0.4405	1.52648	16
17	3.588	0.3778428	1.62216	0.9845	0.4585	1.51055	17
18	3.64	0.3912637	1.60874	0.9854	0.4748	1.496	18
19	3.689	0.4034969	1.5965	0.9862	0.4898	1.48264	19
20	3.735	0.4146988	1.5853	0.9869	0.5036	1.4703	20
21	3.778	0.4249338	1.57507	0.9876	0.5163	1.45888	21
22	3.819	0.4344855	1.56551	0.9882	0.5281	1.44825	22
23	3.858	0.4433126	1.55669	0.9887	0.5391	1.43834	23
24	3.895	0.4515276	1.54847	0.9892	0.5493	1.42906	24
25	3.931	0.4593742	1.54063	0.9896	0.5589	1.42035	25
26	3.964	0.466873	1.53313	0.9901	0.568	1.41215	26
27	3.997	0.4734404	1.52656	0.9904	0.5765	1.40442	27
28	4.027	0.4794597	1.52054	0.9908	0.5845	1.39711	28
29	4.057	0.4853382	1.51466	0.9911	0.592	1.39018	29
30	4.086	0.4909552	1.50904	0.9914	0.5992	1.38361	30
n	d_2	D_3	D_4	C_4	B_3	B_4	n

Source: Matthew Barsalou. Used with permission.

Table 13.3 Calculating a moving range.

Sample	Value	Moving range
1	31.91	
2	32.29	0.38
3	30.44	1.85
4	29.74	0.70
5	32.90	3.16
Mean		1.5225

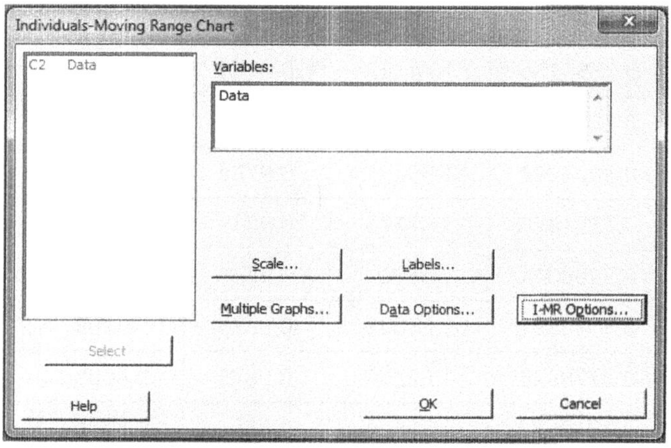

Figure 13.4 Dialog box for an I-MR chart.

The centerline and control limits for the moving range chart are:

$$\text{Centerline} = \overline{MR} = 1.523$$
$$\text{LCL} = 0$$
$$\text{UCL} = 3.267 * \overline{MR} = 3.267 * 1.523 = 4.976$$

In Minitab: Go to **Stat > Control Charts > Variables Charts for Individuals > I-MR** and enter the column containing the data as shown in Figure 13.4. Then click **OK**.

The resulting Minitab I-MR chart is shown in Figure 13.5.

Xbar-S Chart

An Xbar-S chart is used for data with subgroups containing more than eight measurements each. The Xbar is used to monitor stability in the location of the data, and the S chart is used to monitor stability in the variation inherent to the process. For the Xbar chart, the means of each subgroup are plotted in time order.

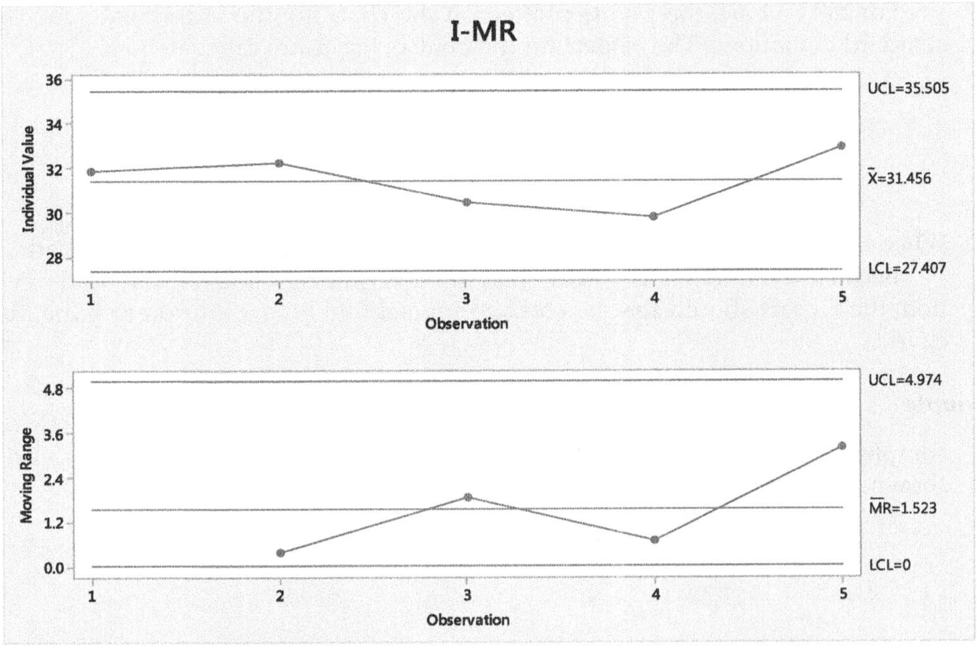

Figure 13.5 Graph for an I-MR chart.

The centerline is equal to $\bar{X} = \frac{\Sigma X_i}{n}$ and the estimate of within-subgroup standard deviation is the average subgroup standard deviation, unbiased using constant c_4:

$$\bar{S} = \frac{\Sigma \dfrac{\left(\dfrac{c_4(n_i)^2}{1-c_4(n_i)^2}\right)s_i}{c_4(n_i)}}{\Sigma\left(\dfrac{c_4(n_i)^2}{1-c_4(n_i)^2}\right)} \text{ when subgroup size is not constant}$$

$$\bar{S} = \frac{\Sigma s_i}{M * c_4(n_i)} \text{ when subgroup size is constant}$$

In the formulas, n_i is the number of observations in each subgroup and M is the total number of subgroups. The control limits for the Xbar are:

$$\text{LCL} = \bar{X} - 3\frac{\bar{S}}{\sqrt{n_i}}$$

$$\text{UCL} = \bar{X} + 3\frac{\bar{S}}{\sqrt{n_i}}$$

Because the control limits are dependent on the subgroup size, when subgroups do not have the same number of observations the limits will vary for each plotted point.

For the S chart, the points plotted on the chart are the individual subgroup standard deviations. The centerline and control limits are calculated as:

$$\text{Centerline} = c_4(n_i)\overline{S}$$

$$\text{LCL} = B_3(n_i)\overline{S}$$

$$\text{UCL} = B_4(n_i)\overline{S}$$

When the LCL of the S chart is negative, a bound at zero may be used instead.

Because the limits of the Xbar chart are determined using the standard deviation, the S chart should first be checked for stability before interpreting the Xbar chart.

Example

Samples consisting of sample size 10 are taken once per shift for seven shifts as shown in Table 13.4.

Then $\overline{X}$ and $\overline{S}$ are calculated as follows:

$$\overline{X} = \frac{\Sigma X_i}{n} = 17.53$$

$$\overline{S} = \frac{\Sigma s_i}{M * c_4(10)} = \frac{0.576 + 0.416 + 0.626 + 0.792 + 0.893 + 1 + 0.497}{7 * 0.973} = 0.705$$

Table 13.4 Seven subgroups.

	Subgroup 1	Subgroup 2	Subgroup 3	Subgroup 4	Subgroup 5	Subgroup 6	Subgroup 7
	17.2	18.1	17.0	17.2	15.6	15.8	17.9
	18.1	18.0	17.5	19.0	16.8	16.0	17.2
	17.4	17.4	17.9	17.5	17.7	18.3	17.6
	16.6	18.5	16.9	18.1	17.7	18.3	16.6
	18.3	17.8	17.2	16.9	17.7	18.2	17.9
	18.0	17.0	18.4	17.3	18.0	17.4	17.4
	18.3	17.5	18.1	16.9	17.4	17.0	18.4
	18.0	17.5	18.2	17.4	16.6	17.3	17.4
	17.2	17.7	16.5	16.1	18.8	18.7	17.6
	17.3	17.8	17.4	18.0	18.0	16.8	18.0
$\overline{x}_i$	17.64	17.73	17.51	17.44	17.43	17.38	17.6
S_i	0.576	0.416	0.626	0.792	0.893	1	0.497

The centerline and control limits of the Xbar chart are calculated as:

$$\text{Centerline} = \bar{X} = 17.53$$

$$\text{LCL} = \bar{X} - 3\frac{\bar{S}}{\sqrt{n_i}} = 17.53 - 3 * \frac{.705}{\sqrt{10}} = 18.20$$

$$\text{UCL} = \bar{X} + 3\frac{\bar{S}}{\sqrt{n_i}} = 17.53 + 3 * \frac{.705}{\sqrt{10}} = 16.86$$

The centerline and control limits of the S chart are calculated as:

$$\text{Centerline} = c_4(n_i)\bar{S} = 0.973 * 0.705 = 0.686$$

$$\text{LCL} = B_3(n_i)\bar{S} = 0.284 * 0.686 = 0.195$$

$$\text{UCL} = B_4(n_i)\bar{S} = 1.716 * 0.686 = 1.177$$

In Minitab: Go to **Stat > Control Charts > Variables Charts for Subgroups > Xbar-S** and enter the column containing the data as well as the subgroup size. If subgroup sizes are not equal, a column to indicate which subgroup each observation is in can be entered. Alternatively, data may be stored across columns with each row representing a subgroup and the dialog completed accordingly (see Figure 13.6). Click **OK**.

The resulting Minitab Xbar-S is shown in Figure 13.7.

Because no points fell outside of the control limits and there were never nine points in a row on either side of the centerline on either chart, there is no indication that the process is not stable.

Xbar-R Chart

An average and range (Xbar-R) chart is used for a subgroup of up to size 8; however, a common subgroup size is 5. The chart is constructed similarly as with the Xbar-S chart, but the unbiased average subgroup range is used instead to estimate the within-subgroup standard deviation. For the Xbar chart, the means of

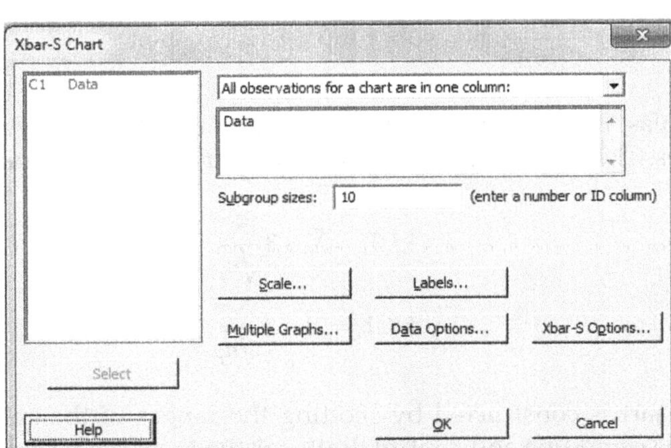

Figure 13.6 Dialog box for an $\bar{X}$ and S chart.

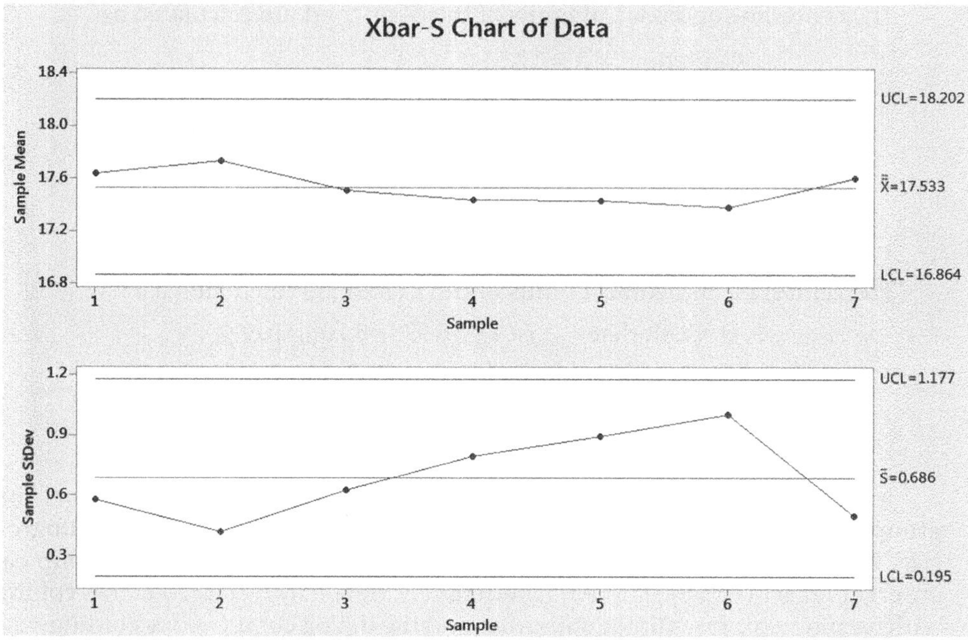

Figure 13.7 Graph for an $\overline{X}$ and S chart.

the subgroups are plotted in time order, the centerline is the overall mean, and the within-subgroup standard deviation estimate is:

$$S_r = \frac{\Sigma\left(\dfrac{\dfrac{d_2(n_i)^2}{d_3(n_i)^2} r_i}{d_2(n_i)}\right)}{\Sigma\left(\dfrac{d_2(n_i)^2}{d_3(n_i)^2}\right)} \text{ when subgroup size is not constant}$$

$$S_r = \frac{\overline{R}}{d_2(n_i)} \text{ when subgroup size is constant}$$

In the formulas, r_i is the range of each subgroup and $\overline{R}$ is the average of the subgroup ranges. The control limits for the Xbar chart are then calculated as:

$$\text{LCL} = \overline{X} - 3\frac{S_r}{\sqrt{n_i}}$$

$$\text{UCL} = \overline{X} + 3\frac{S_r}{\sqrt{n_i}}$$

The R chart is constructed by plotting the ranges of the subgroups in time order and the centerline and control limits for the R chart are calculated as:

$$\text{Centerline} = \bar{R}$$
$$\text{LCL} = D_3\bar{R}$$
$$\text{UCL} = D_4\bar{R}$$

Because the limits of the Xbar chart are determined using the average range, the R chart should first be checked for stability before interpreting the Xbar chart.

Example

The following data was collected by taking five consecutive parts once per shift for seven shifts and recording the weight of each part as shown in Table 13.5.

The overall mean and Sr are calculated as:

$$\bar{X} = \frac{\Sigma X_i}{n} = 17.50$$

$$S_r = \frac{\bar{R}}{d_2(n_i)} = \frac{1.829}{2.326} = 0.786$$

For the Xbar chart, the centerline is the overall mean and the control limits are:

$$\text{LCL} = 17.50 - 3\frac{0.786}{\sqrt{5}} = 16.45$$

$$\text{UCL} = 17.50 + 3\frac{0.786}{\sqrt{5}} = 18.56$$

For the S chart, the centerline and control limits are:

$$\text{Centerline} = \bar{R} = 1.829 = 1.829$$
$$\text{LCL} = D_3\bar{R} = 0 * 1.829 = 0$$
$$\text{UCL} = D_4\bar{R} = 2.114 * 1.829 = 3.866$$

In Minitab: Go to **Stat > Control Charts > Variables Charts for Subgroups > Xbar-R** and enter the column containing the data as well as the subgroup

Table 13.5 Seven subgroups.

Subgroup 1	Subgroup 2	Subgroup 3	Subgroup 4	Subgroup 5	Subgroup 6	Subgroup 7
18.7	17.4	17.5	18.4	16.5	18.6	17.1
18.5	17.6	16.8	17.4	18.0	17.5	17.9
17.7	18.0	17.0	16.3	17.6	17.8	16.6
17.3	17.2	17.5	18.1	16.4	15.1	16.7
18.3	17.7	18.0	17.8	18.6	16.7	18.2

size and click **OK** (see Figure 13.8). If subgroup sizes are not equal, a column indicating which subgroup each observation is in can be entered. Alternatively, data may be stored across columns with each row representing a subgroup and the dialog completed accordingly.

The resulting Minitab Xbar-R chart is shown in Figure 13.9.

Because no points fail the desired tests, the charts indicate that the process is stable.

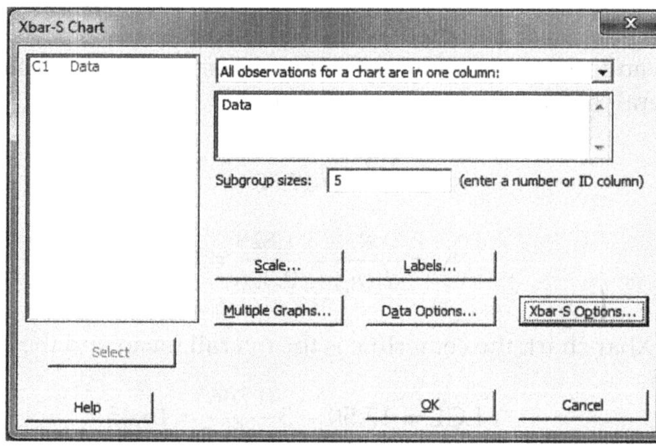

Figure 13.8 Dialog box for an $\overline{X}$ and R chart.

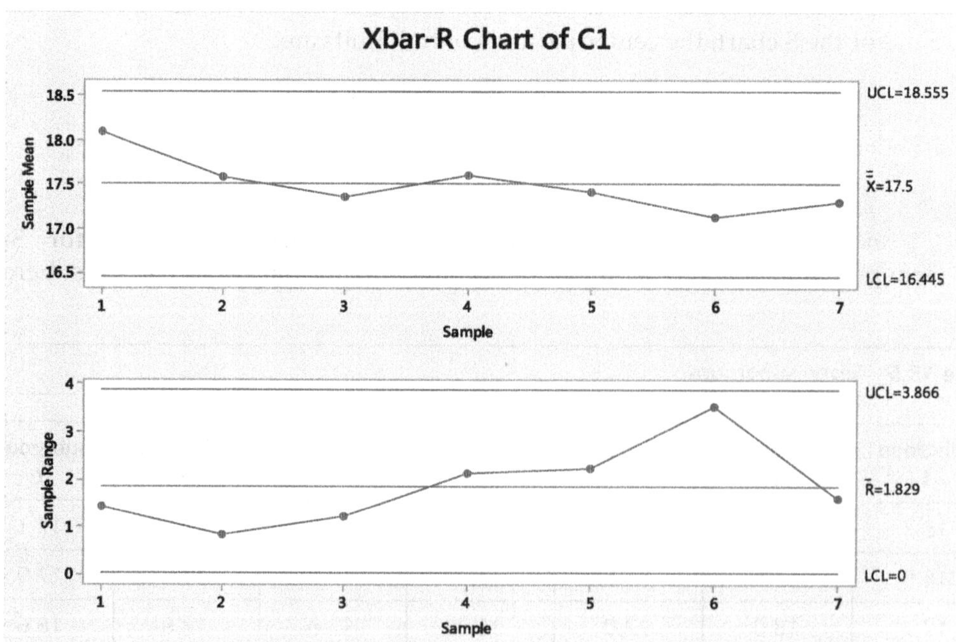

Figure 13.9 Graph for an Xbar-R chart.

CONTINUOUS DATA WITH WEIGHTED VALUES CONTROL CHARTS

CUSUM Chart

Whereas many control charts are intended to detect moderate or large process shifts or deviations (Breyfogle 2003), in some cases it is desirable to detect a small but sustained shift in the location of the data relative to a target value. For example, a circuit board printing process may produce a hole with a target diameter of 1.2 mm. There is an indication that a part is beginning to fail and needs to be replaced if the diameter begins to deviate consistently from that measurement by even a small amount.

The points on a cumulative sum (CUSUM) chart are based on the cumulative sum of the differences between each point and the target value. In other words, if in the example described the first hole measured is 1.18 mm, then the difference between that and the target of 1.2 mm is –0.02 mm. If the second hole measured is 1.19 mm, then the second point uses the sum of that deviation (–0.01) and the previous of –0.02, for a cumulative sum of –0.03. This continues for each point plotted, where that point contains the sum of all deviations from the target for it and every point preceding it.

If the process drifts away from the target value in either direction, the plotted points will gradually drift further away from the target until reaching a control limit and signaling the process has deviated. Unlike the previously described control charts, the CUSUM chart plots two points for each subgroup or value i:

$$CL_i = min\left\{0, CL_{i-1} + \bar{x}_i - \left(T - 0.5\frac{\sigma}{\sqrt{m}}\right)\right\}$$

$$CU_i = max\left\{0, CU_{i-1} + \bar{x}_i - \left(T - 0.5\frac{\sigma}{\sqrt{m}}\right)\right\}$$

CL_{i-1} and Cu_{i-1} are the previously plotted points, $\bar{x}_i$ is the mean of the subgroup or the individual value if subgroups are not used, T is the target, σ is the standard deviation, and m is the subgroup size. The standard deviation when subgroups are not used is $\hat{\sigma} = \frac{\overline{MR}}{d_2(2)}$ as used in a moving range chart, and when subgroups are used it is $\hat{\sigma} = \frac{s_{pooled}}{c_4(d+1)}$ as used in an S chart.

The centerline and control limits for the CUSUM chart are calculated as:

$$\text{Centerline} = 0$$

$$LCL = -4\frac{\hat{\sigma}}{\sqrt{m}}$$

$$UCL = 4\frac{\hat{\sigma}}{\sqrt{m}}$$

The only test used to detect that a process has deviated from the target is a point beyond one of the control limits.

Example

Data are collected on the circuit board holes described previously and recorded in Table 13.6.

In Minitab: Go to **Stat > Control Charts > Time-Weighted Charts > CUSUM**, complete the dialog including the subgroup size of one and the target of 1.2 (as shown in Figure 13.10), and click **OK**.

The resulting chart is shown in Figure 13.11.

Starting at point 27, the chart signals that the process is deviating from the target of 1.2 mm. It appears that the problem may have started after point 20 if we examine where the points began to steadily move away from the centerline.

As a means of comparison, an individuals chart (I chart) the same data is shown in Figure 13.12.

There has not yet been a signal that the process is out of control in spite of some upward trend that might appear to occur on the I chart. This example illustrates how sensitive the CUSUM chart is for detection of sustained shifts. Because of this sensitivity, it is important that the chart be used for situations in which it is believed the process can be centered precisely on target. Otherwise, even a perfectly stable process will eventually provide a signal that it is off-target.

Table 13.6 Circuit board hole diameters.

Diameter (mm)	
1.178	1.202
1.207	1.194
1.203	1.211
1.193	1.209
1.184	1.199
1.208	1.214
1.223	1.222
1.194	1.204
1.225	1.232
1.165	1.212
1.198	1.219
1.210	1.234
1.223	1.242
1.202	1.228
1.190	1.212

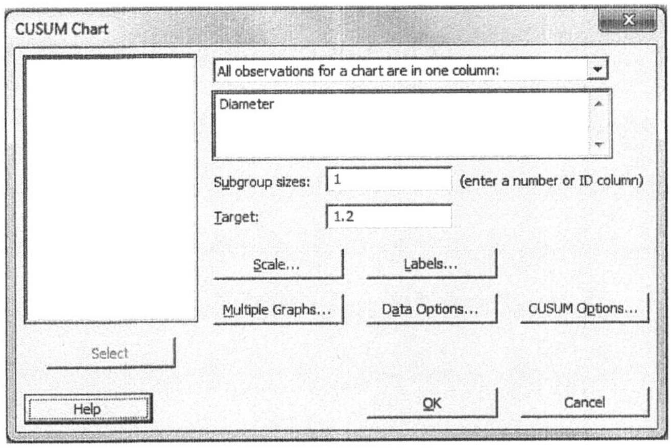

Figure 13.10 Minitab dialog box for CUSUM chart.

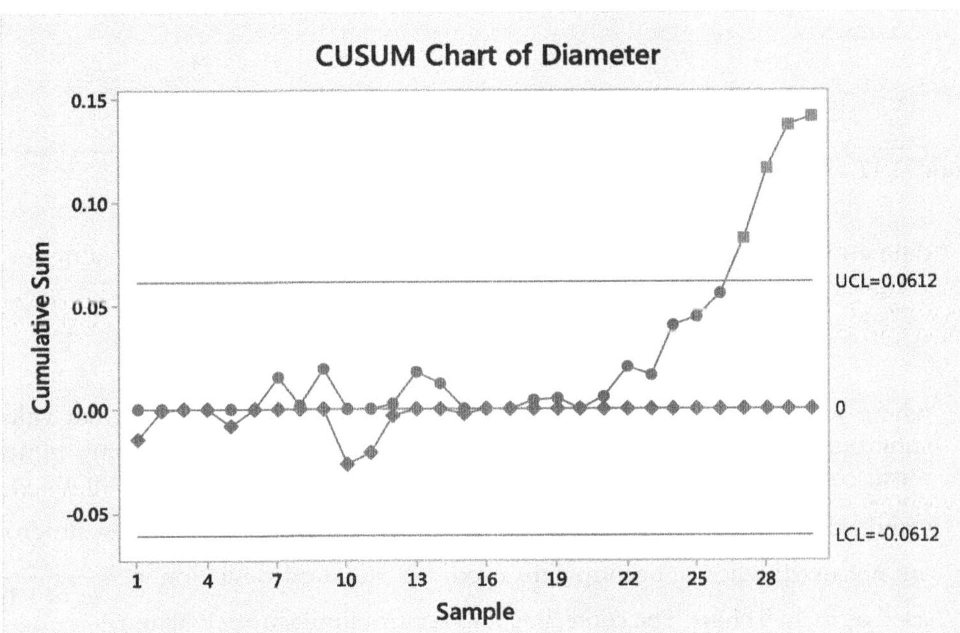

Figure 13.11 CUSUM chart for hole diameters.

EWMA Chart

Whereas the CUSUM chart is intended to detect small, sustained shifts away from a target value, the EWMA chart is intended to detect such shifts from the current location of the process (Breyfogle 2003). It is similar to an I or Xbar chart, but with the intention of detecting small, sustained shifts rather than moderate or large shifts or unusual points. Another difference is that the points plotted are based on a weighted average of the most recent data values and the preceding points.

The points plotted are referred to as the *exponentially weighted moving averages* (EWMAs). Before creating the chart, the initial average is set as the average of all

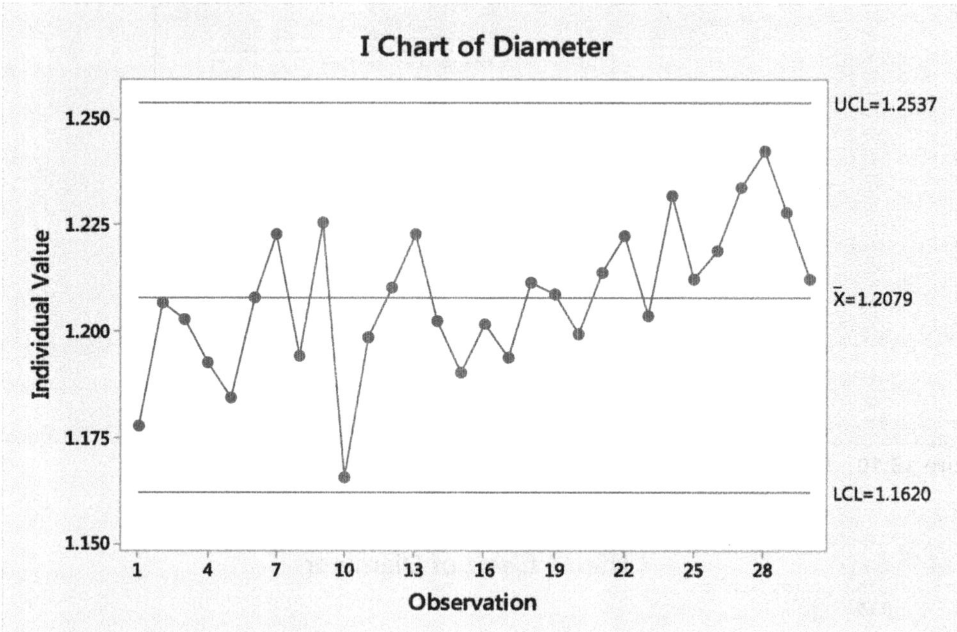

Figure 13.12 Individuals chart of hole diameter.

data and a weight for the most recent value is chosen. A common value for the weight is 0.2. The EWMA z_i for any point i is then calculated as

$$z_i = w\overline{x}_i + (1-w)z_{i-1},$$

where w is the weight and $\overline{x}_i$ is the subgroup mean or the individual value if subgroups are not used. Higher values of weight will make the current subgroup mean count more and lower values will make it count less. The estimate of the standard deviation is $\hat{\sigma} = \dfrac{\overline{MR}}{d_2(2)}$ as used in a moving range chart when subgroups are not used; when subgroups are used, the standard deviation $\hat{\sigma} = \dfrac{s_{pooled}}{c_4(d+1)}$ is used as in an S chart. The centerline and control limits are calculated as:

$$Centerline = \overline{x}$$

$$LCL = \overline{x} - 3\frac{\hat{\sigma} * w}{\sqrt{n}} \text{ for } i = 1$$

$$LCL = \overline{x} - 3\frac{\hat{\sigma}}{\sqrt{n}}\sqrt{\left(\frac{w}{2-w}\right)\left[1-(1-w)^{2i}\right]} \text{ for } i > 1$$

$$UCL = \overline{x} + 3\frac{\hat{\sigma} * w}{\sqrt{n}} \text{ for i} = 1$$

$$UCL = \overline{x} + 3\frac{\hat{\sigma}}{\sqrt{n}}\sqrt{\left(\frac{w}{2-w}\right)\left[1-(1-w)^{2i}\right]} \text{ for } i > 1$$

The only test used to determine a lack of stability on an EWMA chart is a point beyond the control limits.

Example

From each lot of incoming plastic beads, five containers are randomly selected and weighed. The mean weight is tracked over time using an EWMA chart to detect if the supplier is consistently providing the same amount of material. The data collected are shown in Table 13.7.

In Minitab: Go to **Stat > Control Charts > Time-Weighted Charts > EWMA**, complete the dialog as shown in Figure 13.13, and click **OK**.

The chart in Figure 13.14 shows that the supplier has been sending less plastic in each container in recent deliveries.

Table 13.7 Plastic bead container weights.

	Container 1	Container 2	Container 3	Container 4	Container 5
1	224.2	225.2	225.6	224.6	221.7
2	224.8	224.1	226.4	226.9	227.8
3	224.0	224.3	227.3	223.0	226.4
4	223.5	227.8	226.0	224.8	224.4
5	226.7	223.8	225.8	227.2	226.9
6	229.2	225.1	221.3	226.1	226.2
7	224.1	223.4	224.4	227.7	224.4
8	220.8	225.5	224.4	224.9	227.9
9	224.6	225.5	222.6	227.7	226.0
10	225.4	225.1	226.3	222.8	223.8
11	225.4	223.6	226.4	224.2	223.8
12	230.0	226.5	225.3	224.6	226.9
13	226.4	223.8	221.7	225.7	224.3
14	221.3	223.4	224.8	224.1	220.4
15	227.7	224.7	225.4	225.5	225.4
16	225.6	224.1	224.5	224.6	224.8
17	228.9	224.4	227.5	228.3	225.0
18	222.4	224.3	223.5	227.1	222.8
19	220.1	220.1	220.0	223.7	224.1
20	219.0	221.2	222.0	219.2	223.3

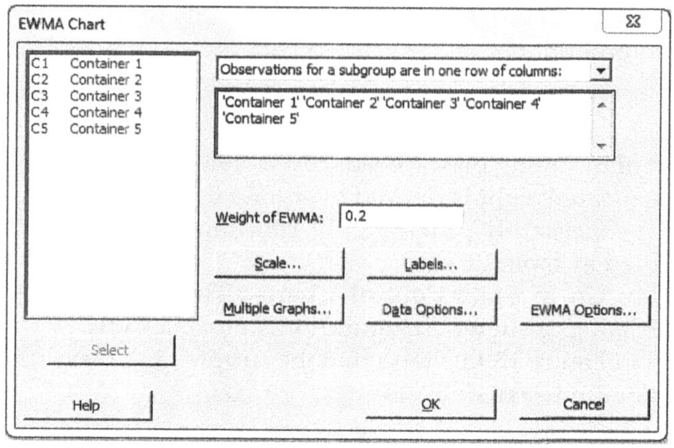

Figure 13.13 Minitab dialog box for EWMA chart.

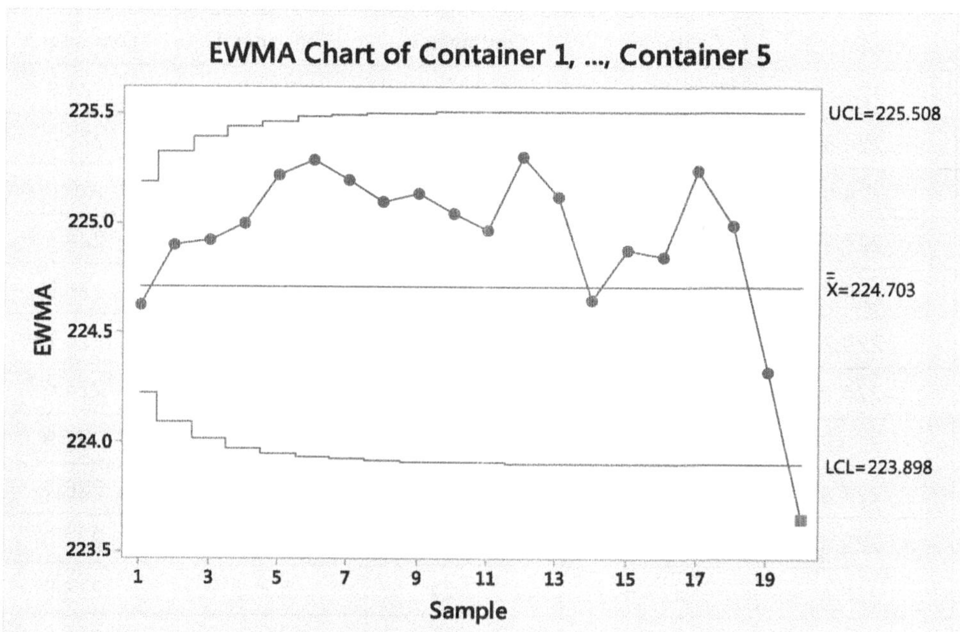

Figure 13.14 EWMA chart for plastic bead container weights.

ATTRIBUTE DATA CONTROL CHARTS

Attribute charts use count data for defects and defectives. A *defect* is a nonconformity such as a scratch or measurement deviation, and a *defective* is a nonconforming unit such as a part with a measurement deviation (Breyfogle 2003). One defective part may have multiple defects, such as one car door with two scratches and one measurement deviation. In this example there is one defective and three defects.

c Chart

A *c* chart is used to monitor the count of defects in each subgroup when subgroup size is constant. Each part could have more than one defect, such as a piece of sheet metal with three scratches. The count of defects for each subgroup is plotted. The centerline ($\bar{c}$) is the average number of defects per subgroup. It and the control limits are calculated as:

$$\text{Centerline} = \bar{c} = \frac{\sum c_i}{M}$$

$$\text{LCL} = \bar{c} - 3\sqrt{\bar{c}}$$

$$\text{UCL} = \bar{c} + 3\sqrt{\bar{c}}$$

In these calculations, c_i is the count of defects in each subgroup and M is the number of subgroups. The centerline and control limits of the *c* chart are based on the Poisson distribution. A property of the Poisson distribution is that the mean is equal to the variance; therefore a single *c* chart monitors both the location and spread of the data.

Example

Table 13.8 presents production data from seven days.

The center line and control limits are:

$$\text{Centerline} = \bar{C} = \frac{8+11+7+9+7+12+10}{7} = \frac{64}{7} = 9.14$$

$$\text{LCL} = \bar{c} - 3\sqrt{\bar{c}} = 9.14 - 3\sqrt{9.14} = 0.07$$

$$\text{UCL} = \bar{c} + 3\sqrt{\bar{c}} = 9.14 + 3\sqrt{9.14} = 18.12$$

In Minitab: Go to **Stat > Control Charts > Attributes Charts > C** and enter the column containing the data as shown in Figure 13.15, and then click **OK**.

The resulting Minitab *c* chart is shown in Figure 13.16.

U Chart

The U chart is used to monitor the count of defects per unit, and the data can be either collected individually or in subgroups. The subgroup sizes can vary. The count of defects per unit is plotted for each subgroup. The centerline and control limits are calculated as:

$$\text{Centerline} = \bar{u} = \frac{\sum c_i}{\sum n_i}$$

$$\text{LCL} = \bar{u} - 3\sqrt{\frac{\bar{u}}{n_i}}$$

$$\text{UCL} = \bar{u} + 3\sqrt{\frac{\bar{u}}{n_i}}$$

Table 13.8 Production data example.

Day	1	2	3	4	5	6	7
Defects	8	11	7	9	7	12	10
Sample Size	50	50	50	50	50	50	50

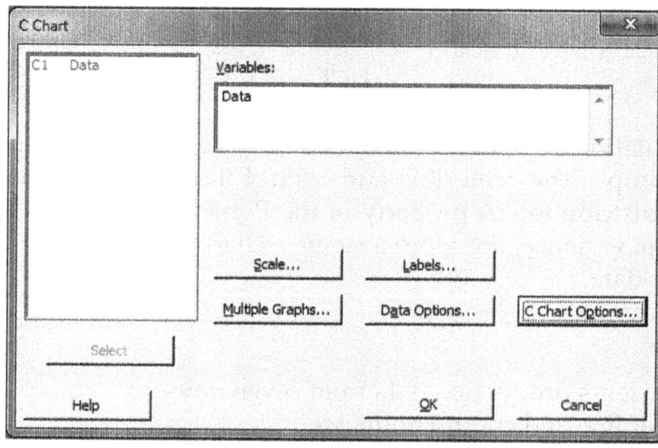

Figure 13.15 Dialog box for a *c* chart.

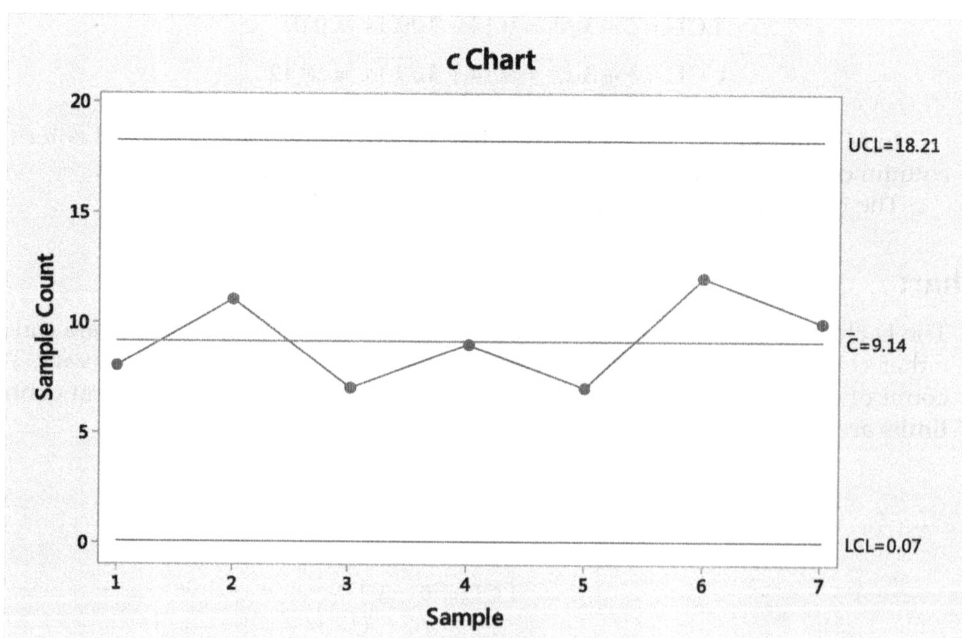

Figure 13.16 Graph for a *c* chart.

As with the c chart, the basis for the chart is the Poisson distribution and the chart monitors both location and spread of the process simultaneously.

Example

Table 13.9 presents the production data from eight days.

The plotted points are the defects/unit from each subgroup. For example:

$$\text{First point} = \frac{32}{115} = 0.278$$

$$\text{Second point} = \frac{26}{109} = 0.239$$

and so on for all the subgroups.

The center line and control limits are:

$$\text{Centerline} = \bar{U} = \frac{32+26+39+34+29+31+34+37}{115+109+119+132+114+118+123+121} = 0.278$$

$$LCL = \bar{u} - 3\sqrt{\frac{\bar{u}}{n_i}} = 0.278 - 3\sqrt{\frac{0.278}{n_i}}$$

$$UCL = \bar{u} + 3\sqrt{\frac{\bar{u}}{n_i}} = 0.278 + 3\sqrt{\frac{0.278}{n_i}}$$

For example, the control limits for the first subgroup are:

$$\text{LCL}_1 = 0.278 - 3\sqrt{\frac{0.278}{115}} = 0.131$$

$$\text{UCL}_1 = 0.278 + 3\sqrt{\frac{0.278}{115}} = 0.426$$

All plotted points and control limits are presented in Table 13.10.

In Minitab: Go to **Stat > Control Charts > Attributes Charts > U** and enter the column containing the data. Then enter the column containing the subgroup sizes and click **OK** (see Figure 13.17).

The resulting Minitab U chart is shown in Figure 13.18.

The chart indicates a stable process with no points outside of the control limits and no patterns that signal the presence of a special cause of variation. The impact

Table 13.9 Eight days of production data.

Day	1	2	3	4	5	6	7	8
Defects	32	26	39	34	29	31	34	37
Sample Size	115	109	119	132	114	118	123	121

of the unequal sample sizes can be seen by the control limits varying in width for each subgroup.

p Chart

A *p* chart is used for the proportion of defective units; the subgroup size can vary. The chart is created by plotting the proportion defective for each subgroup in time order. The centerline is equal to the overall proportion defective in the data set. The centerline and the control limits are calculated as:

$$\text{Centerline} = \bar{p} = \frac{\sum x_i}{\sum n_i}$$

$$\text{LCL} = \bar{p} - 3\sqrt{\frac{\bar{p}(1-\bar{p})}{n_i}}$$

$$\text{UCL} = \bar{p} + 3\sqrt{\frac{\bar{p}(1-\bar{p})}{n_i}}$$

In the calculations, x_i represents the counts of defectives in each subgroup and n_i the subgroup sizes. When $\bar{p}n_i < 0.5$, the false alarm rate for the chart is excessively high and should not be used (Minitab, "Attribute Control Charts"). In this case, an individuals chart of the proportion defective for each subgroup may be a reasonable alternative.

Example

Table 13.11 presents data on the number of parts found to be defective from samples of varying size taken on 12 production days.

The centerline and control limits are calculated as:

Table 13.10 Plotted points and control limits for U chart.

Subgroup	Defects	Sample size	Point plotted	LCL	UCL
1	32	115	0.278	0.130	0.426
2	26	109	0.239	0.126	0.430
3	39	119	0.328	0.133	0.423
4	34	132	0.258	0.140	0.416
5	29	114	0.254	0.130	0.426
6	31	118	0.263	0.132	0.424
7	34	123	0.276	0.135	0.421
8	37	121	0.306	0.134	0.422

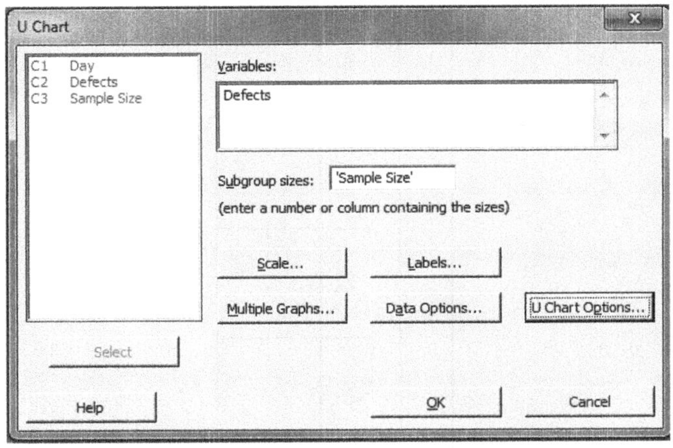

Figure 13.17 Dialog box for a U chart.

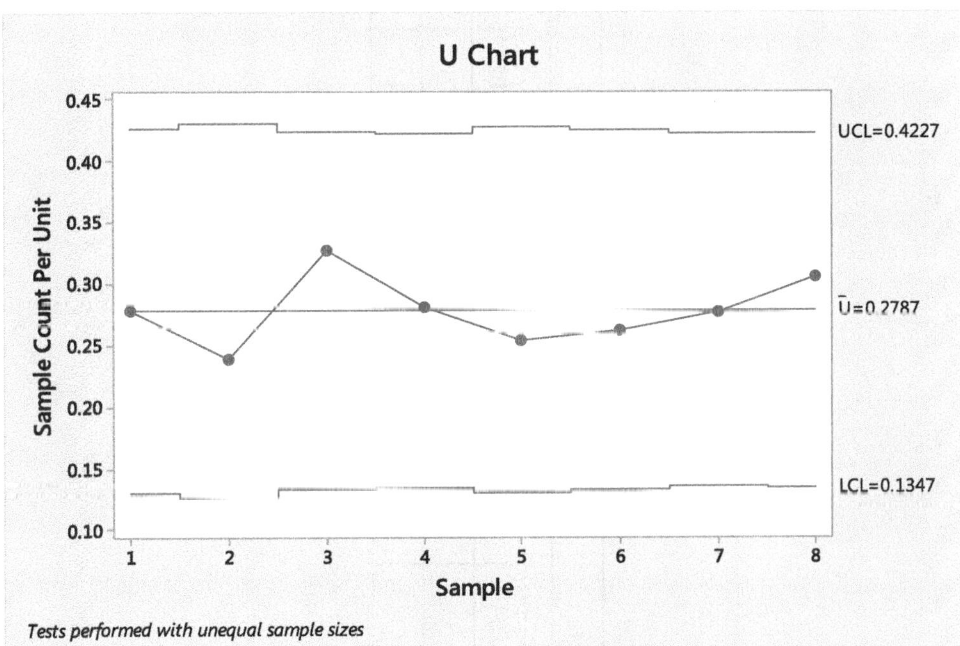

Figure 13.18 Graph for a U chart.

Table 13.11 12 days of production data.

Sample Size	87	94	96	72	84	77	102	76	90	89	78	85
Number defective	12	11	14	9	17	13	15	8	14	16	11	10
Proportion Defective	0.138	0.117	0.146	0.125	0.202	0.169	0.147	0.105	0.156	0.180	0.141	0.118

$$\text{Centerline} = \bar{p} = \frac{\sum x_i}{\sum n_i} = \frac{12+11+14+\cdots+10}{87+94+96+\cdots+85} = \frac{150}{1030} = 0.1456$$

$$\text{LCL} = \bar{p} - 3\sqrt{\frac{\bar{p}(1-p)}{n_i}} = 0.1456 - 3\sqrt{\frac{0.1244}{n_i}}$$

$$\text{UCL} = \bar{p} + 3\sqrt{\frac{\bar{p}(1-p)}{n_i}} = 0.1456 + 3\sqrt{\frac{0.1244}{n_i}}$$

The centerline is constant for all subgroups, but the control limits are dependent on the sample size for each subgroup. For example, the control limits for subgroup 1 are:

$$\text{LCL}_1 = 0.1456 - 3\sqrt{\frac{0.1244}{87}} = 0.032$$

$$\text{UCL}_1 = 0.1456 + 3\sqrt{\frac{0.1244}{87}} = 0.259$$

The control limits for all subgroups are shown in Table 13.12.

In Minitab: Go to **Stat > Control Charts > Attributes Charts > P** and enter the column containing the number of defective units in **Variables**. Then in **Subgroup sizes**, enter the column containing the sample sizes (as shown in Figure 13.19) and click **OK**.

The resulting Minitab p chart is shown in Figure 13.20.

Table 13.12 Control limits for p chart.

N Defective	Sample size	Point plotted	LCL	UCL
12	87	0.138	0.032	0.259
11	94	0.117	0.036	0.255
14	96	0.146	0.038	0.254
9	72	0.125	0.021	0.270
17	84	0.202	0.030	0.261
13	77	0.169	0.025	0.266
15	102	0.147	0.041	0.250
8	76	0.105	0.024	0.267
14	90	0.156	0.034	0.257
16	89	0.180	0.033	0.258
11	78	0.141	0.026	0.265
10	85	0.118	0.031	0.260

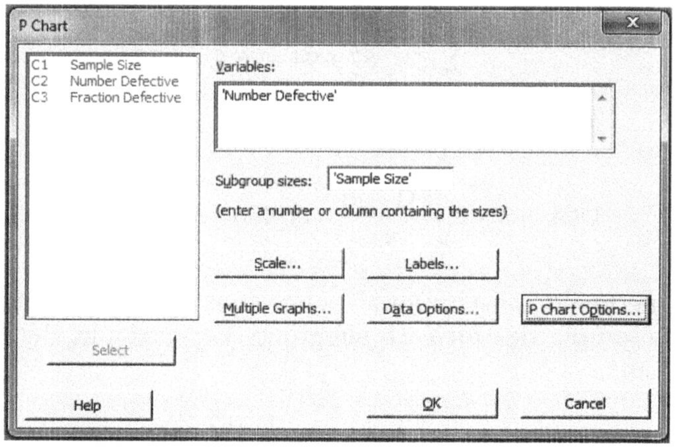

Figure 13.19 Dialog box for a *p* chart.

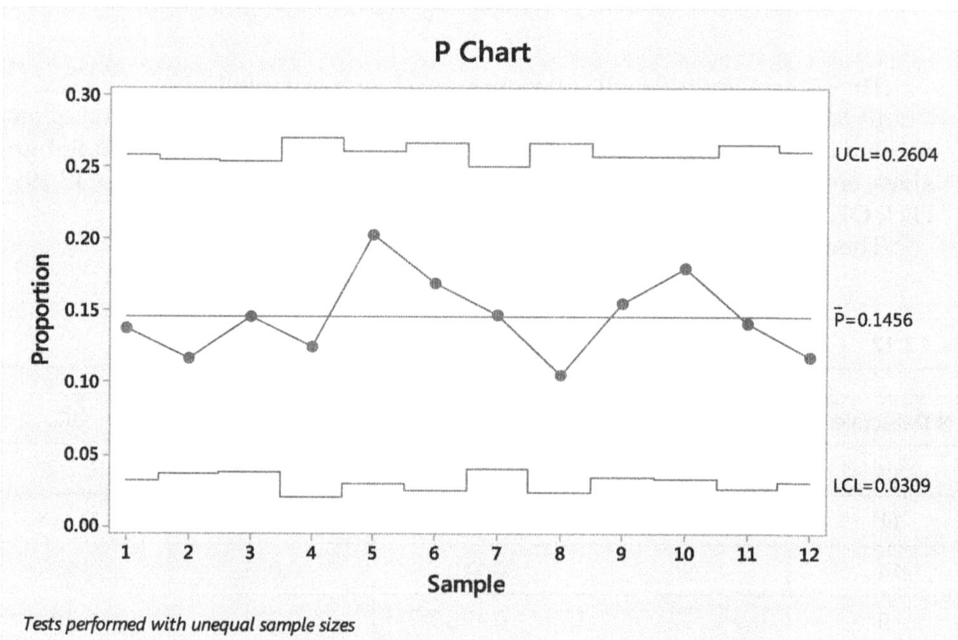

Figure 13.20 Graph for a *p* chart.

The chart indicates the process is stable because no points fall outside the control limits and there are no runs of nine or more points on either side of the center line.

NP Chart

An NP chart is used for the number of defective units when the count of defective units is more intuitive to track than the proportion. The chart is typically used when the subgroup size is constant. Rather than plotting the proportion defective

Table 13.13 15 shifts' worth of data.

Shift	1	2	3	4	5	6	7	8	9	10	11	12	13	14	15
Number Defective	14	16	11	17	16	14	12	15	17	11	15	13	16	11	15
Fraction Defective	0.14	0.16	0.11	0.17	0.16	0.14	0.12	0.15	0.17	0.11	0.15	0.13	0.16	0.11	0.15

for each subgroup as is done on a p chart, the NP chart plots the number defectives in each subgroup. The centerline and control limits are calculated as:

$$\bar{p} = \frac{\sum x_i}{\sum n_i}$$

$$\text{Centerline} = n_i\bar{p} = n_i\frac{\sum x_i}{\sum n_i}$$

$$\text{LCL} = n_i\bar{p} - 3\sqrt{n_i\bar{p}(1-\bar{p})}$$

$$\text{UCL} = n_i\bar{p} + 3\sqrt{n_i\bar{p}(1-\bar{p})}$$

When subgroup sizes are not equal, the centerline and control limits vary between subgroups. Otherwise, they are constant.

Example

A sample of 100 parts is checked every shift for 15 shifts (see Table 13.13).

The overall proportion, centerline, and control limits are calculated as:

$$\bar{p} = \frac{\sum x_i}{\sum n_i} = \frac{14+16+11+\cdots+15}{1500} = 0.142$$

$$\text{Centerline} = n_i\bar{p} = 100 * 0.142 = 14.2$$

$$\text{LCL} = n_i\bar{p} - 3\sqrt{n_i\bar{p}(1-\bar{p})} = 14.2 - 3\sqrt{100 * 0.142 * (1-0.142)} = 3.73$$

$$\text{UCL} = n_i\bar{p} - 3\sqrt{n_i\bar{p}(1-\bar{p})} = 14.2 + 3\sqrt{100 * 0.142 * (1-0.142)} = 24.67$$

In Minitab: Go to **Stat > Control Charts > Attributes Charts > NP** and enter the column containing the data as well as the column containing the subgroup size, as shown in Figure 13.21, and click **OK**.

The resulting Minitab NP chart is shown in Figure 13.22.

Because no points fail the desired tests, the process is considered to be stable.

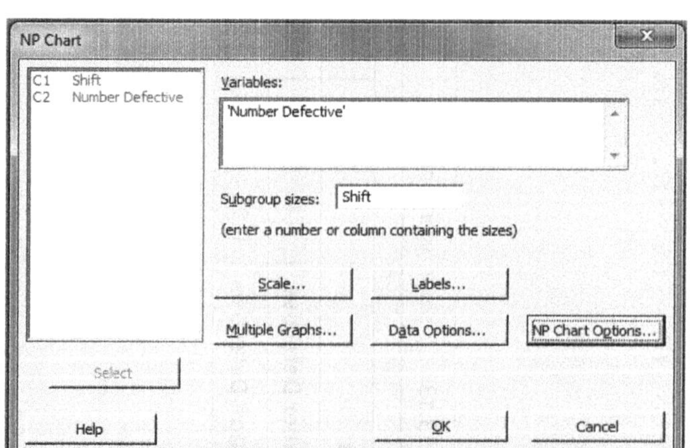

Figure 13.21 Dialog box for an NP chart.

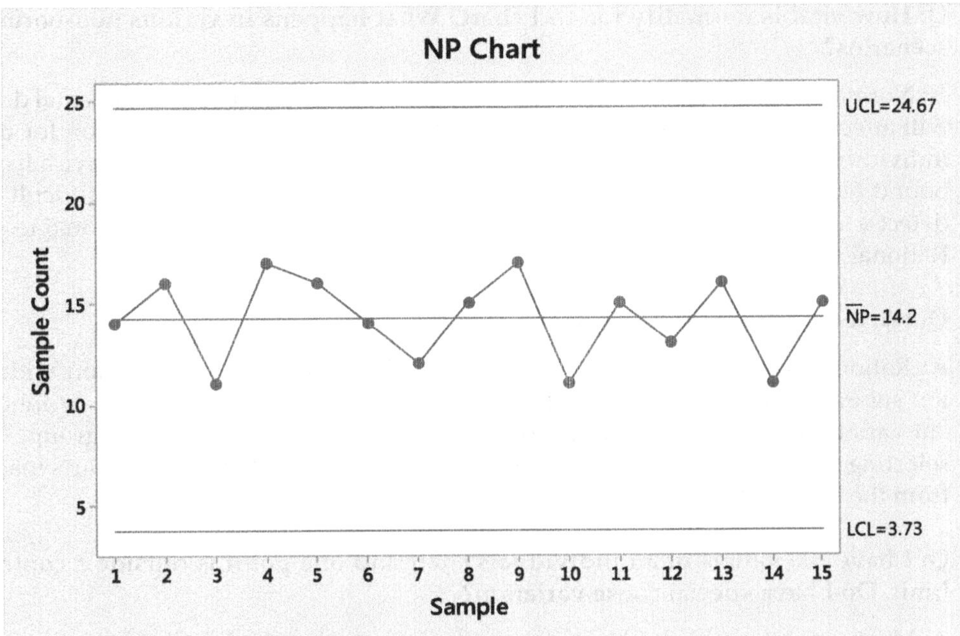

Figure 13.22 Graph for an NP chart.

QUESTIONS AND ANSWERS

Q: How much data do I need before I set my control limits as constant?

A: At least 20 points should be taken before setting control limits for future points on a chart with subgroups, and at least 100 points on an individuals chart. Collecting more points helps ensure the limits are set at appropriate values.

Q: My data are not normally distributed. Can I use an Xbar-R or I-MR chart?

A: Normality is not a formal assumption of a control chart, and the rules associated with the charts were not formed based on a distributional assumption. Control charts that use subgroups are generally very robust to non-normality, and there is no need to formally check for distribution before creating the chart. Control charts that plot individuals are often robust as well, but not in every case. For data bounded at zero or some other bound and highly skewed, the chart may not be able to detect shifts in one direction without using additional rules, and the chart may provide higher false alarm rates than desired.

Q: Do I need to update my SPC control limits every time I get new data, or can I set the limits and keep them as they are?

A: You should set the control limits once you have sufficient data so that you can detect deviations from the initial condition.

Q: How vital is normality for an I chart? What happens in various non-normal scenarios?

A: Normality is not a requirement for any control chart; however, non-normal data will affect the false alarm rate and ability to detect shifts in one direction for the individuals chart in certain circumstances. Specifically, when the data have a hard bound (such a 0 for time data) and is highly skewed, it may be very difficult to detect a shift in the direction of the bound and will likely require additional tests. Rational subgroups are a method for avoiding this when possible.

Q: What are rational subgroups?

A: Rational subgroups are those taken in such a way that the variation within the subgroups does not include long-term process drifting and trends or inherent variation between the subgroups. For example, one might take subgroups by selecting consecutively produced parts every few hours or by selecting parts made from the same batch of raw material.

Q: I have 200 values in an individuals chart and one point is outside a control limit. Do I have special cause variation?

A: Maybe, but it could also be a false positive due to the number of values. Investigate to determine if special cause variation may be present before making process changes.

Q: Why is every point that is out of control limits on an individuals chart red if 2.7 of 1000 points have a statistical chance of being out of the control limits with a stable process that is normally distributed?

A: There is a false alarm rate with any control chart that results in points occasionally falling outside of the control limits, even for a perfectly stable process. However, there is no way to differentiate points that are false alarms from those that are due to special causes, so all points are highlighted. Some judgment should be used for charts containing a large number of points.

Chapter 14
Design of Experiments

Design of experiments (DOE) are planned experiments with controlled conditions used to evaluate the effects of multiple input factors on an output. The results of analyzing the designed experiment show likely causation—not simply correlation—between the input factors and the output variable by controlling the conditions and randomizing the order of those conditions. The analysis can be used to determine the optimal conditions and the resulting impact on the output is expected to be consistent with the results predicted statistically.

Additionally, performing a DOE is more economical than One-Factor-at-a-Time (OFaT) testing. In OFaT testing, each factor of interest is changed without changing any others and then immediately evaluated for impact on the response. Performing OFaT testing is very inefficient compared to DOE because many more observations must be collected in order to estimate the effects of all factors. Furthermore, performing a designed experiment allows for interactions between input factors to be estimated (Barrentine 1999).

In DOE, the output variable being investigated is called the *response*. The process conditions being investigated that potentially affect the output are called *factors*. For example, in a heat treating process, the factors might be temperature and cooling time and the response might be the hardness of the material. The temperature levels considered might be 210°, 550°, and 1300° and the cooling times considered might be 30 seconds, 2 minutes, and 5 minutes. A designed experiment might require nine runs, each of which would involve setting the temperature at one of those levels, cooling the piece for one of the cooling time levels, and then measuring hardness on the Rockwell hardness scale C.

There are seven steps to performing a DOE. The problem needs to be recognized and a problem statement created. The factors, levels, and response variable need to be identified. The factors or response may be identified first depending on the problem encountered. The type of DOE needs to be selected and then performed. After completion, the data need to be analyzed and conclusions and recommendations should be formed (Montgomery 1997).

There are many types of designed experiments, and the primary criteria for determining the best type of design is the mathematical model for the response that is desired to be estimated. The most common terms considered for inclusion in such a model are main effects of the factors of interest, quadratic (squared) terms, and interactions between main effects.

For a *categorical factor*, the main effect is the estimated impact that each level of the factor has on the response. For a *continuous factor*, the main effect is akin to the coefficient in regression and represents the impact on the response of increasing the factor by one unit. In the example shown in Figure 14.1, the mean of the response goes down as factor A is changed from a low level to a high level.

Quadratic (squared) terms may be of interest for continuous factors in certain cases where a quadratic effect makes practical sense. Quadratic terms commonly occur when the relationship between the X variable and the Y variable is not a straight line as described by a main effect. For example, gas consumption generally improves as a car drives faster up until some optimal speed such as 90 kilometers per hour. It then begins to go down as the speed increases above that. In the normal range of driving speeds from 0 to 110 kilometers per hour, a quadratic term—speed squared—is likely needed to adequately model the relationship between speed and gas usage. Examples of quadratic fits are shown in Figures 14.2 and 14.3.

Interactions between main effects occur when the impact that one factor makes on the response is dependent on the level of another factor. For example, the impact of drill speed on hole diameter may be very small for one type of drill bit but much larger for a second type of drill bit. As a result, there would be an interaction between drill speed and bit type. In the example in Figure 14.4, the second row of the first column of graphs illustrates an interaction between factors A and C. The dashed line demonstrates that when factor C is at its low level, the effect of moving from the low level of factor A to the high level is a moderate decrease in

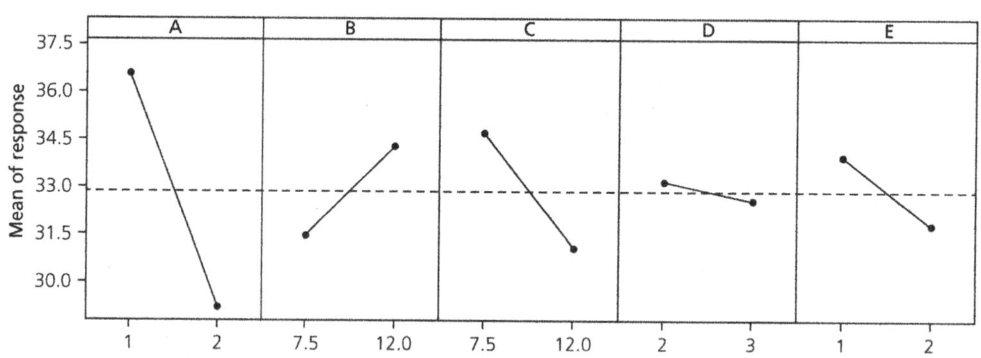

Figure 14.1 Main effects plot.

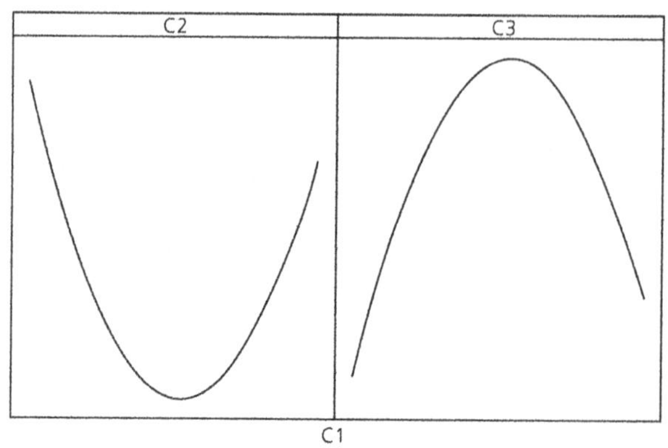

Figure 14.2 Quadratic fit for C2 and C3 vs. C1.

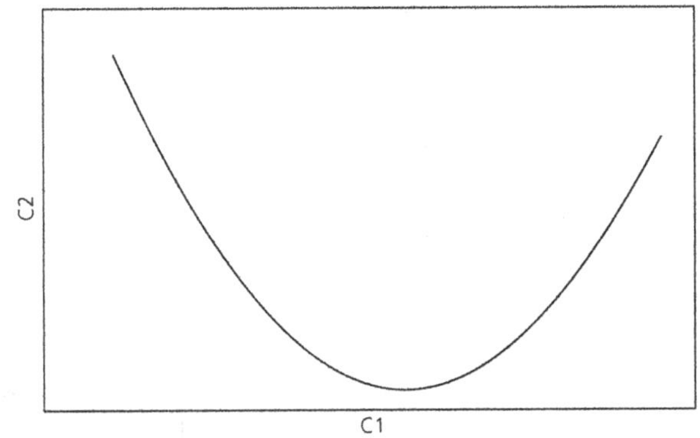

Figure 14.3 Quadratic fit for C2 vs. C1.

the response. However, the solid line illustrates that at the high level of factor C, the effect of the same move is a much larger decrease. Therefore, the effect of factor A is dependent on the level of factor C and an interaction exists. Parallel lines are an indication that there is no interaction, and lines that are far from parallel indicate a stronger interaction.

The example describes an interaction between two main effects, known as a *2-way interaction* because it is the interaction between two inputs. Higher-order interactions are very uncommon but can occur. For example, there might be a *3-way interaction* between factors A, C, and D.

Generally, designed experiments are intended to either identify statistically significant main effects or to evaluate and estimate all terms of interest. A designed experiment may even have both objectives. Screening designs have the goal of identifying which main effects are statistically significant. These designs typically require very few runs but are often not capable of estimating quadratic terms or interactions. Modeling designs have the goal of evaluating and estimating all terms of interest in the final model, including quadratic terms and interactions. These designs require more runs but fulfill the ultimate goal of identifying a model that can be used for optimization.

One experiment may not be sufficient to determine the final model. Many potential main effects of interest and/or runs are expensive, so it may be better to first perform a screening experiment to identify the main effects and then subsequently use a larger modeling design. In other cases, when the input factors can take a wide range of settings a smaller design may allow the region in which the optimal setting exists to be identified, and then a more extensive modeling design may be performed using settings in that region to find the optimal settings with more precision. No more than 25% of the budget for experiments should be spent on the first experiment (Box, Hunter, and Hunter 2005).

When performing the experiment, each experimental unit should be the same as the others to reduce variability. In some cases, it may not be possible to obtain enough homogenous units to perform the experiment on. For example, completing the experiment may require performing runs during two different shifts and the units experimented on cannot be assumed to be the same because they are produced under different conditions. *Blocking* is a technique in which units are separated into homogenous (comparable) groupings and the variation between

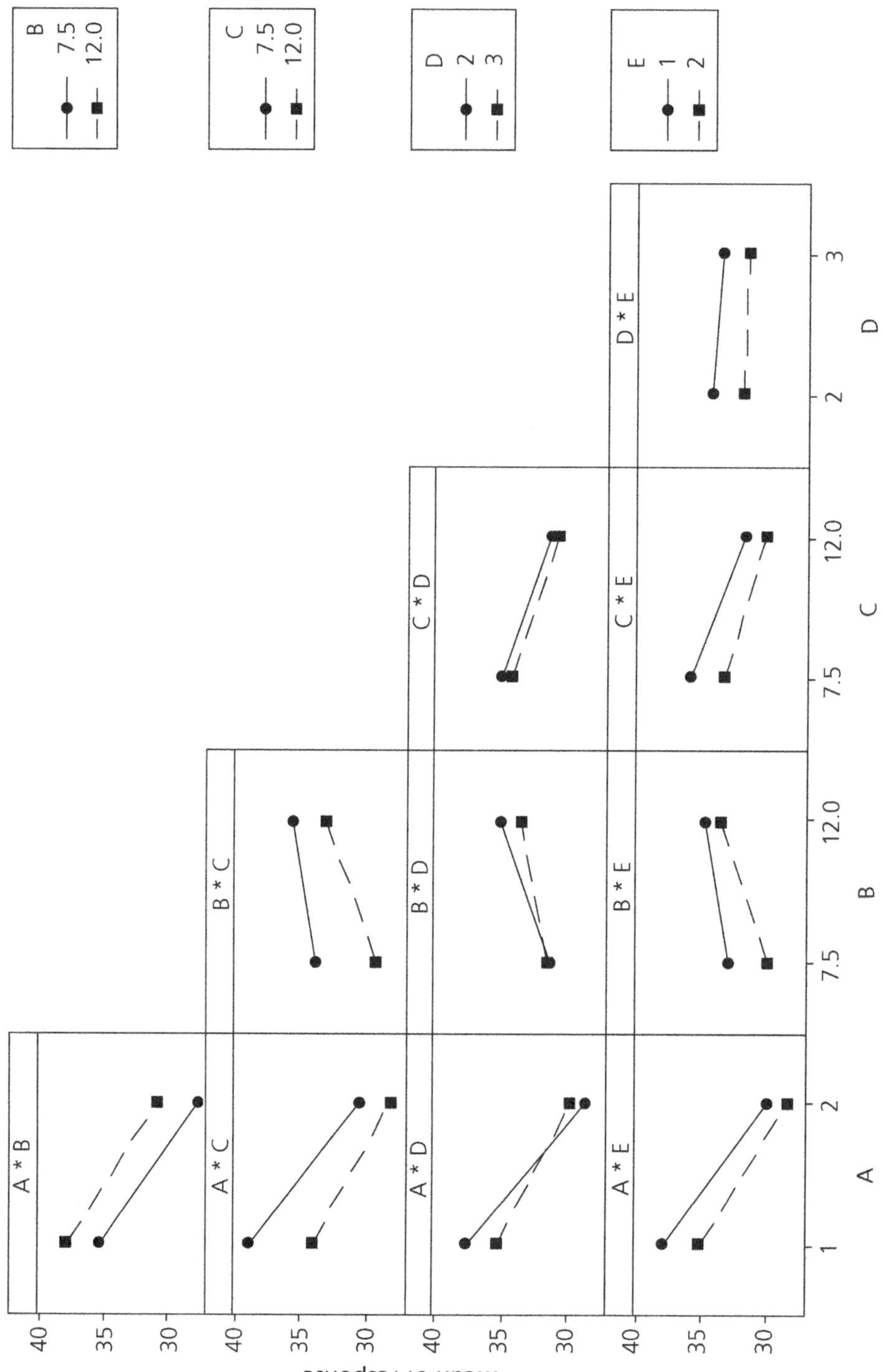

Figure 14.4 Interaction plot.

Table 14.1 Common resolutions.

Resolution	Description
Resolution III	At least some main effects are confounded with 2-way interactions
Resolution IV	2-way interactions are aliased with other 2-way interactions and/or main effects are confounded with 3-way interactions
Resolution V	2-way interactions are aliased with 3-way interactions and/or main effects are aliased with 4-way interactions

groups is accounted for in the analysis. For example, material from two different sources may be separated into blocks.

A critical aspect of DOE is the randomization of the runs within blocks or across the entire experiment when blocking is not needed. Randomization helps protect the practitioner from confounding effects with outside factors or changing conditions and allows us to statistically test whether effects seen are likely due to the factors tested or simply due to random variation. Box, Hunter and Hunter tell us to "block what you can and randomize what you cannot" (2005, 93). Minitab adds randomization by default.

Resolution is a measure of *aliasing,* also known as *confounding,* between the factors. When confounding is present in the design, certain main effects or interactions cannot be mathematically differentiated from others and cannot be estimated separately. Resolution is reported in Roman numerals as the sum of the lowest order terms that are confounded. Common resolutions are explained in Table 14.1.

In a screening design, typically only main effects are of interest and a resolution III design may be fine. In modeling designs, it is often desirable to test 2-way interactions for statistical significance. In that case, carefully consider which 2-way interactions are possible and choose a design accordingly. When few are possible, a resolution IV design may be performed in which the possible 2-way interactions are only aliased with those that seem impossible. Otherwise, a resolution V or higher design is advisable.

Another important aspect of DOE is the inclusion of replication. *Replication* is a technique in which one or more of the factor level combinations are used for more than one run in the experiment. Replication allows for the estimation of pure error, which is the variation among runs that were performed at the same factor level combinations. When replication does not exist in an experiment, the error term in the model is based on the concept known as *sparsity of effects.* This concept is the idea that higher-order interactions are unlikely to exist and thus the residual variation in a model that does not include them is due to unexplained error and not the existence of those terms. When replication is used, the error term in the analysis can be separated into pure error as well as "lack of fit," the error explained by higher-order terms not in the model.

The idea of replication should not be confused with *repeat measures,* where multiple items are measured in each run of the experiment. Replication involves repeating the same factor level combinations but randomizing the order of the replicated runs along with all runs. Repeat measures, on the other hand, represent a single run of the experiment and should be averaged to create a single response value for that run. If it is inexpensive to increase the size of an experiment, replication is almost always preferable to repeat measures.

Once the experiment has been designed, run, and the resulting response values recorded, a model is formed. The model formed is similar to that of ANOVA or regression but with multiple terms. As with ANOVA, the analysis of a designed experiment produces an ANOVA table, R-squared values, and residual plots. Many aspects of the model, including p-values for the terms in the model, would be very cumbersome to calculate by hand and should be done using statistical software.

As with regression it is important to examine the residuals from the model formed by collecting and analyzing the data from a designed experiment. Residual plots can help identify whether any runs were potentially done incorrectly or recorded incorrectly, as well as whether there are unusual patterns over time or fits. Variation in residuals should appear random over time and when plotted against the fitted values and should be approximately normal if confidence or prediction intervals will be made using the model.

FACTORIAL DESIGNS

Factorial designs are the most common designed experiments. There are many different types of factorial designs and the choice of which to use requires careful consideration of cost, factors, and the model of interest. Factorial designs can estimate main effects, and some designs can estimate interactions as well as test for the presence of curvature. The presence of curvature is an indication that one or more quadratic terms may be needed.

Most factorial designs analyze the effect on the response when continuous factors are changed between a low and a high level or when a categorical factor is changed from one level to another. These designs are generally called *2-level factorial designs*.

A full factorial design is one in which at least one run is performed for every possible combination of factor levels. For example, Table 14.2 shows a full factorial design for 3 factors with 2 levels each that has not yet been randomized.

Full factorial designs allow all main effects and every possible interaction to be estimated with no confounding. However, they require many runs when the number of factors is high and may be expensive to perform. A design in which all terms are uncorrelated is known as an *orthogonal design*. A full factorial design with k factors requires 2^k runs. There are often better designs that can be used, based on the terms of interest.

When there are at least three factors of interest, a *fractional factorial design* is an alternative that uses less runs but includes some confounding. A fractional factorial design is created by dividing a 2-level, full factorial design into a fraction of the original design so that less runs are required. This results in confounding between terms, so it is important to consider the resolution of the design and which terms are confounded. In a fractional factorial design, all terms are either not confounded or completely confounded with other terms. There is no partial confounding, where two or more terms are correlated with one another but not completely confounded. Figure 14.5 shows the resolution of full factorial designs and fractional factorial designs for 2–15 factors.

Table 14.2 An unrandomized 3 factor, 2 level full factorial design.

Run	Factor A	Factor B	Factor C
1	Low	Low	Low
2	Low	Low	High
3	Low	High	Low
4	Low	High	High
5	High	Low	Low
6	High	Low	High
7	High	High	Low
8	High	High	High

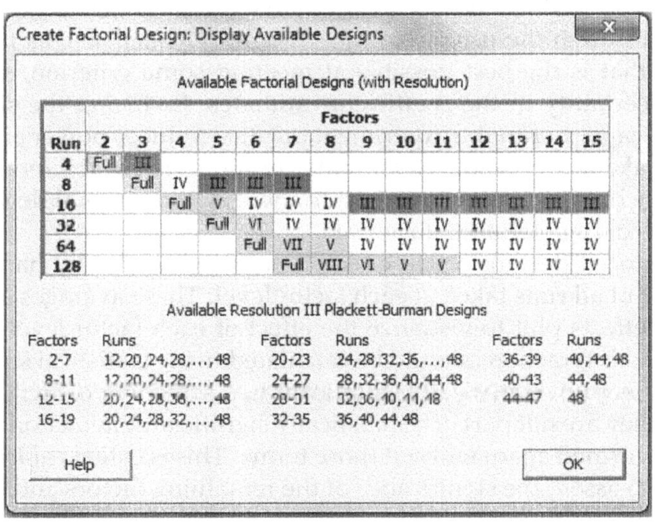

Figure 14.5 Number of runs, factors, and the possible resolutions available in Minitab.

For example, a full factorial design for 5 factors requires $2^5 = 32$ runs. A half fractional factorial requiring 16 runs has resolution V and, in most cases, is still a very good design. A quarter fractional factorial requiring just 8 runs, however, has resolution III and in most cases is only a good design for screening and not modeling.

Center points can be added to the design when one or more factors are continuous and can be set at the midpoint between the two levels used. Center points allow for the presence of curvature to be detected and can provide an estimate of pure error when replicated. Curvature exists when the relationship between the factor and response is not linear between the low and high values. For example, if the fit is quadratic, then curvature exists. Pure error is the variation when multiple observations are taken under the exact same conditions. Five or more center points should be used to provide a reasonably precise estimate of pure error.

Aside from fractional factorial designs, another common screening factorial design is the *Plackett-Burman design*. Plackett-Burman designs are not orthogonal and terms may be correlated, especially main effects with 2-way interactions. Because of this, the designs should only be used as screening designs. If the number of factors is a power of 2, then the smallest Plackett-Burman design is the same as a fractional factorial design.

When runs are expensive, Plackett-Burman designs may require fewer runs than a fractional factorial in order to screen for significant main effects. However, fractional factorial designs can be more easily augmented with additional runs to increase the resolution and allow for modeling. For screening, both designs should be considered.

When one or more categorical factors have three or more levels of interest, a general full factorial design may be used. This design uses all combinations of factor levels in the design and, like 2-level factorial designs, can be replicated and utilize blocks when appropriate. Even without replication, general full factorial designs can be very large and expensive. For example, a design with 6 factors of 3 levels each requires a total of 729 runs.

One way to reduce the size of very large designs such as general full factorial designs is through the use of optimal designs. *Optimal designs* find the design of some size that is the best possible at meeting some criterion, such as minimizing the uncertainty in the coefficient estimates. Reducing the size of the design makes it cheaper to run the experiment but also reduces power and, if reduced too far, may make it impossible to estimate certain terms in the model. Furthermore, software such as Minitab is required to identify the optimal design from a larger design such as a general full factorial.

For factorial designs, main effects can be calculated by hand by calculating the average of all runs taken at each factor level. These averages are often graphed on a main effects plot to visualize the effect of each factor level on the response. P-values and interactions should be calculated using statistical software. Terms for factors that are not statistically significant in a screening design may be removed as long as they are not part of a statistically significant interaction. The analysis can then be performed again without those terms. This is called *reducing the model*, and the ability to assess the significance of the remaining factors and estimate residual error may improve after doing so.

Example

Suppose we wanted to perform a DOE to optimize the quality of a product produced by a particular machine. First, we must determine what a high-quality product is and how to measure it. Subject matter experts may tell us what attribute to measure, and with this information we can clearly define the response variable. The next step would be identifying the factors that potentially have the highest influence on the response variable. In this case, process experts identify the following factors and settings that encompass possible operating conditions: the setting of knob A to 200 and 250, the setting of knob B to 40 and 60, and the raw material as either basic or premium.

Because there are only three factors of interest and each of run of the experiment is fairly cheap, a screening design is not necessary and a full factorial design can be used. The knobs cannot be set to values between the chosen settings, so curvature will not be tested. Further, to increase the power of the design, three replicates will be performed. Table 14.3 shows the factor combinations for each

Table 14.3 All experimental runs and results.

Run	Knob A	Knob B	Material	Response
1	200	40	Basic	30.2
2	250	40	Basic	43.1
3	200	60	Basic	27.3
4	250	60	Basic	40.3
5	200	40	Premium	31.6
6	250	40	Premium	47.5
7	200	60	Premium	31.6
8	250	60	Premium	46.0
9	200	40	Basic	33.1
10	250	40	Basic	46.0
11	200	60	Basic	24.4
12	250	60	Basic	41.7
13	200	40	Premium	37.4
14	250	40	Premium	44.6
15	200	60	Premium	28.8
16	250	60	Premium	47.5
17	200	40	Basic	30.2
18	250	40	Basic	40.3
19	200	60	Basic	27.3
20	250	60	Basic	38.8
21	200	40	Premium	36.0
22	250	40	Premium	44.6
23	200	60	Premium	28.8
24	250	60	Premium	47.5

run and the resulting response value. Note that while the order of the runs was randomized to run the experiment, they are shown in a sorted order in the table.

We can calculate the main effects by determining the mean of each factor at high and low and calculating the difference. For example, the main effect for knob A is:

Knob A high:

$$\frac{(43.1+40.3+47.5+46.0+46.0+41.7+44.6+47.5+40.3+38.8+44.6+47.5)}{12}=43.99$$

Knob A low:

$$\frac{(30.2+27.3+31.6+31.6+33.1+24.4+37.4+28.8+30.2+27.3+36.0+28.8)}{12}=30.56$$

Knob A main effect $= 43.99 - 30.56 = 13.43 = 13.43$

This means that the effect on the response of moving from a setting of 200 to 250 on knob A is an increase of 13.43. This is then repeated for the other factors. Table 14.4 shows the main effects for all three factors.

Then create a graph of the main effects (Figure 14.6).

Although this illustrates the main effects for each factor, a p-value should be calculated to determine if there is evidence that each factor actually influences the

Table 14.4 Main effects for all factors.

	Mean at high level	Mean at low level	Main effect
Knob A	43.99	30.56	13.43
Knob B	35.83	38.72	–2.89
Material	39.33	35.23	4.10

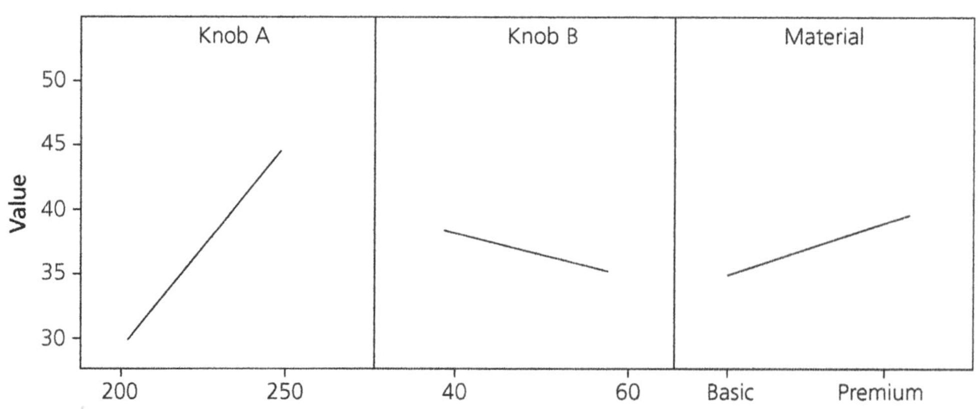

Figure 14.6 Main effects plot.

response. Further, 2-way interactions should be evaluated if they make intuitive sense.

In Minitab: Go to **Stat > DOE > Factorial > Create Factorial Design** and select the number of factors as shown in Figure 14.7. Then click on **Designs** and select the desired design (see Figure 14.8). Click on **Factors** and name the factors (as depicted in Figure 14.9), and then click **OK.**

Minitab will then generate a worksheet such as the one in Figure 14.10. The cell C8 can then be relabeled as "Response" and the results entered in this column after completion of the experiment.

In Minitab: Go to **Stat > DOE > Factorial > Analyze Factorial Design** and select the column containing the data as shown in Figure 14.11.

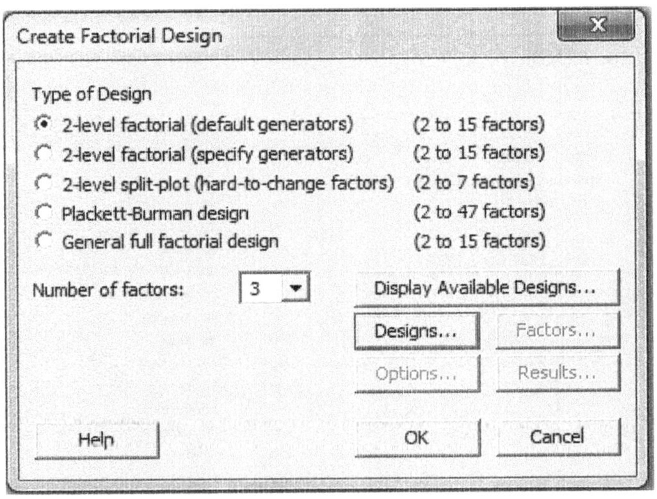

Figure 14.7 Dialog box for a DOE.

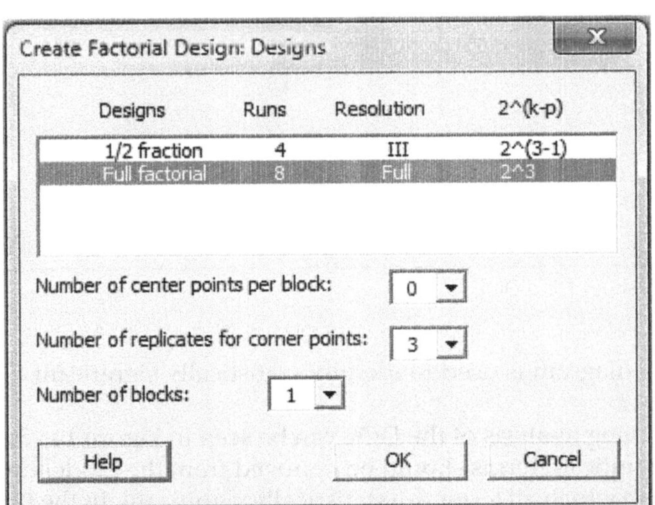

Figure 14.8 Designs for a DOE.

Create Factorial Design: Factors

Factor	Name	Type	Low	High
A	Knob A	Numeric	200	250
B	Knob B	Numeric	40	60
C	Material	Text	Basic	Premium

Help OK Cancel

Figure 14.9 Factors for a DOE.

Worksheet 1 ***

↓	C1	C2	C3	C4	C5	C6	C7-T	C8
	StdOrder	RunOrder	CenterPt	Blocks	Knob A	Knob B	Material	
1	6	1	1	1	250	40	Premium	
2	2	2	1	1	250	40	Basic	
3	17	3	1	1	200	40	Basic	
4	18	4	1	1	250	40	Basic	
5	5	5	1	1	200	40	Premium	
6	21	6	1	1	200	40	Premium	
7	9	7	1	1	200	40	Basic	
8	14	8	1	1	250	40	Premium	
9	12	9	1	1	250	60	Basic	
10	15	10	1	1	200	60	Premium	
11	11	11	1	1	200	60	Basic	
12	10	12	1	1	250	40	Basic	
13	19	13	1	1	200	60	Basic	
14	24	14	1	1	250	60	Premium	
15	22	15	1	1	250	40	Premium	
16	13	16	1	1	200	40	Premium	
17	23	17	1	1	200	60	Premium	
18	8	18	1	1	250	60	Premium	
19	1	19	1	1	200	40	Basic	
20	16	20	1	1	250	60	Premium	
21	4	21	1	1	250	60	Basic	
22	20	22	1	1	250	60	Basic	
23	7	23	1	1	200	60	Premium	
24	3	24	1	1	200	60	Basic	
25								
26								

Figure 14.10 Worksheet for a DOE.

A Pareto diagram is used to identify statistically significant factors, as shown in Figure 14.12.

The resulting analysis of the DOE can be seen in Figure 14.13.

The insignificant terms should be removed from the model until a final model is identified in which all terms are statistically significant. In the **Graphs** subdialog of **Analyze Factorial Design**, "Four in one" should be chosen to provide residual

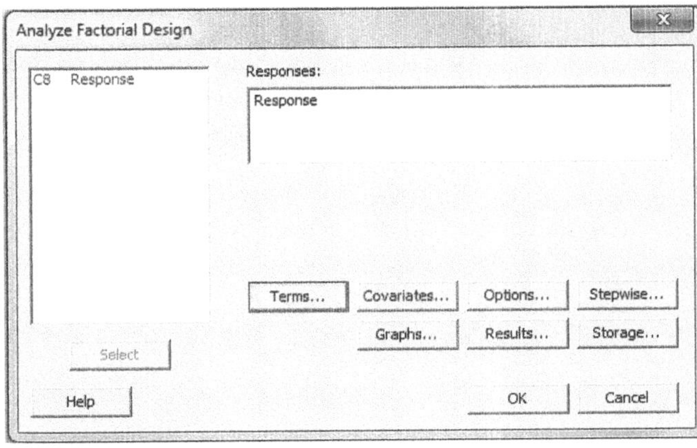

Figure 14.11 Analyze factorial design.

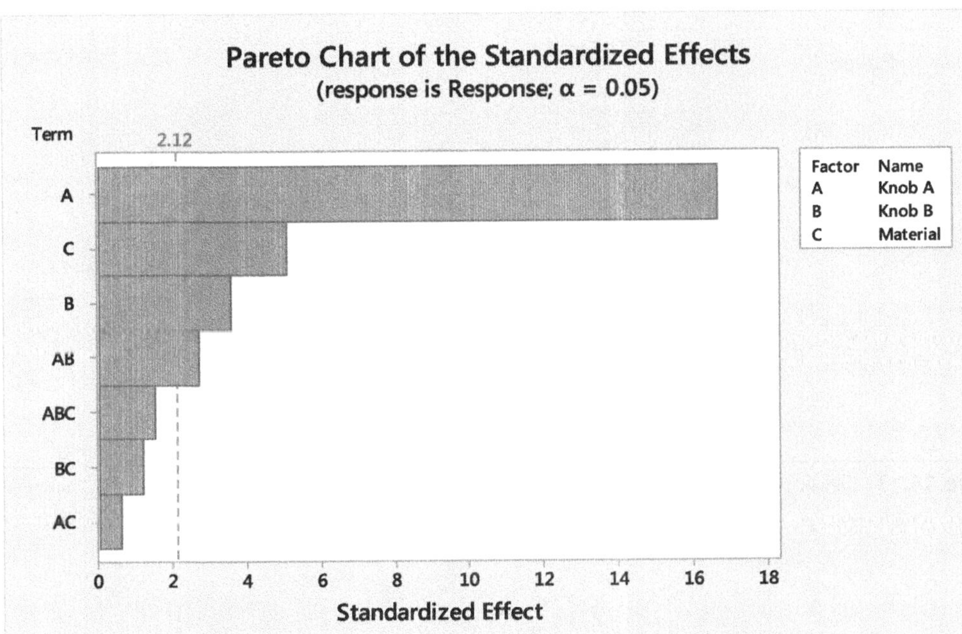

Figure 14.12 Pareto chart of main effects.

plots, as shown in Figure 14.14. The resulting output is shown in Figure 14.15 and the residual plots in Figure 14.16.

Go to **Stat > DOE > Factorial > Factorial Plots** and select the factors to analyze. Minitab will then provide a main effects plot (see Figure 14.17) and interaction plot (see Figure 14.18).

The analysis demonstrates that all three main effects are statistically significant, as is an interaction between the settings of knob A and knob B. The interaction plot in Figure 14.18 illustrates that at the high level of knob A, the setting of knob B has little impact. However, at the low setting of knob A, the response is much higher when knob B is at the low setting. The residual plots demonstrated

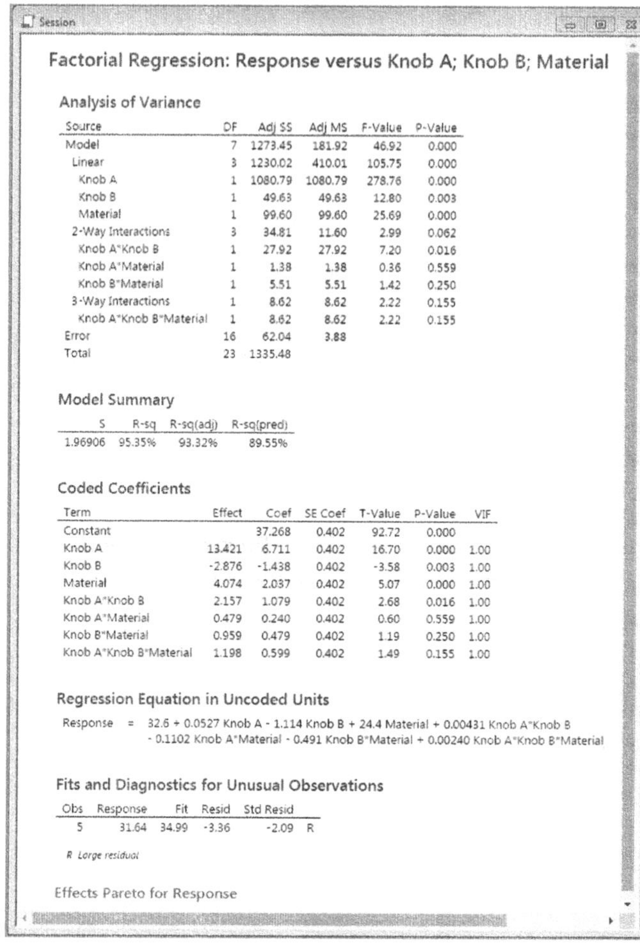

Session

Factorial Regression: Response versus Knob A; Knob B; Material

Analysis of Variance

Source	DF	Adj SS	Adj MS	F-Value	P-Value
Model	7	1273.45	181.92	46.92	0.000
Linear	3	1230.02	410.01	105.75	0.000
Knob A	1	1080.79	1080.79	278.76	0.000
Knob B	1	49.63	49.63	12.80	0.003
Material	1	99.60	99.60	25.69	0.000
2-Way Interactions	3	34.81	11.60	2.99	0.062
Knob A*Knob B	1	27.92	27.92	7.20	0.016
Knob A*Material	1	1.38	1.38	0.36	0.559
Knob B*Material	1	5.51	5.51	1.42	0.250
3-Way Interactions	1	8.62	8.62	2.22	0.155
Knob A*Knob B*Material	1	8.62	8.62	2.22	0.155
Error	16	62.04	3.88		
Total	23	1335.48			

Model Summary

S	R-sq	R-sq(adj)	R-sq(pred)
1.96906	95.35%	93.32%	89.55%

Coded Coefficients

Term	Effect	Coef	SE Coef	T-Value	P-Value	VIF
Constant		37.268	0.402	92.72	0.000	
Knob A	13.421	6.711	0.402	16.70	0.000	1.00
Knob B	-2.876	-1.438	0.402	-3.58	0.003	1.00
Material	4.074	2.037	0.402	5.07	0.000	1.00
Knob A*Knob B	2.157	1.079	0.402	2.68	0.016	1.00
Knob A*Material	0.479	0.240	0.402	0.60	0.559	1.00
Knob B*Material	0.959	0.479	0.402	1.19	0.250	1.00
Knob A*Knob B*Material	1.198	0.599	0.402	1.49	0.155	1.00

Regression Equation in Uncoded Units

Response = 32.6 + 0.0527 Knob A - 1.114 Knob B + 24.4 Material + 0.00431 Knob A*Knob B
- 0.1102 Knob A*Material - 0.491 Knob B*Material + 0.00240 Knob A*Knob B*Material

Fits and Diagnostics for Unusual Observations

Obs	Response	Fit	Resid	Std Resid	
5	31.64	34.99	-3.36	-2.09	R

R Large residual

Effects Pareto for Response

Figure 14.13 Session window for a DOE.

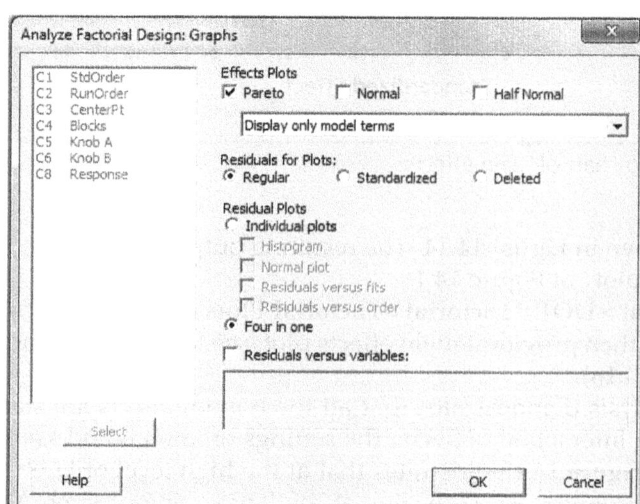

Figure 14.14 Minitab dialog box for DOE graphs.

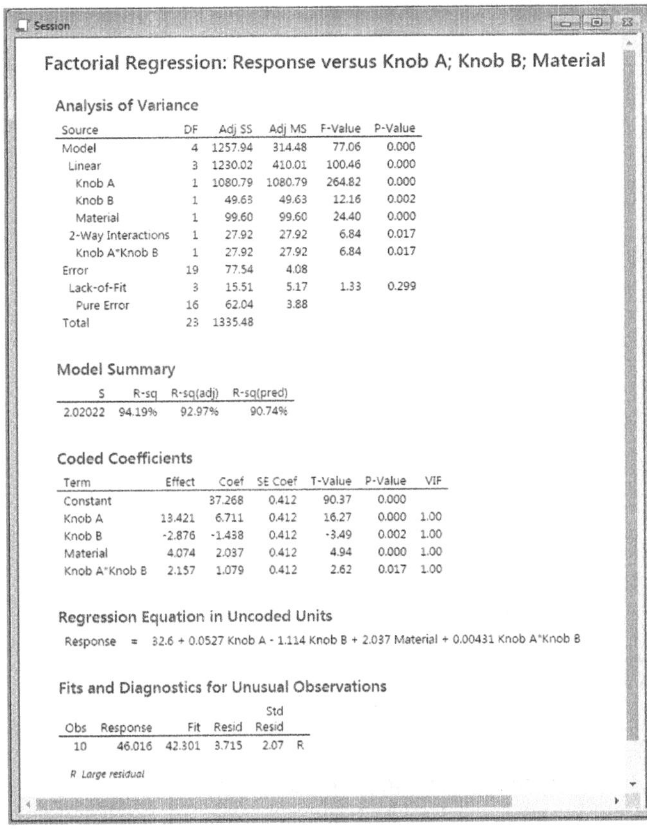

Session

Factorial Regression: Response versus Knob A; Knob B; Material

Analysis of Variance

Source	DF	Adj SS	Adj MS	F-Value	P-Value
Model	4	1257.94	314.48	77.06	0.000
Linear	3	1230.02	410.01	100.46	0.000
Knob A	1	1080.79	1080.79	264.82	0.000
Knob B	1	49.63	49.63	12.16	0.002
Material	1	99.60	99.60	24.40	0.000
2-Way Interactions	1	27.92	27.92	6.84	0.017
Knob A*Knob B	1	27.92	27.92	6.84	0.017
Error	19	77.54	4.08		
Lack-of-Fit	3	15.51	5.17	1.33	0.299
Pure Error	16	62.04	3.88		
Total	23	1335.48			

Model Summary

S	R-sq	R-sq(adj)	R-sq(pred)
2.02022	94.19%	92.97%	90.74%

Coded Coefficients

Term	Effect	Coef	SE Coef	T-Value	P-Value	VIF
Constant		37.268	0.412	90.37	0.000	
Knob A	13.421	6.711	0.412	16.27	0.000	1.00
Knob B	-2.876	-1.438	0.412	-3.49	0.002	1.00
Material	4.074	2.037	0.412	4.94	0.000	1.00
Knob A*Knob B	2.157	1.079	0.412	2.62	0.017	1.00

Regression Equation in Uncoded Units

Response = 32.6 + 0.0527 Knob A - 1.114 Knob B + 2.037 Material + 0.00431 Knob A*Knob B

Fits and Diagnostics for Unusual Observations

Obs	Response	Fit	Resid	Std Resid	
10	46.016	42.301	3.715	2.07	R

R Large residual

Figure 14.15 Output for the reduced model.

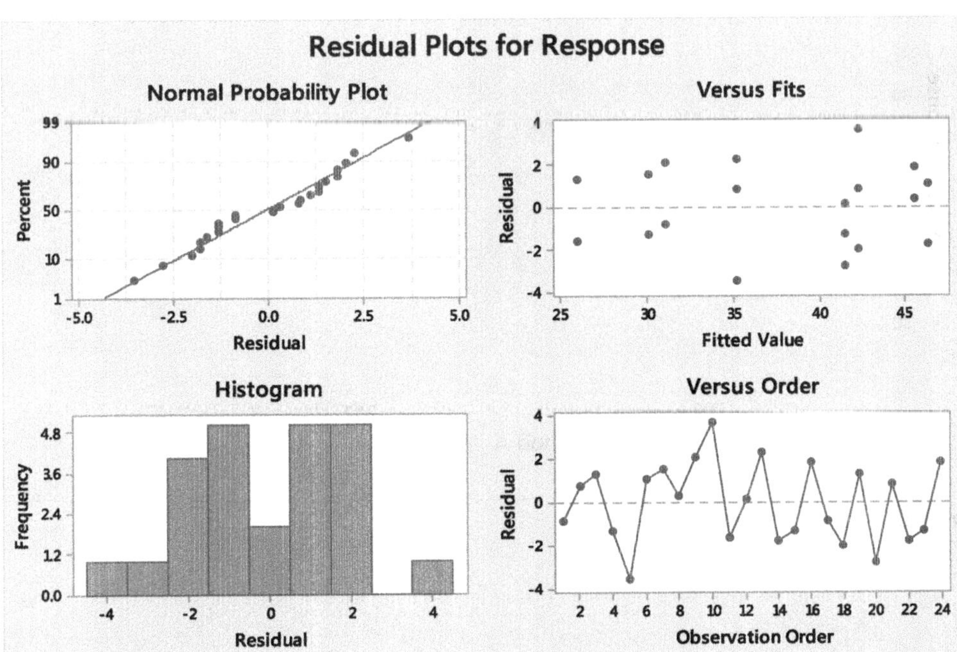

Figure 14.16 Residual plots.

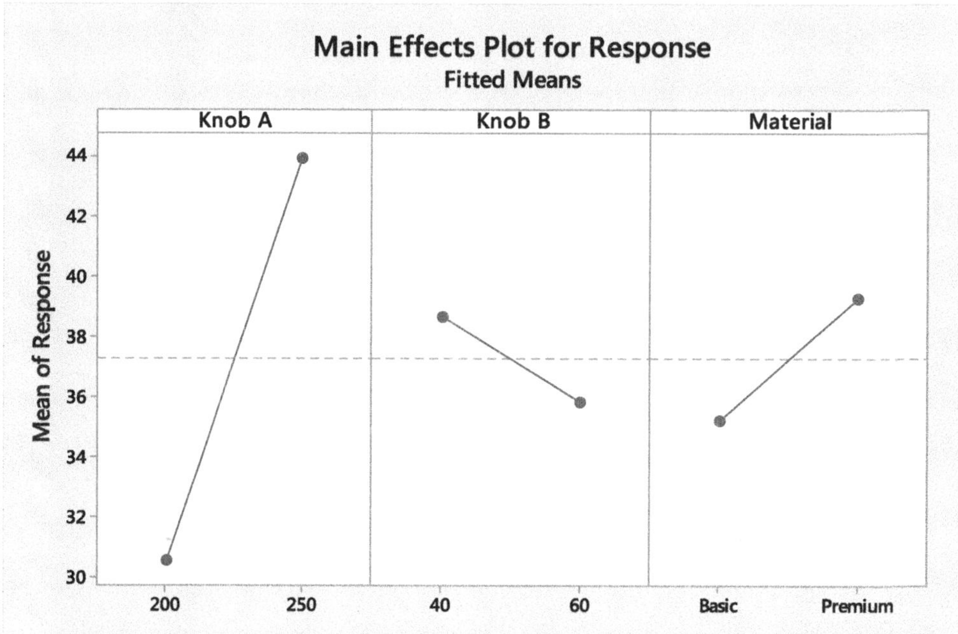

Figure 14.17 Main effects plot.

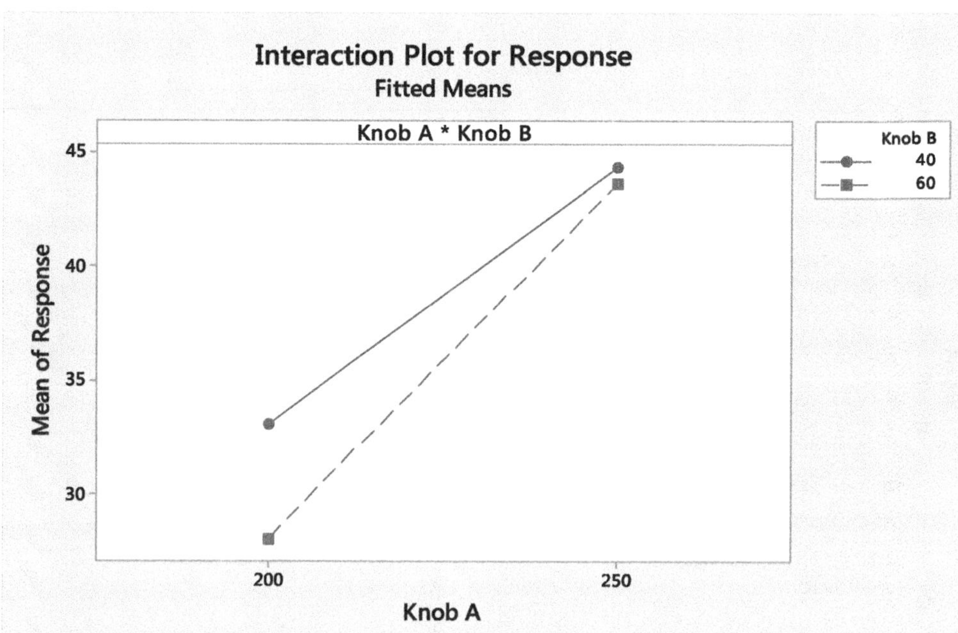

Figure 14.18 Interaction plot.

that there were no unusual observations or patterns, and the residuals are approximately normally distributed.

Higher values of the response are desired for a better quality product. To find the optimal settings, in Minitab go to **Stat > DOE > Factorial > Response Optimizer**. Complete the dialog as shown in Figure 14.19.

The resulting optimization plot is shown in Figure 14.20.

The plot predicts that at the optimal settings of knob A = 250, knob B = 40, and material = premium, the response will equal approximately 40.4. Output that includes more information about this prediction appears in the session window and is shown in Figure 14.21.

The 95% confidence interval shown indicates that we are 95% sure that the mean response at the optimal factor settings is between 44.446 and 48.305. The 95% prediction interval shown indicates that we expect 95% of observations to be between 41.728 and 51.023 when the optimal settings are used.

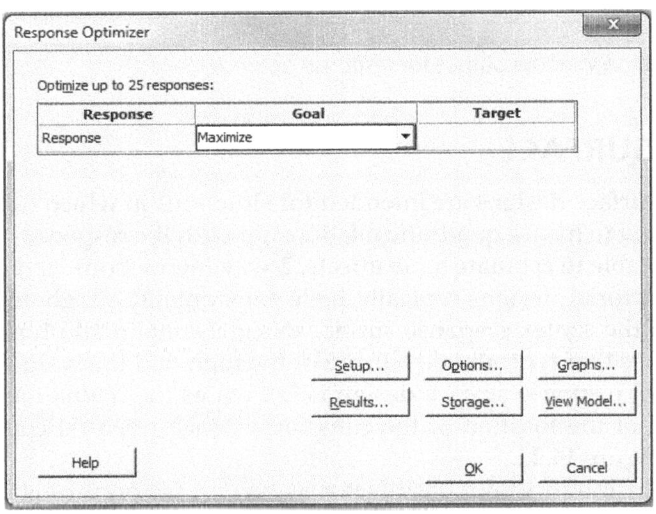

Figure 14.19 Response optimizer dialog box.

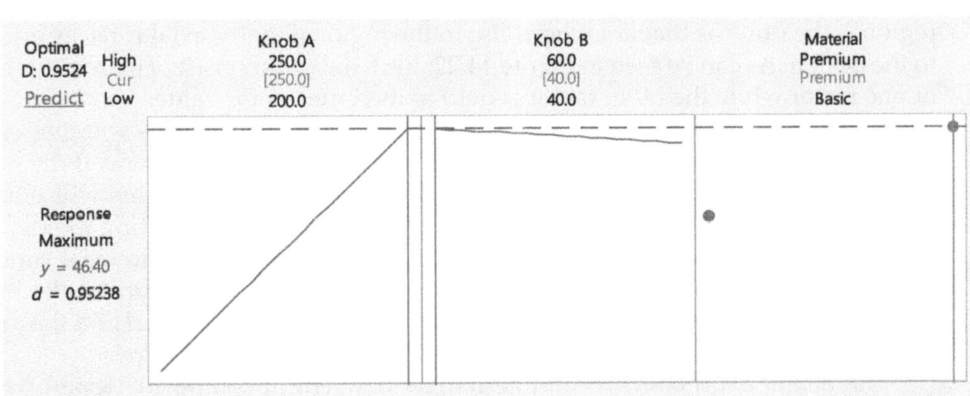

Figure 14.20 Optimization plot.

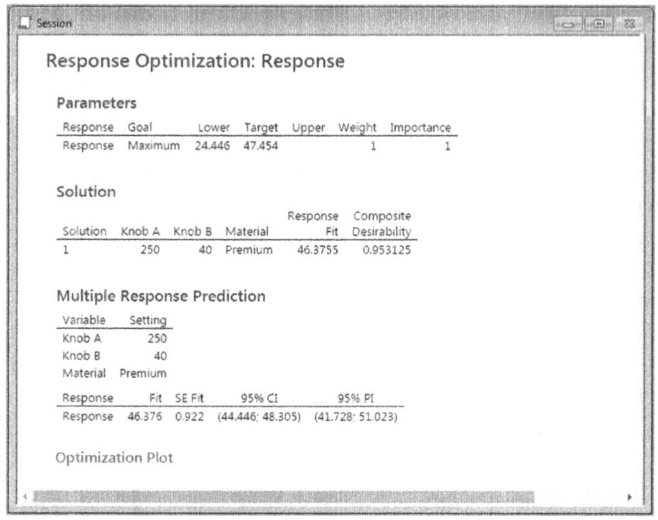

Session

Response Optimization: Response

Parameters

Response	Goal	Lower	Target	Upper	Weight	Importance
Response	Maximum	24.446	47.454		1	1

Solution

Solution	Knob A	Knob B	Material	Response Fit	Composite Desirability
1	250	40	Premium	46.3755	0.953125

Multiple Response Prediction

Variable	Setting
Knob A	250
Knob B	40
Material	Premium

Response	Fit	SE Fit	95% CI	95% PI
Response	46.376	0.922	(44.446; 48.305)	(41.728; 51.023)

Optimization Plot

Figure 14.21 Session window output for response optimizer.

RESPONSE SURFACE

Response surface designs are intended for situations in which two or more factors are suspected to have a quadratic relationship with the response. Response surface designs are able to estimate main effects, 2-way interactions, and quadratic effects. Whereas factorial designs typically have runs only at a high and low value and possible in the center, response surface designs contain additional points known as *axial points* that typically lie outside of the high and low values. The most common type of response surface design is known as the *central composite design*. An illustration of the location of the runs for a 2-factor central composite design is shown in Figure 14.22.

The illustration shows each factor as having factorial runs at a low value of –1 and a high value of 1. Additionally, the design has center runs. Often a central composite design will not include replication but will use at least 5 center points. In doing so, an assumption is made that the pure error at the center of the design's region is the same as that anywhere else in the region. Finally, axial runs are added to the design. As can be seen in Figure 14.22, the axial runs consist of extreme values of one factor while the other factor is held at its center point value.

One consideration when creating a response surface design is whether each factor can take on the more extreme values required for the axial runs. If the high and low values for a factor are physical limitations, then axial runs will not be possible outside of those values. In that case, two options for creating the design are to specify high and low values closer to the center that result in axial runs at the physical limits or to set both the factorial runs and the axial runs at the high and low values. A response surface design in which both the factorial and the axial runs are set at the high and low values is known as a *face-centered design*.

The *Box-Behnken design* is an alternative to a central composite design when three or more factors are of interest. Box-Behnken designs utilize runs where some factors are tested at their high or low level while others are held at their

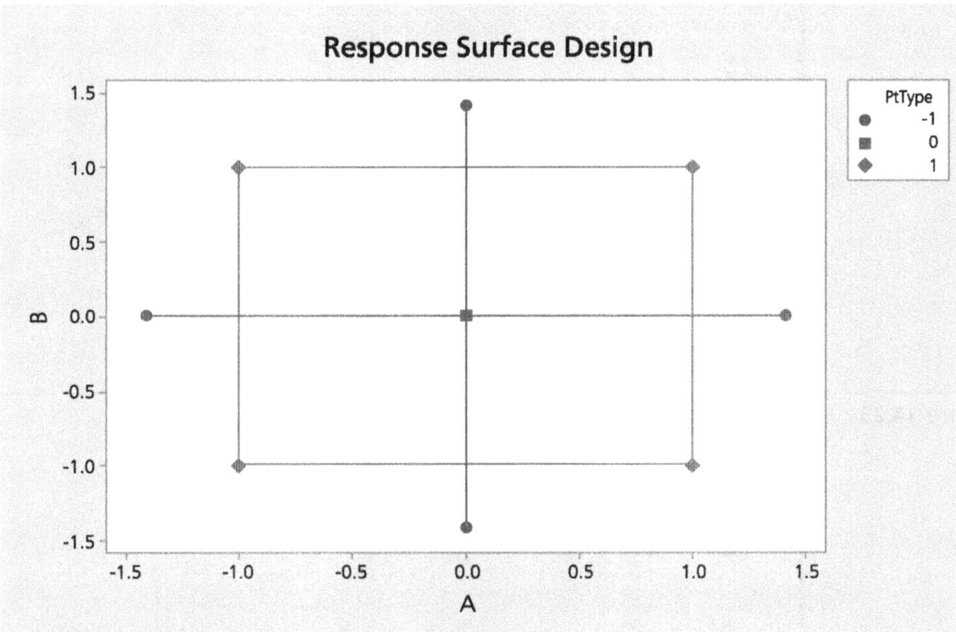

Figure 14.22 2-factor central composite design.

midpoint. They require less runs than a central composite design but do not contain an embedded factorial design and therefore cannot be used for sequential experimentation. They are useful when axial runs fall outside of safe or reasonable operating conditions or when it is not feasible to simultaneously set all factors at their low and/or high levels. For example, a chemical reaction vessel may not be safely operated at high temperature, high pressure, high amount of catalyst, and low mixing speed, but a Box-Behnken design would ensure that in any run only two of those factors would be set to the extreme levels.

Example

A commercial bakery would like to determine the optimal conditions in which to transport bread in order to preserve it as long as possible. The two critical factors are known to be temperature and humidity, and additionally the bakery wants to test whether white and wheat bread require different conditions. The delivery trucks typically use temperatures between 15°C and 25°C and keep humidity between 40–60%, but in both cases they are capable of values well outside of the normal operating range.

In Minitab: Go to **Stat > DOE > Response Surface > Create Response Surface Design** and complete the dialog as shown in Figure 14.23 and click **OK**.

Next, click on **Designs** and complete the dialog as shown in Figure 14.24 and click **OK**. Note that if desired, a face-centered design could have been selected in the Designs subdialog.

Click on **Factors** and complete the dialog as shown in Figure 14.25, and then click **OK** in each dialog.

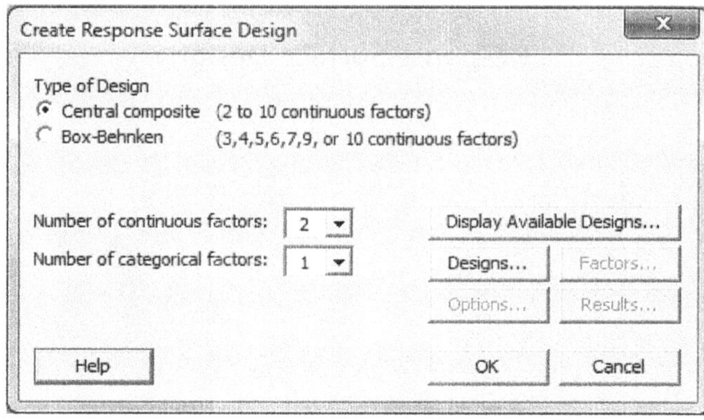

Figure 14.23 Minitab dialog box to create a response surface design.

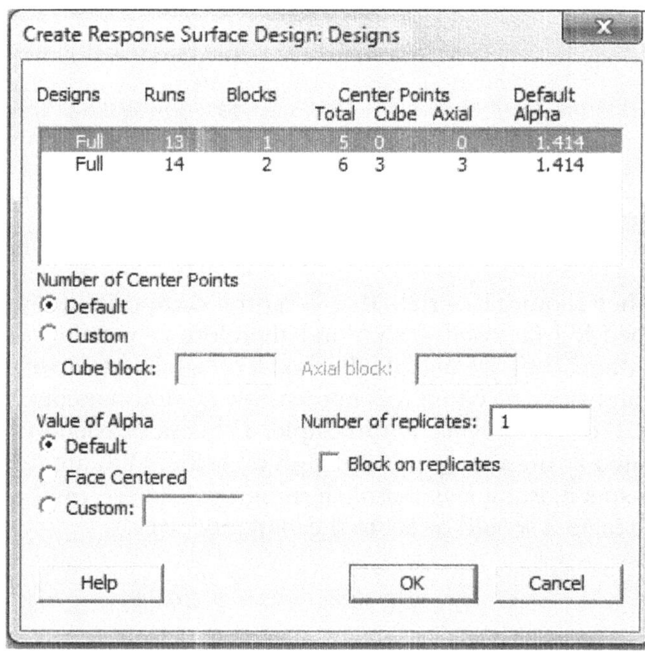

Figure 14.24 Design subdialog box to a create response surface design.

The design for the bread freshness example is shown in Figure 14.26 and has been randomized.

Data are collected by putting a loaf of bread under the specified conditions for each run and measuring the number of days until the loaf has spoiled. The unrandomized design and collected response values are found in Table 14.5.

In Minitab: Go to **Stat > DOE > Response Surface > Analyze Response Surface Design** and complete the dialog as shown in Figure 14.27; click **OK**.

The resulting one-way ANOVA table in the session window is shown in Figure 14.28.

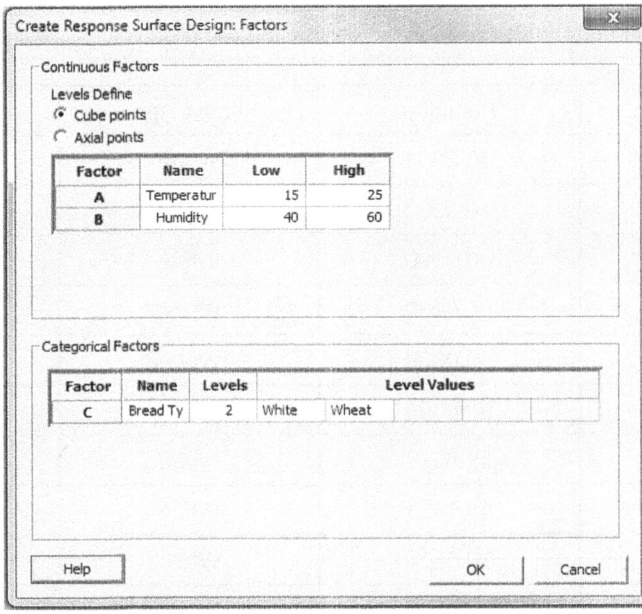

Figure 14.25 Factors subdialog box to create a response surface design.

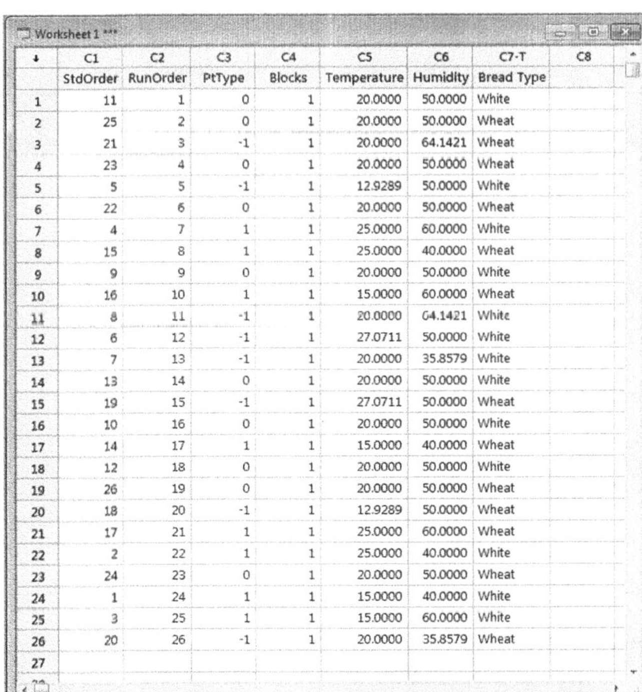

Worksheet 1 ***

↓	C1	C2	C3	C4	C5	C6	C7-T	C8
	StdOrder	RunOrder	PtType	Blocks	Temperature	Humidity	Bread Type	
1	11	1	0	1	20.0000	50.0000	White	
2	25	2	0	1	20.0000	50.0000	Wheat	
3	21	3	-1	1	20.0000	64.1421	Wheat	
4	23	4	0	1	20.0000	50.0000	Wheat	
5	5	5	-1	1	12.9289	50.0000	White	
6	22	6	0	1	20.0000	50.0000	Wheat	
7	4	7	1	1	25.0000	60.0000	White	
8	15	8	1	1	25.0000	40.0000	Wheat	
9	9	9	0	1	20.0000	50.0000	White	
10	16	10	1	1	15.0000	60.0000	Wheat	
11	8	11	-1	1	20.0000	64.1421	White	
12	6	12	-1	1	27.0711	50.0000	White	
13	7	13	-1	1	20.0000	35.8579	White	
14	13	14	0	1	20.0000	50.0000	White	
15	19	15	-1	1	27.0711	50.0000	Wheat	
16	10	16	0	1	20.0000	50.0000	White	
17	14	17	1	1	15.0000	40.0000	Wheat	
18	12	18	0	1	20.0000	50.0000	White	
19	26	19	0	1	20.0000	50.0000	Wheat	
20	18	20	-1	1	12.9289	50.0000	Wheat	
21	17	21	1	1	25.0000	60.0000	Wheat	
22	2	22	1	1	25.0000	40.0000	White	
23	24	23	0	1	20.0000	50.0000	Wheat	
24	1	24	1	1	15.0000	40.0000	White	
25	3	25	1	1	15.0000	60.0000	White	
26	20	26	-1	1	20.0000	35.8579	Wheat	
27								

Figure 14.26 Randomized central composite design for bread freshness.

Table 14.5 Unrandomized response surface design and response values.

Temperature	Humidity	Bread type	Days
15.0000	40.0000	White	6.1
25.0000	40.0000	White	7.1
15.0000	60.0000	White	5.4
25.0000	60.0000	White	5.1
12.9289	50.0000	White	8.2
27.0711	50.0000	White	8.0
20.0000	35.8579	White	6.0
20.0000	64.1421	White	3.4
20.0000	50.0000	White	8.7
20.0000	50.0000	White	7.7
20.0000	50.0000	White	8.8
20.0000	50.0000	White	7.8
20.0000	50.0000	White	9.3
15.0000	40.0000	Wheat	7.0
25.0000	40.0000	Wheat	8.2
15.0000	60.0000	Wheat	5.5
25.0000	60.0000	Wheat	4.2
12.9289	50.0000	Wheat	7.3
27.0711	50.0000	Wheat	8.1
20.0000	35.8579	Wheat	6.6
20.0000	64.1421	Wheat	4.1
20.0000	50.0000	Wheat	9.0
20.0000	50.0000	Wheat	8.1
20.0000	50.0000	Wheat	9.4
20.0000	50.0000	Wheat	9.1
20.0000	50.0000	Wheat	8.9

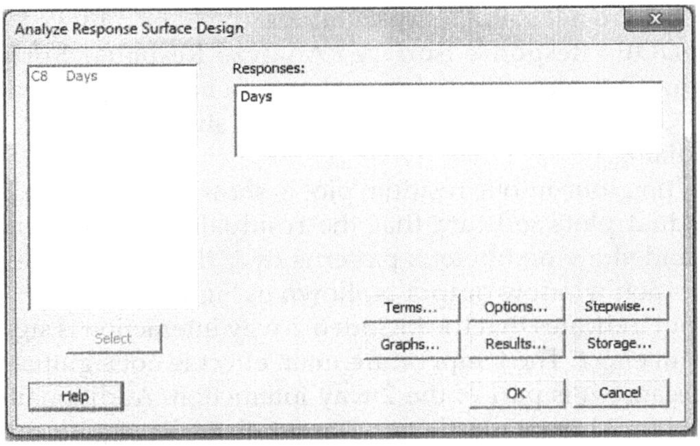

Figure 14.27 Minitab dialog box to analyze a response surface design.

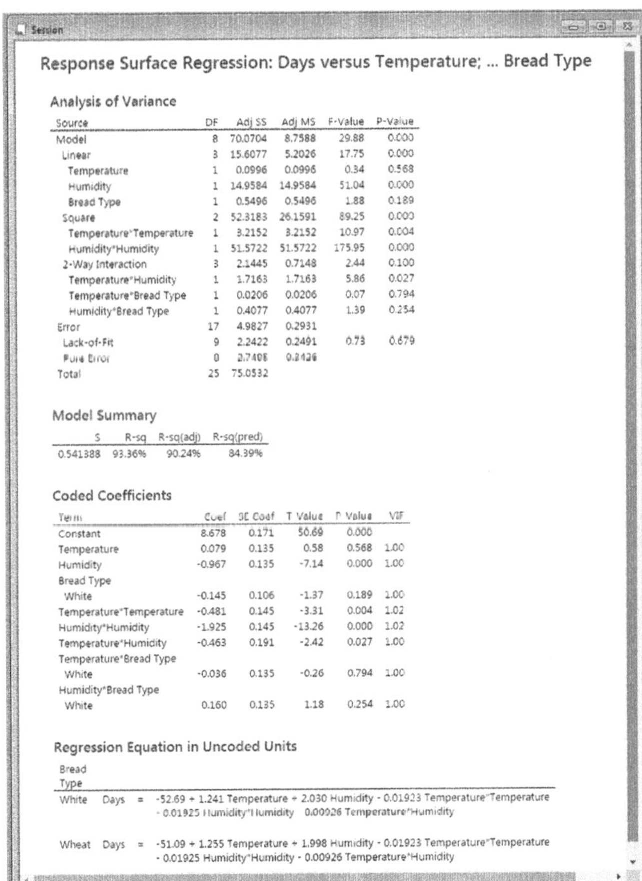

Session

Response Surface Regression: Days versus Temperature; ... Bread Type

Analysis of Variance

Source	DF	Adj SS	Adj MS	F-Value	P-Value
Model	8	70.0704	8.7588	29.88	0.000
Linear	3	15.6077	5.2026	17.75	0.000
Temperature	1	0.0996	0.0996	0.34	0.568
Humidity	1	14.9584	14.9584	51.04	0.000
Bread Type	1	0.5496	0.5496	1.88	0.189
Square	2	52.3183	26.1591	89.25	0.000
Temperature*Temperature	1	3.2152	3.2152	10.97	0.004
Humidity*Humidity	1	51.5722	51.5722	175.95	0.000
2-Way Interaction	3	2.1445	0.7148	2.44	0.100
Temperature*Humidity	1	1.7163	1.7163	5.86	0.027
Temperature*Bread Type	1	0.0206	0.0206	0.07	0.794
Humidity*Bread Type	1	0.4077	0.4077	1.39	0.254
Error	17	4.9827	0.2931		
Lack-of-Fit	9	2.2422	0.2491	0.73	0.679
Pure Error	8	2.7405	0.3426		
Total	25	75.0532			

Model Summary

S	R-sq	R-sq(adj)	R-sq(pred)
0.541388	93.36%	90.24%	84.39%

Coded Coefficients

Term	Coef	SE Coef	T Value	P Value	VIF
Constant	8.678	0.171	50.69	0.000	
Temperature	0.079	0.135	0.58	0.568	1.00
Humidity	-0.967	0.135	-7.14	0.000	1.00
Bread Type					
White	-0.145	0.106	-1.37	0.189	1.00
Temperature*Temperature	-0.481	0.145	-3.31	0.004	1.02
Humidity*Humidity	-1.925	0.145	-13.26	0.000	1.02
Temperature*Humidity	-0.463	0.191	-2.42	0.027	1.00
Temperature*Bread Type					
White	-0.036	0.135	-0.26	0.794	1.00
Humidity*Bread Type					
White	0.160	0.135	1.18	0.254	1.00

Regression Equation in Uncoded Units

Bread Type			
White	Days	=	-52.69 + 1.241 Temperature + 2.030 Humidity - 0.01923 Temperature*Temperature - 0.01925 Humidity*Humidity 0.00926 Temperature*Humidity
Wheat	Days	=	-51.09 + 1.255 Temperature + 1.998 Humidity - 0.01923 Temperature*Temperature - 0.01925 Humidity*Humidity - 0.00926 Temperature*Humidity

Figure 14.28 ANOVA table for unreduced response surface model.

Factors that are not statistically significant must be removed from the model. Go to **Stat > DOE > Response Surface > Analyze Response Surface Design** and click on **Terms**. Complete the dialog as shown in Figure 14.29 and click **OK**.

Click on **Graphs** and complete the dialog as shown in Figure 14.30. Then click **OK** in each dialog.

The resulting four-in-one residual plot is shown in Figure 14.31.

The residual plots indicate that the residuals are approximately normally distributed and show no obvious patterns over the fitted values or observation order. The session window output is shown in Figure 14.32.

The output indicates that the included 2-way interaction is significant, as is the humidity main effect. The temperature main effect is not significant but is kept in the model because it is part of the 2-way interaction. Additionally, both the temperature and humidity quadratic terms are statistically significant.

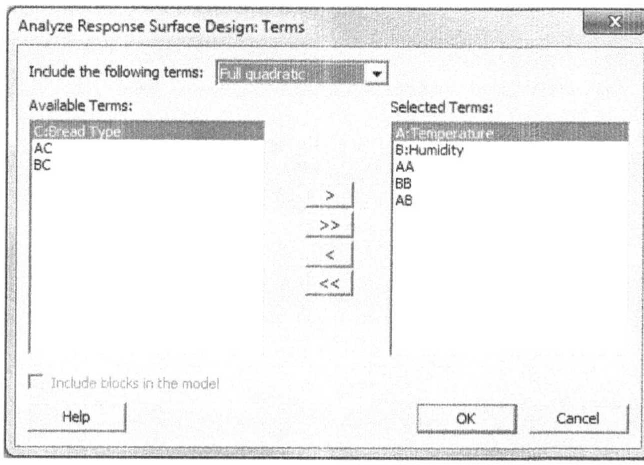

Figure 14.29 Terms subdialog box to analyze a response surface design.

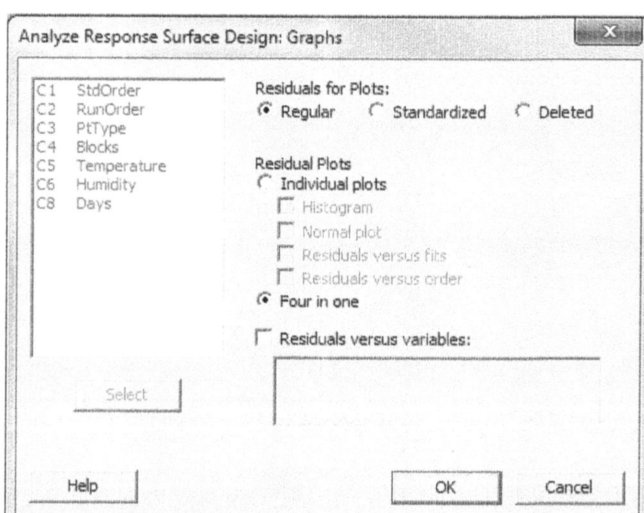

Figure 14.30 Graphs subdialog box to analyze a response surface design.

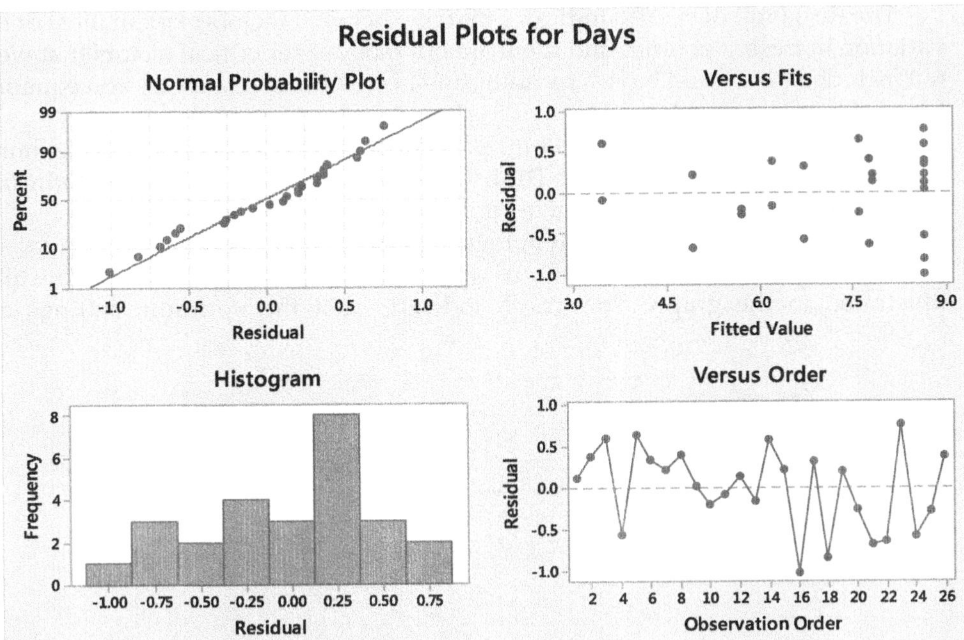

Figure 14.31 Residual plot for response surface model.

Session

Response Surface Regression: Days versus Temperature; Humidity

Analysis of Variance

Source	DF	Adj SS	Adj MS	F-Value	P-Value
Model	5	69.0926	13.8185	46.37	0.000
Linear	2	15.0580	7.5290	25.26	0.000
Temperature	1	0.0996	0.0996	0.33	0.570
Humidity	1	14.9584	14.9584	50.19	0.000
Square	2	52.3183	26.1591	87.77	0.000
Temperature*Temperature	1	3.2152	3.2152	10.79	0.004
Humidity*Humidity	1	51.5722	51.5722	173.04	0.000
2-Way Interaction	1	1.7163	1.7163	5.76	0.026
Temperature*Humidity	1	1.7163	1.7163	5.76	0.026
Error	20	5.9606	0.2980		
Lack-of-Fit	12	3.2200	0.2683	0.78	0.661
Pure Error	8	2.7405	0.3426		
Total	25	75.0532			

Model Summary

S	R-sq	R-sq(adj)	R-sq(pred)
0.545921	92.06%	90.07%	87.01%

Coded Coefficients

Term	Coef	SE Coef	T-Value	P-Value	VIF
Constant	8.678	0.173	50.27	0.000	
Temperature	0.079	0.136	0.58	0.570	1.00
Humidity	-0.967	0.136	-7.08	0.000	1.00
Temperature*Temperature	-0.481	0.146	-3.28	0.004	1.02
Humidity*Humidity	-1.925	0.146	-13.15	0.000	1.02
Temperature*Humidity	-0.463	0.193	-2.40	0.026	1.00

Regression Equation in Uncoded Units

Days = -51.89 + 1.248 Temperature + 2.014 Humidity - 0.01923 Temperature*Temperature
- 0.01925 Humidity*Humidity - 0.00926 Temperature*Humidity

Figure 14.32 Session window output for response surface model.

The R-sq (adj) of 90.07% indicates that the included factors explain most of the variation in freshness times and there are not likely other critical factors that were not included but would have a meaningful effect. If desired, a regression equation for predicting the number of days of freshness is shown in the output.

To find the settings of temperature and humidity that are predicted to maximize the days of freshness, go to **Stat>DOE>Response Surface>Response Optimizer** and complete the dialog as shown in Figure 14.33 and click **OK.**

The resulting graph is shown in Figure 14.34.

The quadratic relationships are apparent from the curved relationships illustrated in the graph. The graph indicates that the optimum settings are

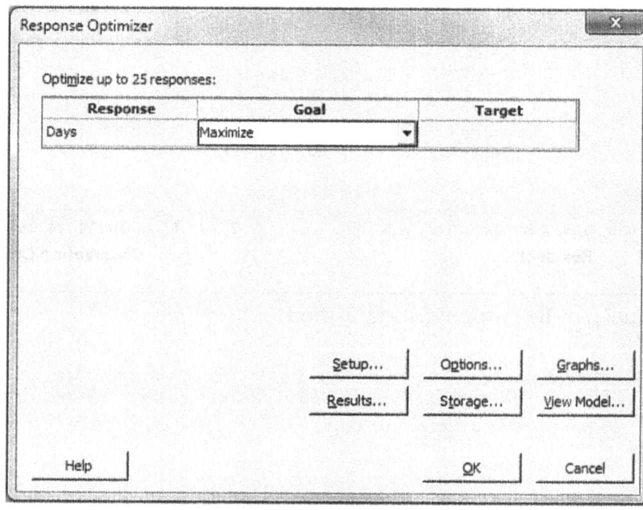

Figure 14.33 Minitab dialog box for response optimizer.

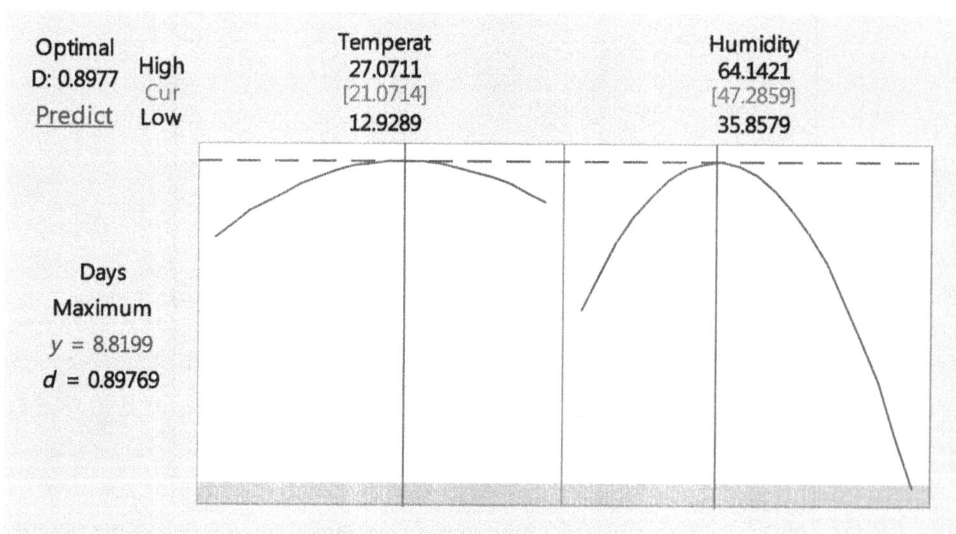

Figure 14.34 Response optimizer graph for response surface design.

Session

Response Optimization: Days

Parameters

Response	Goal	Lower	Target	Upper	Weight	Importance
Days	Maximum	3.37712	9.44026		1	1

Solution

Solution	Temperature	Humidity	Days Fit	Composite Desirability
1	21.0714	47.2859	8.81991	0.897686

Multiple Response Prediction

Variable	Setting
Temperature	21.0714
Humidity	47.2859

Response	Fit	SE Fit	95% CI	95% PI
Days	8.820	0.170	(8.466; 9.174)	(7.627; 10.012)

Optimization Plot

Figure 14.35 Session window output for response optimizer.

estimated to be a temperature of 21.1°C and a humidity of 47.3%. The predicted days of freshness at these settings is 8.8. The session window output is shown in Figure 14.35.

Aside from providing the estimated optimum settings and predicted number of days, the session window output also provides two intervals. A 95% confidence interval on the true mean number of days at the optimum settings is 8.466 to 9.174. This indicates that if these settings are used we are 95% confident that the true mean number of days expected over time falls somewhere in that range. A provided 95% prediction interval of 7.627 to 10.012 indicates that at the optimal settings we expect 95% of loaves to last a number of days within that range.

MIXTURE DESIGNS

Mixture designs are another type of designed experiment in which at least two factors are mixed in some proportion. A special design is required because increasing the amount of one component of the mixture means the other(s) must be decreased. Aside from the components of the mixture, other process factors such as temperature or presence of a catalyst may be included in the design. An example of a situation in which a mixture design would be required is the production of a beverage made with water, high fructose corn syrup, and a flavoring additive. Rather than requiring raw amounts of each component, a mixture design would allow each to be varied between specified high and low proportions while maintaining an overall proportion of 100% when all three ingredients are combined.

QUESTIONS AND ANSWERS

Q: What are the advantages and disadvantages of factorial DOEs when compared to the response surface method?

A: Factorial designs use fewer runs to estimate the coefficients of main effects and interactions and to test for statistical significance. However, unlike response surface designs, they are not appropriate for estimating quadratic terms and therefore may not identify optimal conditions if continuous factors have a relationship with the response that includes curvature.

Q: Is the factorial design the most useful type of DOE?

A: That depends on the desired outcome of the experiment. Factorial designs are efficient at identifying statistically significant factors, including main effects and interactions, but when curvature exists in relationships with continuous factors, factorial designs are not as effective at determining optimal conditions. The appropriate experimental design should be chosen based on the goals of the experiment.

Q: Should I do a Taguchi DOE?

A: No. Taguchi methods use the signal-to-noise ratio in place of ANOVA, and authors such as Box, Bisgaard, and Fung (1988) are critical of this method.

Q: What do I do if I have too many factors and I don't know which ones are relevant?

A: A smaller experimental design known as a screening design can first be performed to determine statistically significant main effects. Then a second experimental design can be created for those significant effects that allows for optimization.

Q: When would I use a response surface design?

A: Response surface designs are desirable when one or more continuous factors are suspected to have quadratic relationships with the response.

Chapter 15
Reliability

Reliability involves the study of when failures occur and can be used to quantify the reliability of parts (Lawson and Erjavec 2001). Typically, this is done by looking at the distribution of the amount of time or some other stress factor until the failure occurs. For example, you might be interested in the time until light bulbs burn out or the kilometers until tire tread wears too thin. Stress factors may be a force applied, temperature, voltage, or some other mechanism known to cause failure during use. Reliability analysis is useful for determining things such as the best warranty duration, service schedules, or recommended replacement interval.

CENSORING

Very often the stress factor of interest is time or some factor requiring a lengthy amount of time. While many parts may be monitored to perform a reliability study on, it can be impractical to wait until all parts have failed to complete the study. The data are called censored when the time to failure is unknown; censored data may be right censored, interval censored, or left censored (O'Connor and Kleyner 2012). To analyze the failures that have happened while not losing the information about those parts that have not yet failed, the remaining parts may be set up as right-censored data.

Right-censored data are data in which we know a left bound but not the true value. For example, we know that the true value is a value greater than 10,000 if an engine belt has been operating continuously for 10,000 hours without failure. The failure time for that belt is treated as a right-censored value. A study of such belts in which some have failed while others have not is shown in Table 15.1.

Another common situation is that the exact failure times are unknown, but they are known to fall within some interval. For example, suppose that every 1000 hours of operation a pump is inspected for O-ring leakage. At 21,000 hours no leakage was detected, but at 22,000 hours it was. While the exact number of hours until the leakage began is not known, it is known to fall somewhere between 21,000 and 22,000 hours. This type of data is known as *arbitrarily censored data*. An example of arbitrarily censored data for O-ring leakage is shown in Table 15.2. Notice that the last row of data indicates right-censoring as the interval contains a lower bound but no upper bound because the test was stopped before a failure occurred.

Table 15.1 Right-censored data.

Failure time (hrs)	Censored?
7374	No
9366	No
6796	No
7782	No
8011	No
7816	No
10000	Yes
10000	Yes
10000	Yes
10000	Yes

Table 15.2 Arbitrarily censored data.

Start time	End time	Count
0	5000	2
5000	10000	13
10000	15000	31
15000	20000	20
20000	25000	11
25000	*	9

FAILURE TYPES

For a part or product, there are three phases of failure over time. The first phase is *early-life or infant mortality,* which refers to failures that occur because the part was not manufactured or assembled correctly. This failure rate is highest when the part is first used, decreases rapidly from there, and typically has a Weibull shape parameter of less than one. The next failure phase is *useful-life or random failure,* and these failures occur at a constant rate over time and typically have a Weibull shape parameter approximately equal to one. They usually have a low rate of occurrence. These failures are typically due to unknown or unexpected reasons. The third phase of failure is *wear-out failures,* and they occur later in the life of the part, have a rate that increases over time, and typically have a Weibull shape parameter greater than one. These failures are due to degradation of the part through use

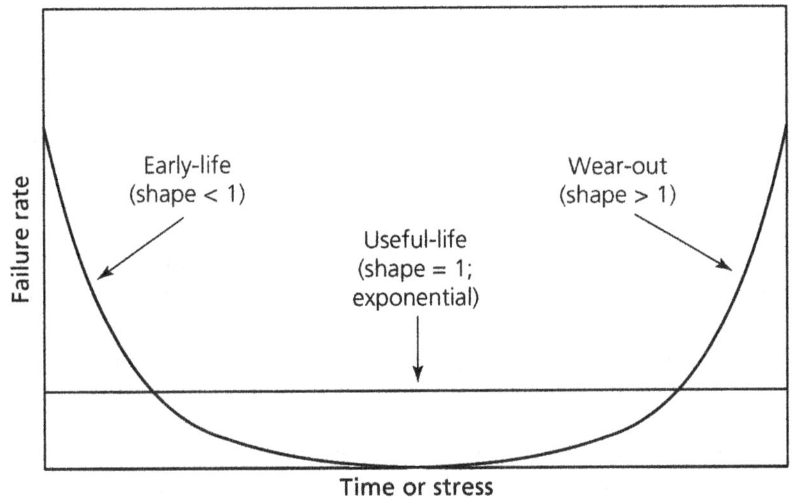

Figure 15.1 Failure rates for each failure type.

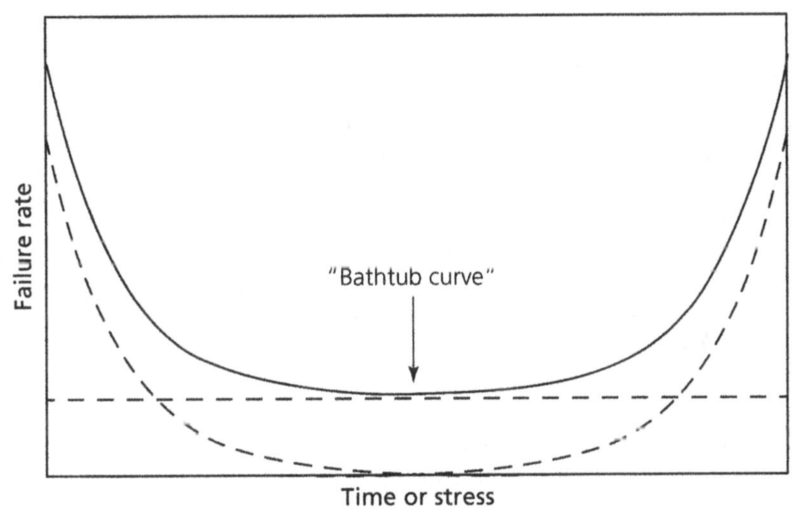

Figure 15.2 The bathtub curve.

(O'Connor and Kleyner 2012). The failure rates over time or stress for each failure type are shown in Figure 15.1.

These failure types occur in some combination for most parts produced but are not equally likely. When combined, the overall failure rate of the part may be described using the bathtub curve as shown in Figure 15.2.

Typically the rate of useful-life failures is low and a phase of constant failure rate may never be observed as infant-mortality failures are quickly replaced with wear-out failures. As the initial quality of a product improves, the rate of infant-mortality failures goes down and for many products wear-out failures may be the dominant source. Therefore, the bathtub curve may be useful for visualizing the failure types, but it does not usually represent the overall failure rate of the part over time.

FAILURE MODES

Certain parts only have a single mechanism in which they fail in use. For example, if light bulbs are tested over time, most likely the only way in which failure would be observed is a burnout where the filament inside fails. This is known as the failure mode. The failure mode is the way in which something fails and not the cause of failure that resulted in the failure mode. One failure mode could have multiple causes.

Other parts may have multiple failure modes that are observed. For example, tread wear-out and blowouts may both be observed as failure modes if tires are observed over time. In both the light bulb case and the tire case, other failure modes are almost certainly possible, but a study would typically be limited to those that are observed under reasonable conditions or have been observed in use by customers.

It is important that the failure mode also be collected if multiple modes occur when data are collected on failure times or stresses. The distribution of failure times is almost certainly different for each failure mode and each must be analyzed separately and then combined into an overall distribution of failure times.

DISTRIBUTION IDENTIFICATION

The most important tool in reliability is distribution analysis, which requires the likely distribution of failure times or stress to be determined. The Weibull distribution is the most commonly used, and in fact some reliability practitioners refer to reliability analysis as Weibull analysis. However, other distributions may occur in certain situations. An overview of common distributions and the situations in which they are most likely to occur is found in Chapter 4. Further, an overview of testing for distribution fit is found in Chapter 7. Both concepts are important in distribution analysis.

Statistical software such as Minitab makes it easy to formally test a data set for how well it fits a large number of distributions. Rather than this approach, the nature of the process should be considered and only those distributions that would make intuitive sense should be examined. Probability plots should be made for each distribution of interest and the best fit should be determined from those for each failure mode. The distribution that makes the most intuitive sense should be assumed if more than one distribution shows a good fit. Weibull is most commonly the best fit for time or stress until failure.

Example

A device that tests the puncture resistance of cans uses increasing pressure. The force necessary to puncture the top of the can is recorded for 20 cans. The data are shown in Table 15.3 and contain no censored values.

In Minitab: **Go to Stat > Reliability/Survival > Distribution Analysis (Right Censoring) > Distribution ID Plot**. When no censoring exists, the dialogs for right-censored data should be used. For most reliability experiments such as this, the most intuitive distributions to examine are Weibull, 3-parameter Weibull, normal, and lognormal. Select the column containing the data, as shown in Figure 15.3, and click **OK**.

Table 15.3 Puncture forces.

Puncture force	
19	63
108	28
10	33
71	27
30	49
24	11
53	66
21	6
67	85
43	70

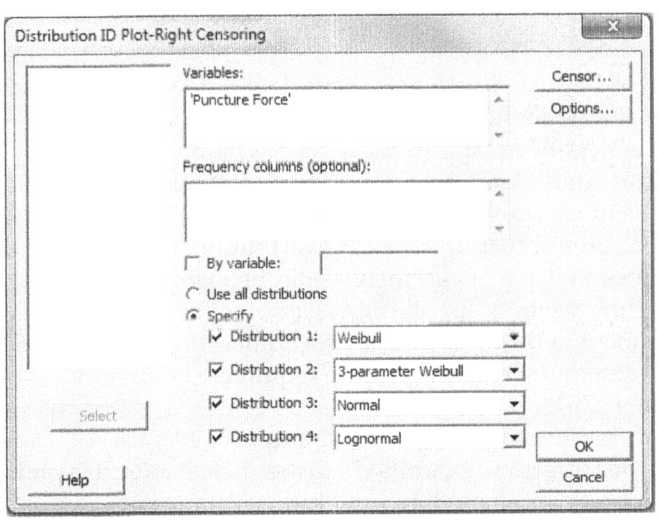

Figure 15.3 Dialog box for distribution ID.

The probability plots for the selected distributions are shown in Figure 15.4.

Visually, the Weibull distribution appears to provide the best fit of the data. Any remaining analysis should assume that the data come from a Weibull-distributed population, since we expect it would be the most likely distribution before the data were collected. In addition, the graph displays the Anderson-Darling (adjusted) value for each distribution. This is the statistic and not the p-value; lower values indicate a better fit. The tabular data given in the session window in Figure 15.5 shows that Weibull was the closest fit based on having the lowest Anderson-Darling statistic and that the 3-parameter Weibull could also

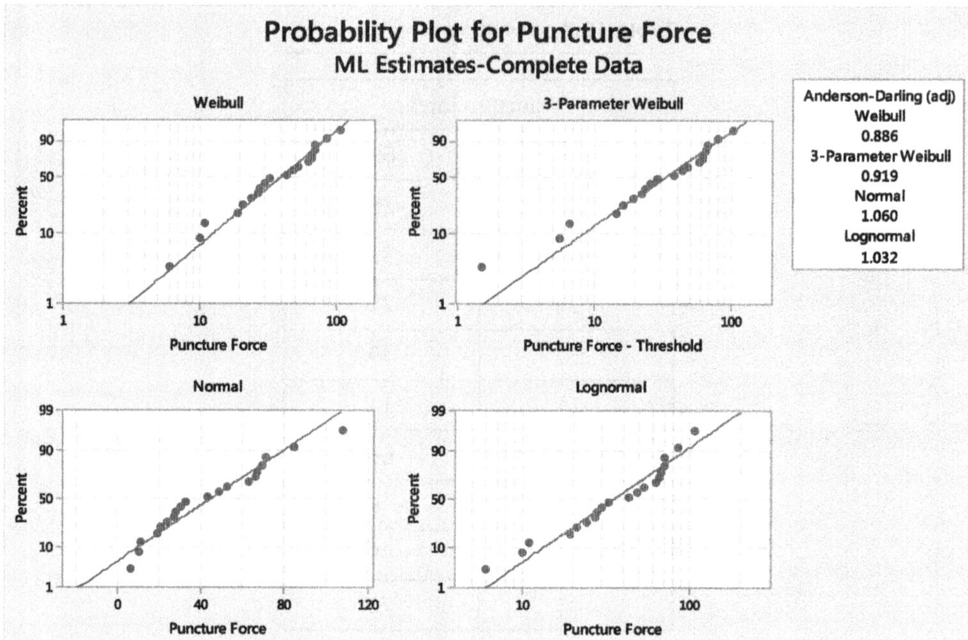

Figure 15.4 Graphical output for distribution ID.

be considered because it has a similarly low value. Minitab also provides further information in the session window.

The goodness-of-fit table in the session window provides the same information as the table of Anderson-Darling (adjusted) statistics on the probability plots. The table of percentiles provides information on which values correspond to a certain percent of failures occurring for each distribution. For example, the first line in the table indicates that 1% of can tops would be expected to puncture under a force of 3.11987 if the Weibull distribution is assumed. It also provides a standard error for this estimate and the 95% confidence interval on this value that indicates we are 95% confident the true mean force required to puncture 1% of cans is between 1.04906 and 9.27842.

This information is provided by default for 1%, 5%, 10%, and 50% of failures and for all distributions examined on the distribution ID plot. The 50% of failures corresponds to the median puncture resistance of cans, since half of all cans will fail at forces less than this value and half will fail at a force higher than this value.

For each distribution considered, the table of MTTF shows the mean force required to puncture a can as well as the standard error of that estimate and a confidence interval. The MTTF is a common reliability metric and stands for *mean time to failure*, although it is used not only for time but any stress used to examine the reliability of a part.

Session

Distribution ID Plot: Puncture Force

Goodness-of-Fit

Distribution	Anderson-Darling (adj)
Weibull	0.886
Lognormal	1.032
Exponential	1.870
Normal	1.060

Table of Percentiles

Distribution	Percent	Percentile	Standard Error	95% Normal CI Lower	Upper
Weibull	1	3.11987	1.73490	1.04906	9.27842
Lognormal	1	5.97157	1.94703	3.15177	11.3142
Exponential	1	0.444225	0.0993317	0.286595	0.688553
Normal	1	-19.0133	11.6968	-41.9387	3.91205
Weibull	5	8.30823	3.22479	3.88257	17.7786
Lognormal	5	10.0062	2.59952	6.01359	16.6496
Exponential	5	2.26716	0.506953	1.46268	3.51413
Normal	5	-0.495247	9.31985	-18.7618	17.7713
Weibull	10	12.8045	4.05374	6.88467	23.8145
Lognormal	10	13.1758	3.01156	8.41826	20.6222
Exponential	10	4.65693	1.04132	3.00446	7.21829
Normal	10	9.37668	8.19967	-6.69438	25.4477
Weibull	50	39.7187	6.29117	29.1186	54.1775
Lognormal	50	34.7816	5.89094	24.9564	48.4749
Exponential	50	30.6371	6.85067	19.7657	47.4878
Normal	50	44.2	6.07602	32.2912	56.1088

Table of MTTF

Distribution	Mean	Standard Error	95% Normal CI Lower	Upper
Weibull	44.2380	6.09793	33.7647	57.9600
Lognormal	46.3373	8.90291	31.7971	67.5265
Exponential	44.2000	9.88342	28.5159	68.5104
Normal	44.2000	6.07602	32.2912	56.1088

Figure 15.5 Session window output for distribution ID.

DISTRIBUTION OVERVIEW

Distribution analysis uses the fitted distribution of failure data to examine when items should be expected to fail. This allows us to make decisions on when to repair or replace parts and also tells us when to expect customers to experience failures. Additionally, warranty duration on products might be determined based on when a certain percentage of parts are expected to have failed.

Recall from Chapter 4 that the *probability density function* (PDF) f(t) of a distribution provides the relative frequencies of values occurring for a distribution. This enables us to see, for example, which range of values we expect to find most data between and how unusual certain values would be. Additionally, the *cumulative distribution function* (CDF) F(t) provides the probability of all values less than or equal to a value of interest. For example, the CDF would enable us to determine the probability of a tire blowout occurring in the first 20,000 miles of use.

When examining reliability, the inverse of the CDF is often of interest and is known as the *survival function* R(t). The formula is simply:

$$R(t)=1-F(t)$$

While the CDF represents the proportion of items expected to fail up to a certain time or stress, the survival function represents the proportion of items that have not yet failed at that point. Using the previous example, this would tell us the percentage of tires that do *not* experience a blowout in the first 20,000 miles.

Another useful measure of reliability is known as the *hazard function*. The hazard function is calculated as:

$$h(t)=\frac{f(t)}{R(t)}$$

The hazard function indicates when an item is more or less likely to fail over time. When plotted, a decreasing hazard function indicates the likelihood of failure decreases over time. A horizontal function indicates constant failure rate over time and an increasing function indicates an increasing rate over time. The curves shown for the three failure phases in Figure 15.1 from the beginning of this chapter are hazard functions. For example, infant mortality is described by a decreasing hazard function, indicating that over time a failure due to infant mortality becomes less likely.

The Weibull distribution exhibits a relationship between the shape parameter and the hazard function that can be useful in making quick assessments of the failure type a particular failure mode falls into. The relationship is outlined in Table 15.4.

Example

You wish to further examine the failure rates of the cans based on the puncture force data found in Table 15.3. Recall that the Weibull distribution was found to provide a good fit for the data.

In Minitab: Go to **Stat > Reliability/Survival > Distribution Analysis (Right Censoring) > Distribution Overview Plot**. Complete the dialog as shown in Figure 15.6 and click **OK**.

The output is shown in Figure 15.7.

The output includes a probability density function plot, Weibull plot, survival function plot, and a hazard function plot. In Figure 15.7, the probability density

Table 15.4 Relationship between shape parameter and hazard.

Shape parameter	Hazard function	Failure type
<1	Decreasing	Infant mortality
1	Constant	Random
>1	Increasing	Wear-out

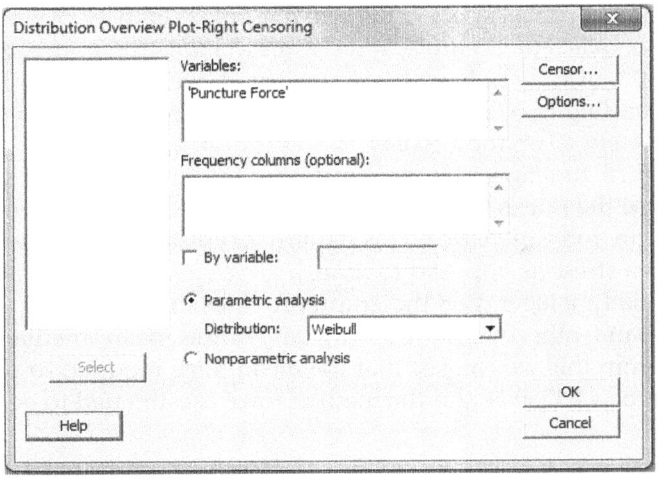

Figure 15.6 Dialog box for distribution overview plot.

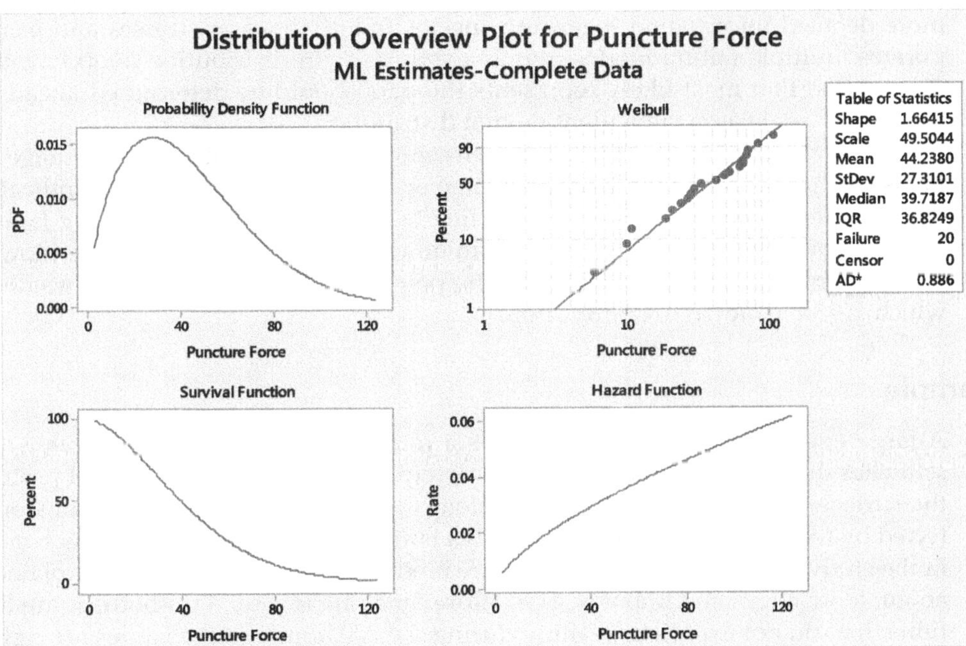

Figure 15.7 Output for distribution overview plot.

function plot illustrates the expected distribution of puncture forces required to puncture the can based on the Weibull distribution fit to the data. It shows that cans will most likely puncture with a force of somewhere between 0 and about 80 Newtons (N), as most of the area under the curve is covered by this area. However, the right tail of the distribution shows that some cans could require much greater force.

The Weibull plot is a probability plot to check for the distribution fit. This plot confirms that the data are well-fit by the Weibull distribution as the points fall along the fitted line with no obvious pattern. The survival function plot shows the percentage of parts expected to have not failed on the y-axis and the puncture force on the x-axis. For example, if we want to know the puncture force in which we expect 50% of cans to have been punctured, we can look for the point on the line corresponding to 50 on the y-axis. This point corresponds to a puncture force of approximately 40 on the x-axis.

The hazard function plot demonstrates the rate of failure versus puncture force. Because the rate is increasing, we can tell that the rate of punctures goes up as the puncture force increases. This pattern is typically the case when the variable of interest is a stress as opposed to time.

Additionally, a legend on the graph provides the estimated parameters of the distribution and other characteristics such as the mean, median, and standard deviation. From this we can see that the mean force required to puncture a can is estimated to be 44.2380 N and the median force is estimated to be 39.7187 N.

DISTRIBUTION ANALYSIS

A distribution analysis is similar to a distribution overview; however, it provides more detailed information regarding specific failure times or stresses and incorporates multiple failure modes into analysis. As with distribution overview, the distribution that most likely represents the data should be determined ahead of time as the analysis is dependent on that distribution.

In distribution analysis, the term *percent* is used to indicate the percentage of items that have failed by a particular time or stress. The term *percentile* indicates the estimated time or stress corresponding to the percent of items having failed. For example, one might want to determine when 1% of fuses will have blown; using distribution analysis, you find the percentile estimated to be the point in which 1% have blown to be 1832 hours.

Example

A large equipment manufacturer uses a plastic film to protect surfaces from scratches during loading and wants to determine how long the film will protect the surfaces before failing. A 200-hour-long study is performed on 50 items protected by the film, and the time in which a failure occurs is recorded for each item in the study. There are two known failure modes that customers have complained about: loose edges and bubbles. The failure times are recorded for both, as are the films that did not experience failure during the 200 hours, so the values are right-censored data. The data are shown in Table 15.5.

In Minitab: Go to **Stat > Reliability/Survival > Distribution Analysis (Right Censoring) > Parametric Distribution Analysis.** Complete the main dialog as shown in Figure 15.8 and click **OK.**

Click on **Censor**, complete the dialog as shown in Figure 15.9, and click **OK.**

Click on **FMode**, complete the dialog as shown in Figure 15.10, and click **OK.**

In the dialog for parametric distribution analysis (Figure 15.8), click on **Graphs**, then check **Survival plot** and **Hazard plot** (as shown in Figure 15.11) and click **OK** in each dialog.

The sessions window output contains a table of percentiles as shown in Figure 15.12.

Table 15.5 Data on protective film failures.

Failure time	Mode	Censored	Failure time	Mode	Censored	Failure time	Mode	Censored	Failure time	Mode	Censored
3	Loose Edge	No	36	Loose Edge	No	72	Loose Edge	No	153	Bubbles	No
6	Loose Edge	No	41	Bubbles	No	76	Loose Edge	No	157	Bubbles	No
6	Loose Edge	No	42	Bubbles	No	81	Loose Edge	No	165	Bubbles	No
6	Loose Edge	No	42	Loose Edge	No	93	Bubbles	No	169	Bubbles	No
6	Loose Edge	No	44	Loose Edge	No	101	Bubbles	No	179	Loose Edge	No
11	Bubbles	No	44	Loose Edge	No	101	Loose Edge	No	181	Bubbles	No
12	Loose Edge	No	45	Loose Edge	No	105	Bubbles	No	188	Bubbles	No
14	Loose Edge	No	46	Loose Edge	No	105	Loose Edge	No	190	Loose Edge	No
15	Bubbles	No	47	Loose Edge	No	106	Loose Edge	No	200	Bubbles	Yes
16	Loose Edge	No	50	Loose Edge	No	107	Loose Edge	No	200	Bubbles	Yes
17	Bubbles	No	55	Loose Edge	No	108	Bubbles	No	200	Bubbles	Yes
17	Loose Edge	No	56	Bubbles	No	109	Loose Edge	No	200	Bubbles	Yes
17	Loose Edge	No	57	Loose Edge	No	110	Bubbles	No	200	Bubbles	Yes
19	Bubbles	No	59	Bubbles	No	111	Loose Edge	No	200	Bubbles	Yes
21	Loose Edge	No	59	Loose Edge	No	118	Loose Edge	No	200	Bubbles	Yes
22	Loose Edge	No	60	Bubbles	No	122	Bubbles	No	200	Bubbles	Yes

Table 15.5 Data on protective film failures. (Continued)

Failure time	Mode	Censored	Failure time	Mode	Censored	Failure time	Mode	Censored	Failure time	Mode	Censored
23	Bubbles	No	60	Loose Edge	No	122	Bubbles	No	200	Bubbles	Yes
24	Bubbles	No	64	Bubbles	No	123	Bubbles	No	200	Bubbles	Yes
24	Loose Edge	No	65	Bubbles	No	124	Loose Edge	No	200	Bubbles	Yes
24	Loose Edge	No	65	Bubbles	No	126	Loose Edge	No	200	Bubbles	Yes
25	Bubbles	No	65	Bubbles	No	128	Bubbles	No	200	Bubbles	Yes
26	Bubbles	No	67	Loose Edge	No	130	Loose Edge	No	200	Bubbles	Yes
30	Bubbles	No	68	Loose Edge	No	131	Loose Edge	No	200	Loose Edge	Yes
30	Loose Edge	No	69	Loose Edge	No	138	Bubbles	No	200	Loose Edge	Yes
35	Bubbles	No	72	Bubbles	No	142	Loose Edge	No	200	Loose Edge	Yes

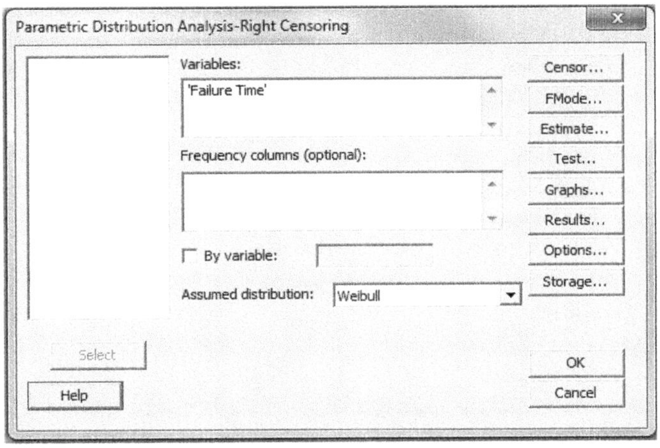

Figure 15.8 Dialog box for parametric distribution analysis: right censoring.

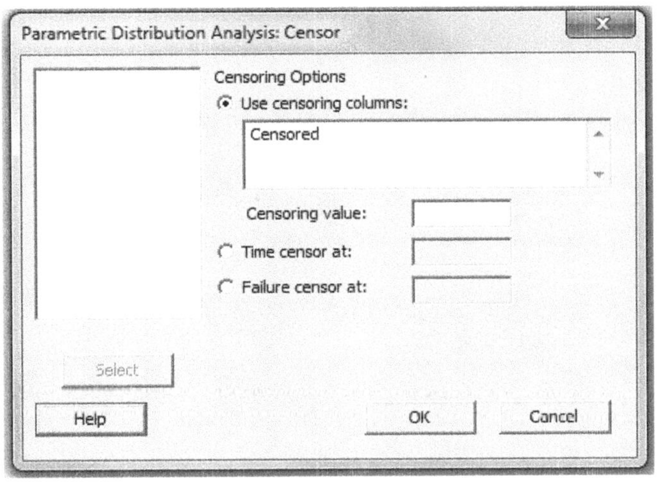

Figure 15.9 Censor subdialog box.

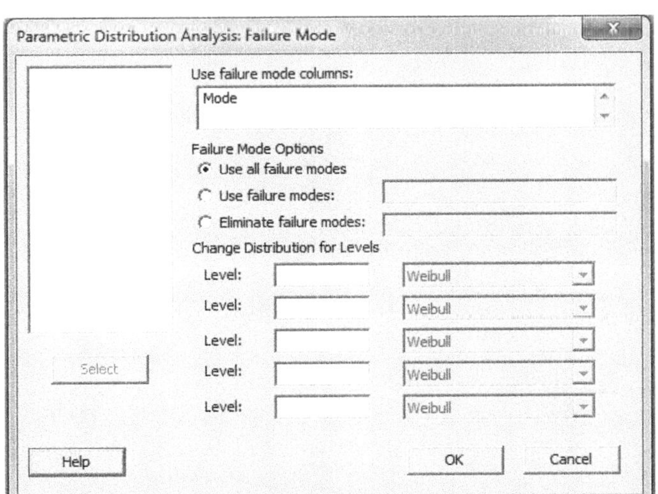

Figure 15.10 Failure mode subdialog box.

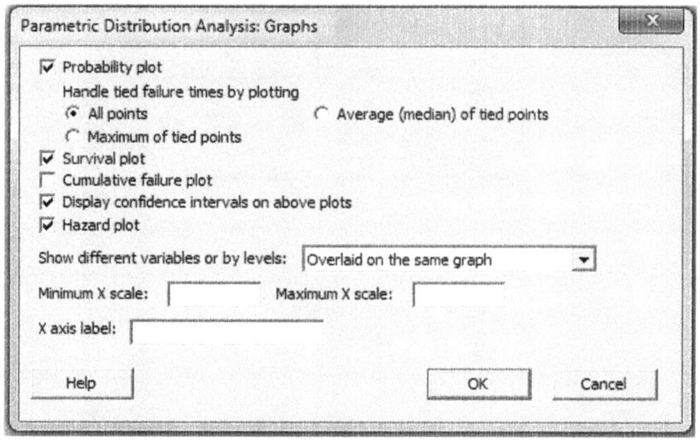

Figure 15.11 Graphs subdialog box.

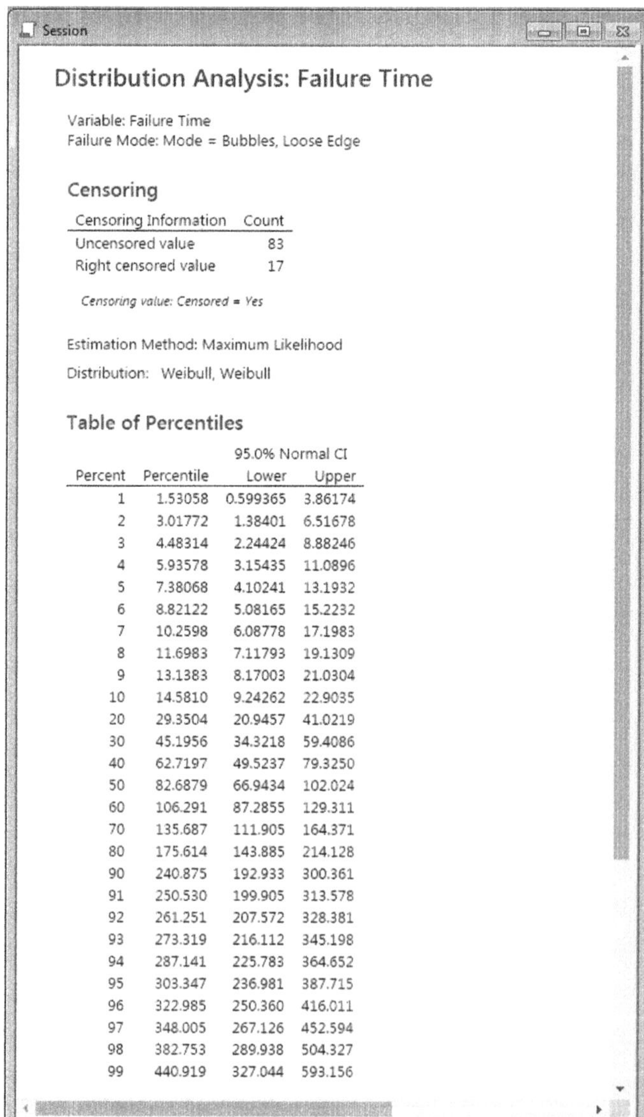

Session

Distribution Analysis: Failure Time

Variable: Failure Time
Failure Mode: Mode = Bubbles, Loose Edge

Censoring

Censoring Information	Count
Uncensored value	83
Right censored value	17

Censoring value: Censored = Yes

Estimation Method: Maximum Likelihood

Distribution: Weibull, Weibull

Table of Percentiles

		95.0% Normal CI	
Percent	Percentile	Lower	Upper
1	1.53058	0.599365	3.86174
2	3.01772	1.38401	6.51678
3	4.48314	2.24424	8.88246
4	5.93578	3.15435	11.0896
5	7.38068	4.10241	13.1932
6	8.82122	5.08165	15.2232
7	10.2598	6.08778	17.1983
8	11.6983	7.11793	19.1309
9	13.1383	8.17003	21.0304
10	14.5810	9.24262	22.9035
20	29.3504	20.9457	41.0219
30	45.1956	34.3218	59.4086
40	62.7197	49.5237	79.3250
50	82.6879	66.9434	102.024
60	106.291	87.2855	129.311
70	135.687	111.905	164.371
80	175.614	143.885	214.128
90	240.875	192.933	300.361
91	250.530	199.905	313.578
92	261.251	207.572	328.381
93	273.319	216.112	345.198
94	287.141	225.783	364.652
95	303.347	236.981	387.715
96	322.985	250.360	416.011
97	348.005	267.126	452.594
98	382.753	289.938	504.327
99	440.919	327.044	593.156

Figure 15.12 Table of percentiles output.

This table shows the time corresponding to commonly desired percents as well as a confidence interval on the true percentile. It is desirable to know how quickly items must be loaded after the film is applied to avoid too many failures. For example, from Table 15.2 we can see that is it estimated 1% of films will fail by 1.53058 hours while the 95% confidence interval indicates the true value is likely somewhere between 0.599365 hours and 3.86174 hours.

The analysis provides graphical output for each failure mode as well as the combined failure rates. Probability plots for the time until loose edges and the time until bubbles are shown in Figure 15.13; an overall probability plot is shown in Figure 15.14.

These plots show the individual failure modes were reasonably well-fit by the Weibull distribution as the points fall close to the centerline without obvious patterns. The combined probability plot on the bottom portion of the graph in Figure 15.14 is based on the Weibull distributions fit to the individual modes and also shows a reasonable fit.

Survival plots for each failure mode are shown in Figure 15.15.

Note that while both plots in Figure 15.15 are showing a similar pattern, the x-axes are different. Films will often go more hours without experiencing loose edges than bubbles. These are combined into an overall survival plot indicating the percentage of parts that did not experience any failure mode by a given time, as shown in Figure 15.16.

For example, looking across from where 50% would fall on the y-axis and then finding the point on the centerline corresponding to this, we can then look at the x-axis value for this point and see that 50% of film will experience some form of failure by about 80 hours.

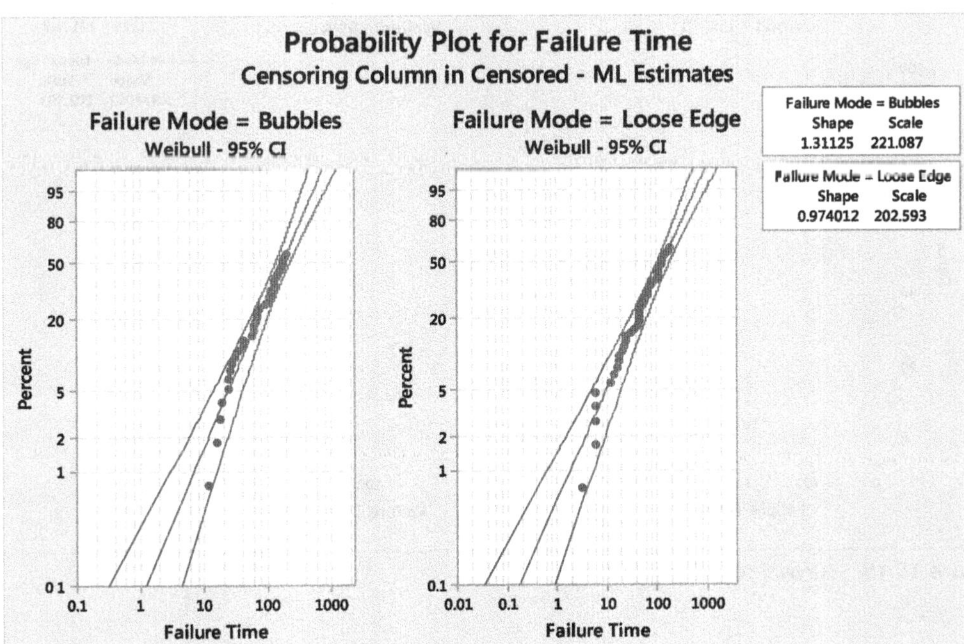

Figure 15.13 Probability plots for bubbles and loose edge.

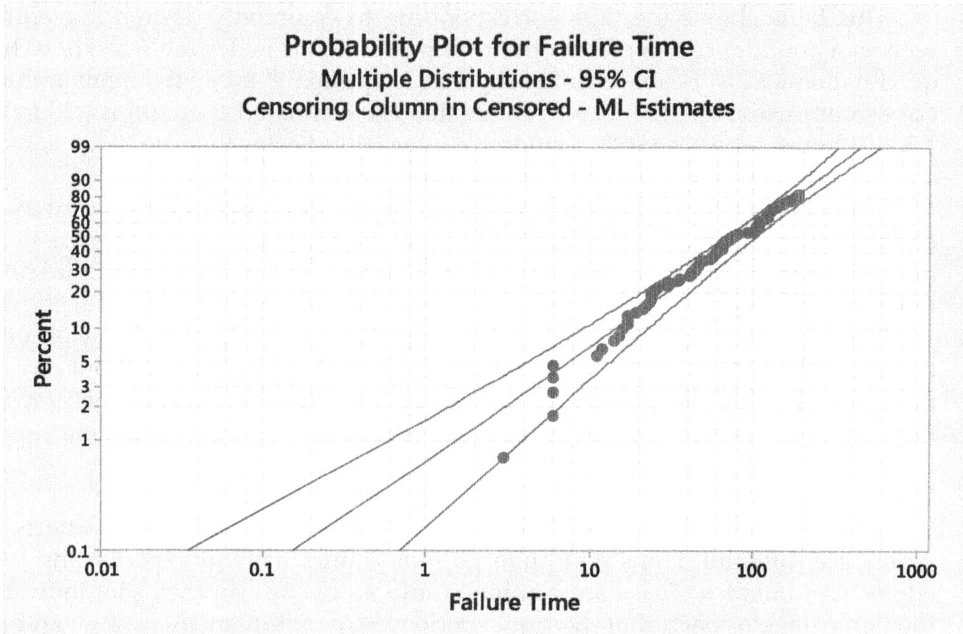

Figure 15.14 Probability plot for film.

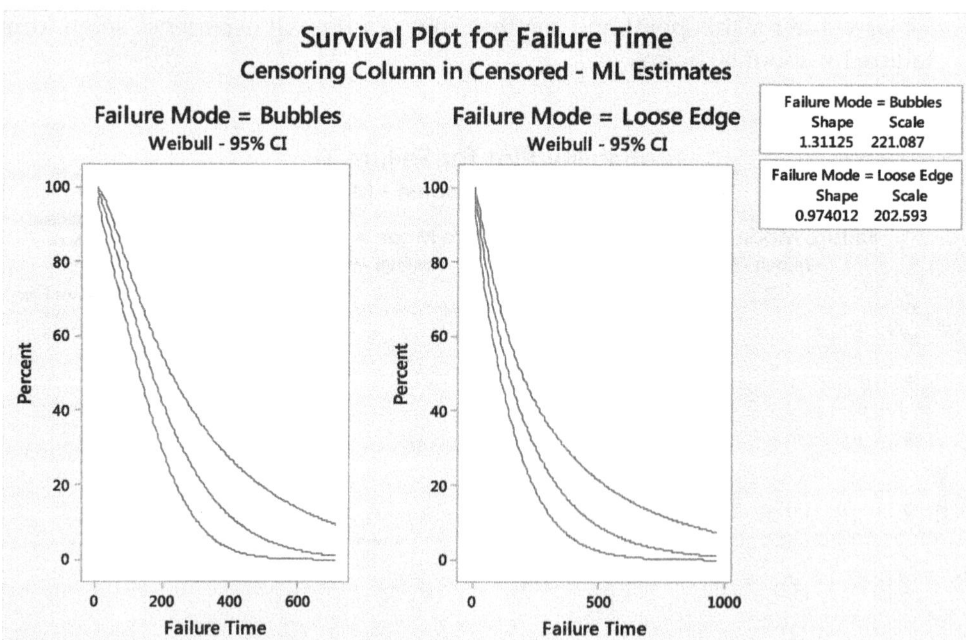

Figure 15.15 Survival plots by failure mode.

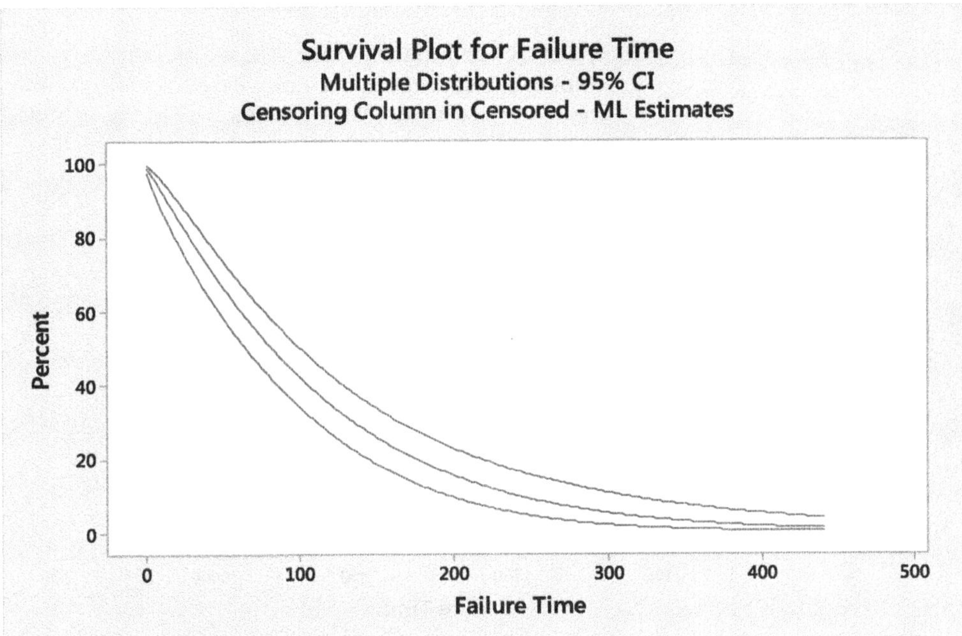

Figure 15.16 Combined survival plot.

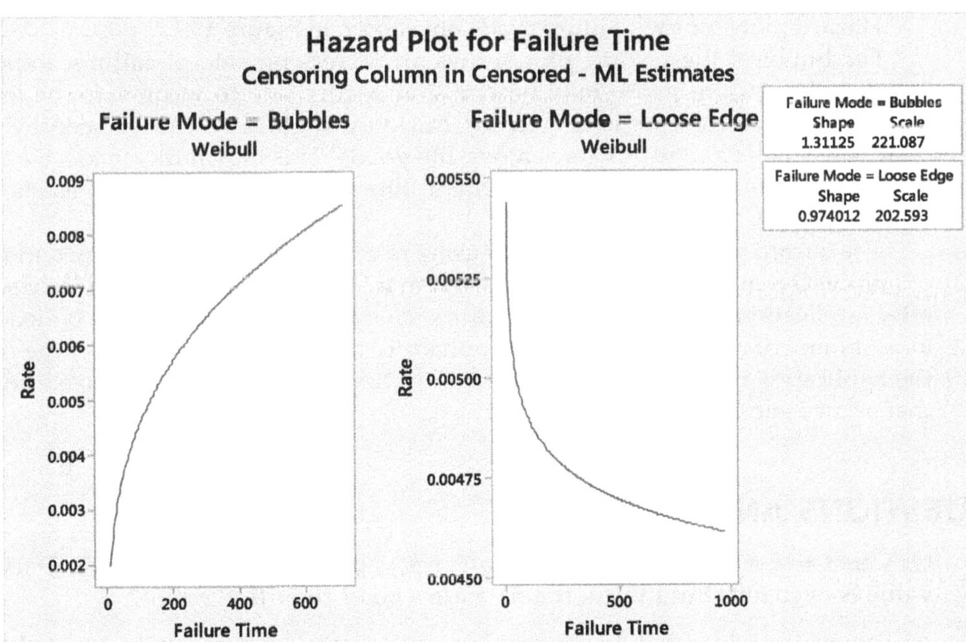

Figure 15.17 Hazard plots by failure mode.

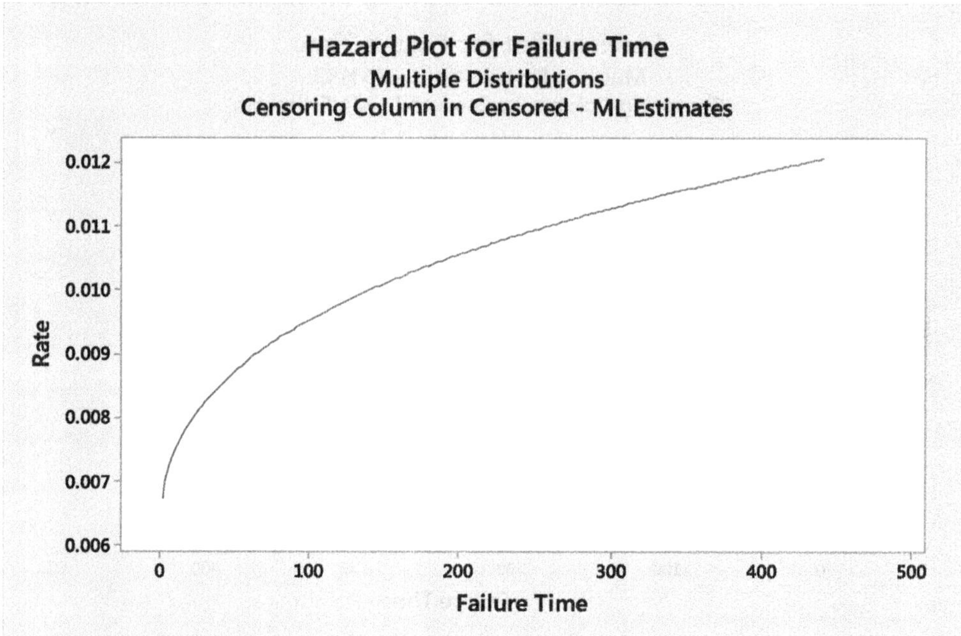

Figure 15.18 Hazard plot for film failure.

Hazard plots for each failure mode are shown in Figure 15.17.

For bubbles, the hazard plot shows an increasing rate of failure, indicating that this failure mode may be one that occurs due to wear-out. For loose edges, however, the rate is decreasing but only slightly, as can be seen by the relatively small change in values along the y-axis. This may indicate loose edges are random failures. These are combined into an overall hazard plot as shown in Figure 15.18.

The hazard plot illustrates an increasing rate of failure for the combined modes of failure. Overall, the study has indicated that failures begin occurring very soon after application and that without process changes, the items need to be loaded almost immediately after the film is applied. To improve the reliability of the film, the application method should be studied to find ways to reduce the loose edges that occur almost immediately.

QUESTIONS AND ANSWERS

Q: Can I use a Weibull distribution for an analysis if the Anderson-Darling value is over 2000 but I think the data are a good fit with the plot?

A: The magnitude of the Anderson-Darling statistic is not easily interpreted and should not be used to determine distribution fit. If the data fall randomly along the fitted line on a probability plot with no obvious patterns, then the distribution is a good fit for the data.

Q: For a 2-parameter Weibull analysis of reliability data, the shape parameter gives some indication of the type of failure mode. For instance, if the slope is >1, then the failure mode is wear-out; if the slope is <1, then the failure mode is "early life." Does the same interpretation apply for 3-parameter Weibull? What if the slope is <1 but the threshold/location is 1000 hours? What assumption might be made about this set of failure data?

A: The same interpretation generally applies for the 3-parameter Weibull as well. The additional parameter "shifts" the lower bound on the distribution from zero to some other value, typically due to a physical constraint. Often times the 3-parameter Weibull fits reliability data based on a stress factor rather than time.

Chapter 16
Other Quality Tools

Many simple yet highly effective quality tools were brought together in the book *Guide to Quality Control* (Ishikawa 1991). These tools can be used for the collection and assessment of data that would later be analyzed statistically as well as for supporting the generation of ideas during brainstorming sessions. These tools can be created on a dry erase board, spreadsheet program, or a presentation program. There are also options available to use some of these tools in Minitab.

CAUSE AND EFFECT DIAGRAM

A cause and effect diagram is also known as an *Ishikawa diagram* or a *fishbone diagram* (Hoerl and Snee 2009). The cause and effect diagram can be used to bring structure to a brainstorming session and to list potential factors that may have caused the problem that is being investigated. Figure 16.1 shows a cause and effect diagram for a diameter that is out of specification. The main branches of the cause and effect diagram represent categories of potential causes; in Figure 16.1, they are measurements, material, personnel, environment, methods, and machines. These are typical main branch labels; however, the labels can be changed based on the problem under investigation. The sub-branches then list the potential causal factors. In this example, the sub-branches under machines are broken down into twigs with further factors.

In Minitab: Enter the defect name and the quantity of items in the Minitab worksheet, as shown in Figure 16.2. Go to **Stat > Quality Tools > Cause-and-Effect** and click in the box under Causes and then select the column containing the items to list. An effect can also be listed (see Figure 16.3). Sublevels can be added by clicking on **Sub** on the right side of the dialog box. The sub-branches can then be added by clicking in Causes and selecting the desired column as shown in Figure 16.4. Click **OK** and the resulting cause and effect diagram is depicted in Figure 16.5.

PARETO CHART

The Pareto chart illustrates the 80/20 principle, where often about 80% of all problems are the result of about 20% of all causes (Juran 2005). The chart plots how often each problem has occurred, sorted from the most frequent to the least frequent. It typically also shows the percentage of all problems that each cause represents and

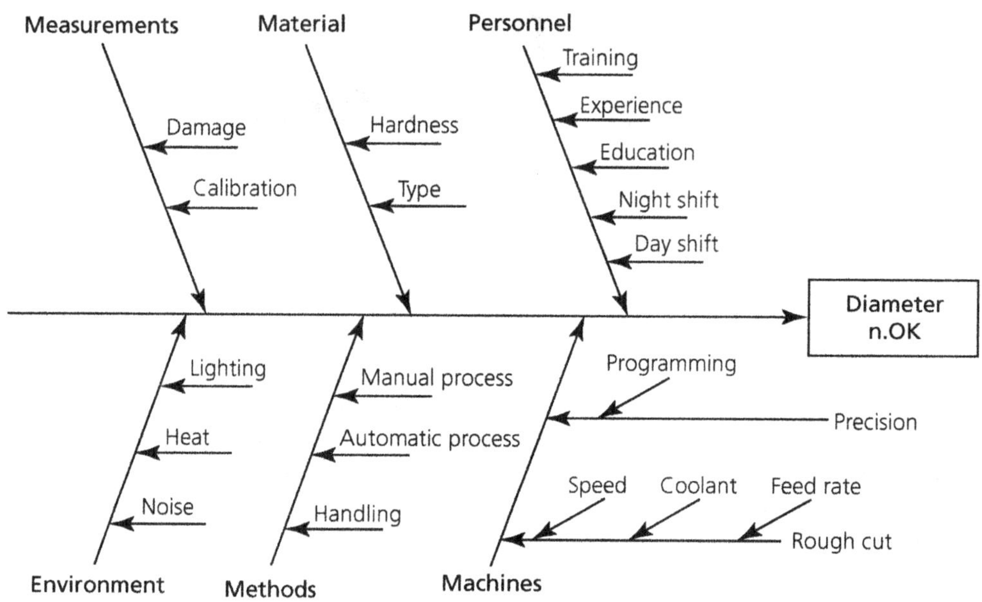

Figure 16.1 Cause and effect diagram.

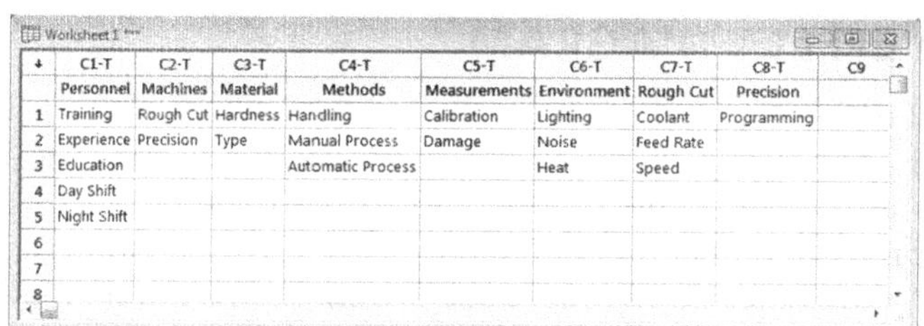

↓	C1-T	C2-T	C3-T	C4-T	C5-T	C6-T	C7-T	C8-T	C9
	Personnel	Machines	Material	Methods	Measurements	Environment	Rough Cut	Precision	
1	Training	Rough Cut	Hardness	Handling	Calibration	Lighting	Coolant	Programming	
2	Experience	Precision	Type	Manual Process	Damage	Noise	Feed Rate		
3	Education			Automatic Process		Heat	Speed		
4	Day Shift								
5	Night Shift								
6									
7									
8									

Figure 16.2 Minitab worksheet for a cause and effect diagram.

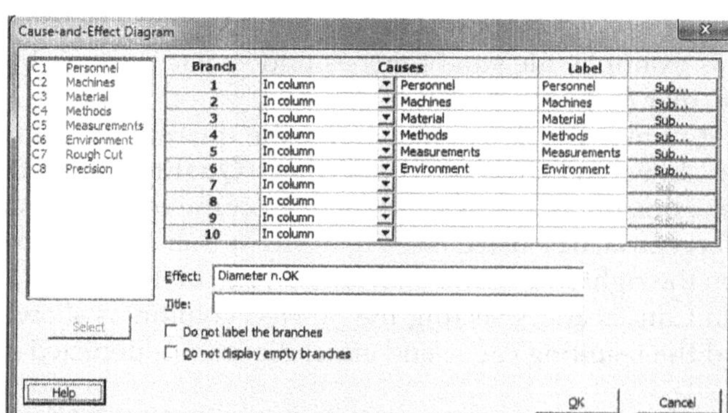

Figure 16.3 Menu for a cause and effect diagram.

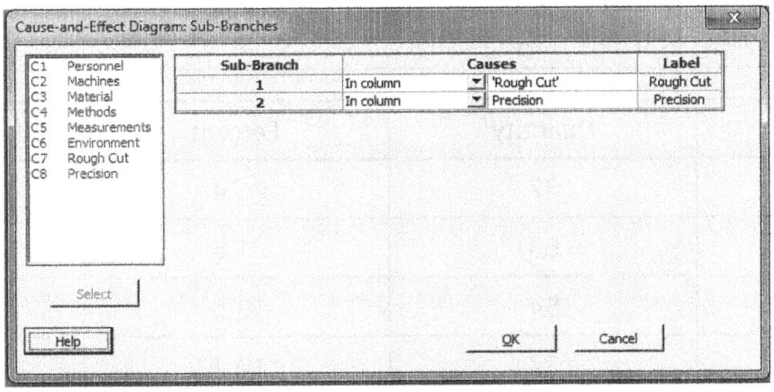

Figure 16.4 Menu for sub-branches in a cause and effect diagram.

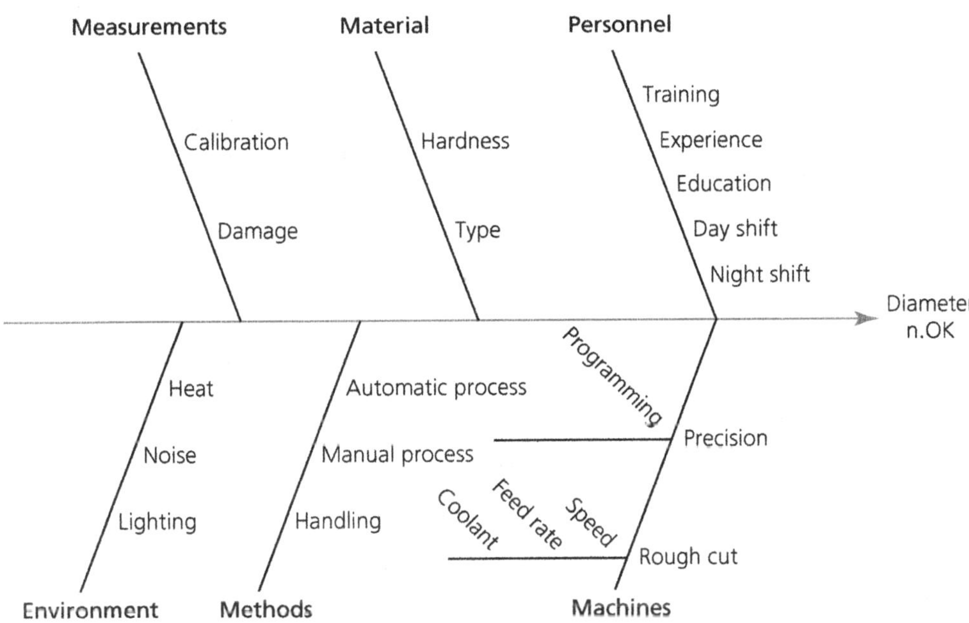

Figure 16.5 Graph for a cause and effect diagram.

the cumulative percentage of problems accounted for by each cause and all more-frequent causes.

The Pareto chart is a useful tool for quality improvement project selection. It is used to identify the causes that, if corrected, will solve the most problems. For example, an organization may find that 4 out of 15 defect types are causing 85% of scrapped or reworked products; therefore, these types would be addressed first. A Pareto chart can also be created based on the costs for problems rather than the counts; this can be useful when a few high-cost items are failing and resulting in a larger financial impact than many lower-cost items.

A Pareto chart is created by first making a table such as the one shown in Table 16.1. The defect types are listed in descending order based on the number of times that defect occurred during the time period of interest. The percent for each defect is then determined by dividing the quantity for the defect by the total

Table 16.1 Table for constructing a Pareto chart.

Defect	Quantity	Percent	Cumulative percent
Blowholes	77	26.4	26.4
Porosities	63	21.6	47.9
Inclusions	54	18.5	66.4
Rust	37	12.7	79.1
Hot cracking	17	5.8	84.9
Hardness n.OK	16	5.5	90.4
Cold shuts	12	4.1	94.5
Residue	9	3.1	97.6
Misruns	7	2.4	100.0

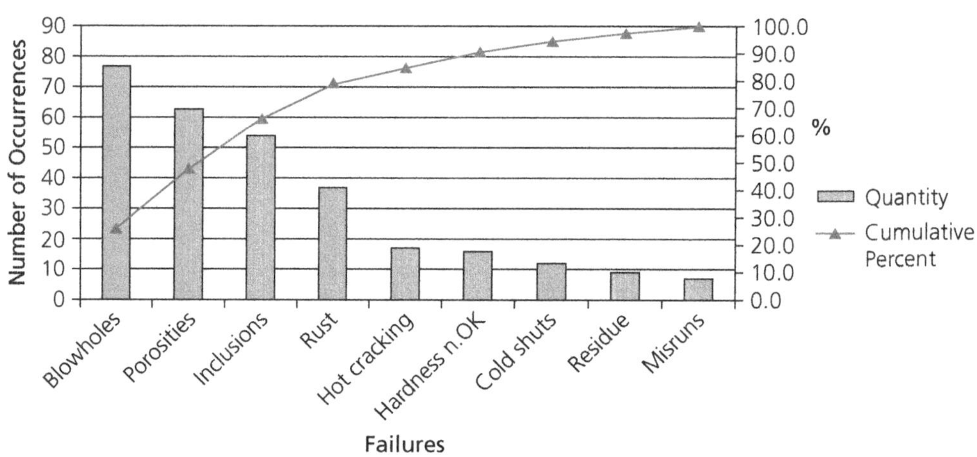

Figure 16.6 Pareto chart.

quantity of all defects. Finally, the cumulative percent is calculated by adding the percent for the first cause to the percent for the second cause and then adding these to the third cause and so on.

The quantity and cumulative percent are then graphed as shown in Figure 16.6.

In Minitab: Enter the defect name and quantity of items in the Minitab worksheet as shown in Figure 16.7. Go to **Stat** > **Quality Tools** > **Pareto Chart** and select the columns containing the defect name and quantities as depicted in Figure 16.8 and then click **OK**. The resulting Pareto chart in Figure 16.9 shows the quantity, percent, and cumulative percent for each item.

The Pareto chart illustrates that blowholes were the most common defect type, accounting for 26% of all defects. Furthermore, the defect types of blowholes, porosities, and inclusions accounted for about two-thirds of all defects. The

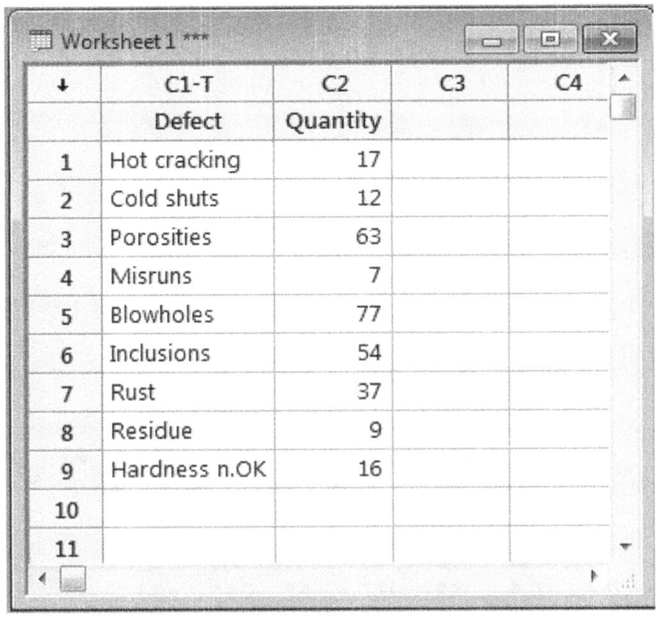

Worksheet 1 ***

↓	C1-T	C2	C3	C4
	Defect	Quantity		
1	Hot cracking	17		
2	Cold shuts	12		
3	Porosities	63		
4	Misruns	7		
5	Blowholes	77		
6	Inclusions	54		
7	Rust	37		
8	Residue	9		
9	Hardness n.OK	16		
10				
11				

Figure 16.7 Minitab worksheet with Pareto chart data.

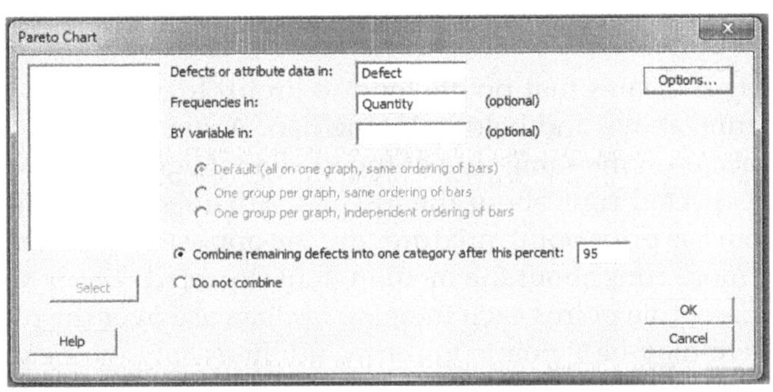

Figure 16.8 Menu for a Pareto chart.

Minitab graph combines together the least frequent causes up to a total combined percentage of 5% into a bar labeled "Other."

RUN CHART

A run chart is used to identify nonrandom patterns in continuous data over time and can be used when there is insufficient data to create a control chart (Tague 2005). A run chart can be plotted on paper or created in a spreadsheet program. The data may be collected in subgroups or individually and plotted in time order, with a centerline at the median. When data are collected in subgroups, the mean or median of each subgroup is plotted. The specific nonrandom patterns that the run chart examines are clusters, mixtures, trends, and oscillations.

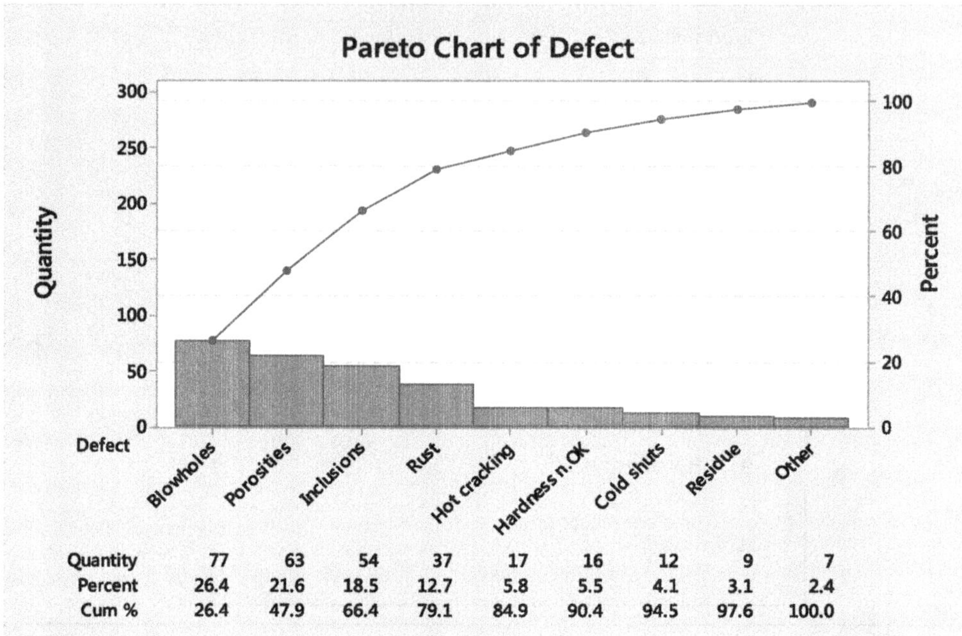

Figure 16.9 Graph for a Pareto chart.

Clustering indicates that points tend to group together; it is measured by the number of runs above and below the median. A run is one or more consecutive points that occur on the same side of the median. High clustering is evidenced by fewer than expected runs about the median. *Mixtures* occur when points tend to fall away from the prior point; mixtures are the opposite of clustering. Mixtures are detected by more runs about the median than expected. *Trends* are patterns where multiple consecutive points each increase or decrease over the previous point. For example, there may be 7 points in a row, all increasing over each other in order. Trends are detected by fewer than expected runs up or down. *Oscillations* are the opposite of trends and describe data that tend to alternate between increasing and decreasing. Oscillation is evidenced by more runs up or down than expected.

The expected number of runs above and below the median is calculated as

$$\text{Runs about median}_{expected} = 1 + \frac{2mn}{N},$$

where m is the number of points above the centerline, n is the number of points less than or equal to the centerline, and N is the total number of points. A Z-statistic can then be calculated as

$$Z = \frac{R - Runs_{expected}}{\sqrt{\dfrac{2mn(2mn - N)}{N^2(N-1)}}},$$

where R is the total number of runs above and below the median. The p-value for clustering can be found by finding the lower tail probability of the

Z-statistic using Tables C.1 and C.2 in Appendix C. The p-value for mixtures is $1 - p_{clustering}$.

The expected number of runs up and down is calculated as

$$\text{Runs up and down}_{expected} = \frac{2N-1}{3},$$

where N is the total number of points. A Z-statistic can then be calculated as

$$Z = \frac{V - Runs_{expected}}{\sqrt{\frac{16N-29}{90}}},$$

where V is the total number of runs up or down. The p-value for trends can be found by finding the lower tail probability of the Z-statistic using Tables C.1 and C.2 in Appendix C. The p-value for oscillations is $1 - p_{trends}$.

Example

Table 16.2 depicts data collected in time order for a run chart that does not have subgroups.

Table 16.2 Data for a run chart.

Delivery time	
5	11
7	10
17	14
9	6
7	8
6	9
7	8
6	10
4	10
10	10
10	10
5	9
6	5
7	6
4	11

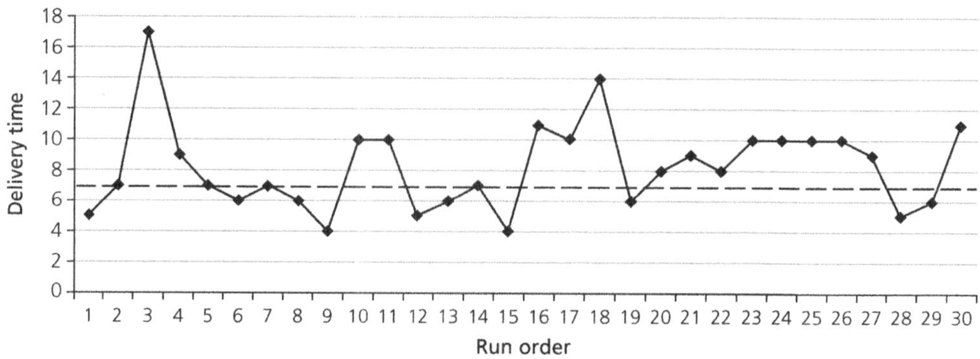

Figure 16.10 Run chart.

To create the chart, first the median of the data is found to be 7. The data are then plotted in time order, with a centerline at the median as shown in Figure 16.10.

The chart shows 12 runs above or below the median, with 14 points above the centerline and 16 below or on the centerline. The expected number of runs about the median and corresponding Z-statistic are:

$$\text{Runs about median}_{expected} = 1 + \frac{2mn}{N} = 1 + \frac{2 * 14 * 16}{30} = 15.93$$

$$Z = \frac{R - Runs_{expected}}{\sqrt{\dfrac{2mn(2mn - N)}{N^2(N-1)}}} = \frac{12 - 15.93}{\sqrt{\dfrac{2 * 14 * 16(2 * 14 * 16 - 30)}{30^2(30-1)}}} = -1.47$$

From Table C.1 in Appendix C, the left tail probability of Z=–1.47 is 0.0708. This is the p-value for clustering, and the p-value for mixtures is 1 – 0.0708 = 0.9292.

The chart in Figure 16.10 shows 17 runs up or down. The expected number of runs up or down and corresponding Z-statistic are:

$$\text{Runs up and down}_{expected} = \frac{2N-1}{3} = \frac{2 * 30 - 1}{3} = 19.67$$

$$Z = \frac{V - Runs_{expected}}{\sqrt{\dfrac{16N - 29}{90}}} = \frac{17 - 19.67}{\sqrt{\dfrac{16 * 30 - 29}{90}}} = -1.19$$

From Table C.1 in Appendix C, the left tail probability of Z=–1.19 is 0.117. This is the p-value for trends, and the p-value for oscillations is 1 – 0.117 = 0.883.

In Minitab: Enter the defect name and quantity of items in the Minitab worksheet as shown in Figure 16.11. Go to **Stat > Quality Tools > Run Chart** and select the column containing the data (see Figure 16.12). Enter the **Subgroup** size or use the column as the entire subgroup and click **OK**. The run chart in Figure 16.13 does not seem to depict any patterns in the data.

All p-values are above the desired alpha level of 0.05, so we do not have evidence that there is clustering, mixtures, trends, or oscillations in the data.

Worksheet 1 ***

↓	C1	C2
	Delivery Time	
1	5	
2	7	
3	17	
4	9	
5	7	
6	6	
7	7	
8	6	
9	4	
10	10	
11	10	
12	5	
13	6	
14	7	
15	4	
16	11	
17	10	
18	14	
19	6	
20	8	
21	9	
22	8	
23	10	
24	10	
25	10	
26	10	
27	9	
28	5	
29	6	
30	11	
31		

Figure 16.11 Minitab worksheet with run chart data.

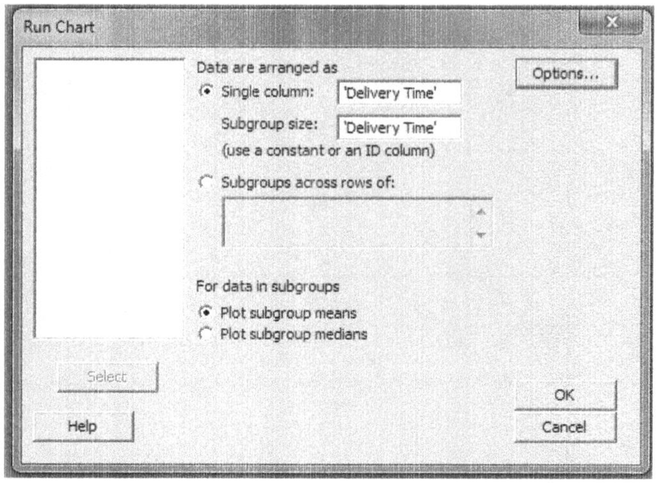

Figure 16.12 Menu for a run chart.

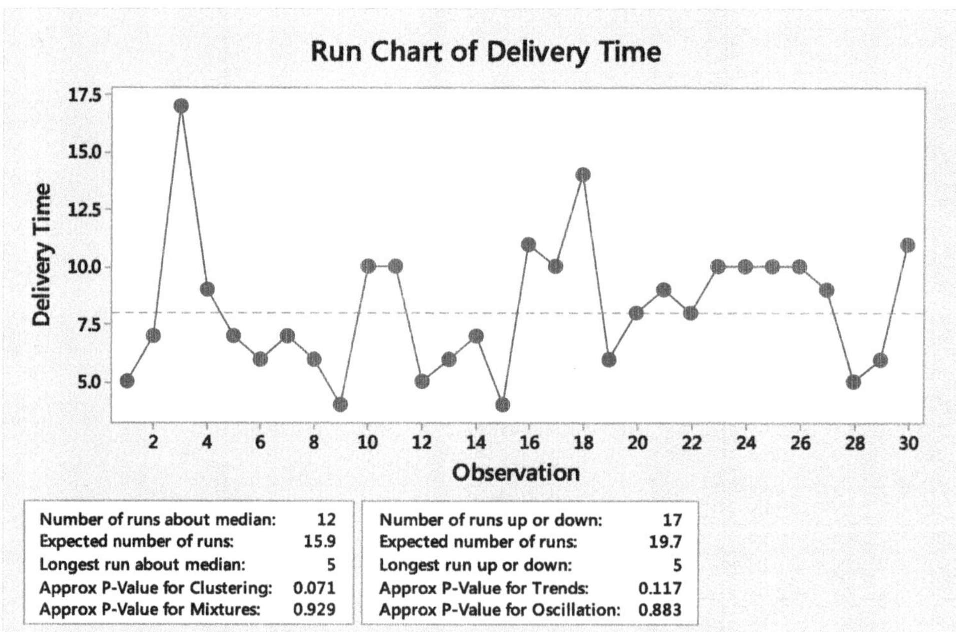

Figure 16.13 Graph for a run chart.

QUESTIONS AND ANSWERS

Q: Do I need to use measurements, material, personnel, environment, methods, and machines as the top branches when creating a cause and effect diagram?

A: Although they are advisable, the top branch labels can be changed to fit individual situations.

Q: What is the difference between a run chart and a control chart? When would I use a run chart instead of a control chart?

A: A run chart checks for specific nonrandom patterns in data, even though the patterns may be stable over time. A control chart, on the other hand, is used to indicate instability in the data based on the presence of special-cause variation.

Q: What is a Pareto chart used for?

A: Pareto charts are used to identify problems that should be addressed first based on the 80/20 principle.

Appendix A
Layman's Statistics Glossary

α error (Type I error)—The risk that you are willing to take that you incorrectly reject the null hypothesis, even though it was in fact true when performing a hypothesis test.

β error (Type II error)—The probability that the null hypothesis is not rejected, even though the alternative hypothesis was true when performing a hypothesis test.

Confidence interval—Provides a range of values that the true value of a parameter, such as the mean, is likely to fall within, as well as the probability that it is inside the interval. Confidence intervals are useful in demonstrating the precision of the estimated parameter.

Descriptive statistics—Measures of the distribution of a data set, such as the central tendency and spread. Common descriptive statistics include the mean, median, and standard deviation.

Nonparametric tests—Hypothesis tests that do not assume a particular distribution of the population(s) and typically are tests of medians rather than means.

One-way ANOVA—Allows a hypothesis test of whether at least one of three or more populations has a different mean than the others.

Parameter—An attribute of a population that is used to describe its distribution. Parameters are unknown but can be estimated using statistics from a sample taken from that population.

Population—A theoretical pool of infinite samples that has unknown parameters. Data collected in a sample are used to estimate the parameters that would describe that population.

p-value—When a hypothesis test is performed the p-value represents the probability of seeing data at least as extreme as the data collected if the null hypothesis is true. When that probability is very small—smaller than the desired alpha level—the conclusion is that the null hypothesis is likely not true and it is rejected in favor of the alternative hypothesis.

Residuals—In a statistical model such as regression, the differences between the observed data and the model's prediction.

Sample—The items selected to collect data on. There are many strategies for selecting a sample, and ultimately each is intended to help ensure the sample represents the population of interest.

Statistic—An estimate of a population parameter taken from sample data. Some statistics are used to directly estimate parameters of a distribution, such as the mean or standard deviation, while others are used to determine p-values in hypothesis testing.

Unbiasing constant—The estimates of short-term or within-subgroup variation used in control charts and capability analysis are biased estimators of the population parameters they represent. In other words, the expected value of each is slightly different from the parameter by a known amount based on sample size. Unbiasing constants adjust for this bias.

Appendix B
Common Hypothesis Tests

Note: ALL hypothesis tests in Table B.1 assume that the data are independent and identically distributed (IID). For more information, review "Other Considerations" in Chapter 5.

Table B.1 Hypothesis tests and data requirements for Y vs. single X.

What to test	One group	Two groups	Three-plus groups
Mean	**1-sample t test (σ unknown):** For highly skewed data, sample size should be at least 30. **1-sample Z test (σ known):** For highly skewed data, sample size should be at least 30.	**2 sample t:** Populations should have similar shapes. If classical test is used, populations should have equal variances; otherwise use Welch's test. **Paired t:** If differences are highly skewed, sample size should be at least 30.	**One-way ANOVA:** For the classical test all populations should be normally distributed and have equal variances. The Welch test typically performs well even when these requirements are not met.
Median	**1-sample sign:** No data requirements need to be met. **1 sample Wilcoxon:** The population must be symmetric.	**Mann-Whitney:** The population distributions must have the same shape. **Mood's median:** No data requirements need to be met.	**Kruskal-Wallis:** The population distributions must have the same shape. **Mood's median:** No data requirements need to be met.
Variance	**1 variance test:** To use the chi-square test population must be normally-distributed; otherwise, use Bonett's test.	**2 variance test:** To use the F test populations must be normally-distributed; otherwise, use Levene's test for sample sizes less than 20 or highly skewed populations and Bonett's test for all other cases.	**Equal variances test:** For heavily skewed or heavy-tailed populations or sample sizes under 20 use Levene's test; otherwise, use Bonett's test.
Proportion	**1 proportion test:** To use the normal approximation, both occurrences and non-occurrences should be at least 5.	**2 proportion test:** To use the normal approximation, both occurrences and non-occurrences should be at least 5 for each proportion.	**One-way Chi-square goodness-of-fit test:** The expected count for each level should be at least 5.

Table B.2 Models and data requirements for Y vs. X.

Type of Xs	Single X	Multiple Xs
Continuous	**Simple regression:** If sample size is less than 15, residuals should be normally distributed or p-values may be inaccurate. Otherwise, if variance of residuals is not constant across fitted values or if residuals are non-normal, prediction intervals may not be accurate. Residuals should be stable over time.	**Multiple regression:** If sample size is less than 15, residuals should be normally distributed or p-values may be inaccurate. Otherwise, if variance of residuals is not constant across fitted values or if residuals are non-normal, prediction intervals may not be accurate. Residuals should be stable over time.
Categorical	**One-way ANOVA:** For the classical test, all populations should be normally distributed and have equal variances. The Welch test typically performs well even when these requirements are not met.	**General Linear model:** If residuals are non-normal or if variance of residuals is not constant across fitted values or levels of factors, prediction intervals may be inaccurate. Residuals should be stable over time.
Both	**N/A**	**General Linear model:** If residuals are non-normal or if variance of residuals is not constant across fitted values or levels of factors, prediction intervals may be inaccurate. Residuals should be stable over time.

Table B.3 Types of data and data requirements for control charts.

Type of data	Control chart
Continuous values collected in subgroups	**Xbar-R:** Use for subgroups of < 8; subgroup size can vary in Minitab. Control limits may not be precise until at least 100 values have been plotted. Normality is not required. **Xbar-S:** Use with subgroups of ≥ 8; subgroup size can vary in Minitab. Control limits may not be precise until at least 100 values have been plotted. Normality is not required.
Continuous values collected individually	**I-MR:** Use for data that are not collected in subgroups. Normality is not required; however, for process that naturally produced highly skewed data that are bounded, the lower control limit may be below the bound. Another chart or additional tests should be used.
Continuous values collected individually or in subgroups	**EWMA or CUSUM:** Use to detect subtle, sustained shifts from the mean or target value over time.
Defectives	***p* chart:** Monitors the proportion of defectives; subgroup size can vary in Minitab. Subgroup sizes should be large enough that the proportion defective times the subgroup size should be at least 0.5 for each subgroup. **NP chart:** Monitors the proportion of defectives when subgroup size is constant and using the count of defectives makes interpretation easier than the proportion. All plotted points should be at least 1 or larger subgroups should be used.
Defects	**U chart:** Defects per unit; subgroup size can vary in Minitab. Subgroups should be large enough that average defect rate per unit times the subgroup size is at least 0.5 for every subgroup. ***c* chart:** Defects per subgroup with multiple defects per unit possible; use when the total number of defects in the subgroup is easier to interpret than defects per unit. Subgroups should be large enough that each data point is at least 1.

Table B.4 Types of design of experiments.

Objective	Design(s)
Screen factors to find important main effects	**Fractional factorial (Resolution III or IV):** Confounding may be an issue but design can be easily augmented with additional runs to estimate interactions between significant main effects. **Plackett-Burman:** May require fewer runs than a fractional factorial but less easily augmented.
Determine a predictive model with 2-level factors and no curvature	**Full factorial:** Allows main effects and all interactions to be estimated independently, but design may be large. **Fractional factorial (Resolution V or higher):** Allows main effects to be estimated independently, and 2-way interactions are only confounded with 3-way or higher-order interactions.
Determine a predictive model with at least one 3-level or higher factor and no curvature	**General full factorial:** Allows main effects and all interactions to be estimated but requires a run for every combination of factor levels, so design may be large.
Determine a predictive model with curvature	**Response surface:** Allows main effects and quadratic terms to be estimated independently as well as interactions depending on resolution.
Determine a predictive model when the factors are proportions in a mixture	**Mixture designs:** Allows main effects and interactions to be estimated when terms are proportions that are codependent; for example, the amount of corn syrup, water, and flavoring in a beverage.
Determine a predictive model when one or more factors are hard to change	**Split-plot designs:** Reduces how often levels of one or more factors are changed during the experiment but has less power to test those factors.

Appendix C
Tables

LOWER-TAIL PROBABILITIES OF THE STANDARD NORMAL DISTRIBUTION

The tables here provide the lower-tail probability of the Z distribution corresponding to values ranging from –3 to 3. For a value of z found in the table, this is equal to $P(Z \leq z)$ or the probability of a Z value less than or equal to z. The whole number and first decimal place are found in the first column, and rows indicate the second decimal place as shown on top of the tables. For example, to find the probability of Z being less than or equal to –1.00, the value of –1.0 is found in the left column and the value of .00 is found in the top row. The cell where the –1.0 row intersects the .00 column provides the probability. In this case $P(Z \leq -1.00)$ is found to be 0.1587. Graphically, this probability is represented as shown in Figure C.1.

To find the upper-tail probability, the lower-tail probability is subtracted from 1.

Table C.1 provides the lower-tail probabilities for negative values of z, and Table C.2 provides those of positive values of z.

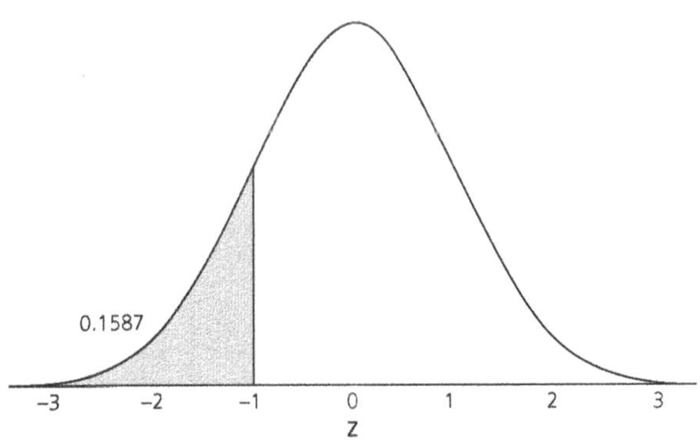

Figure C.1 Graphical illustration of $P(Z \leq -1.00)$.

Table C.1 Lower-tail probabilities for negative values of Z.

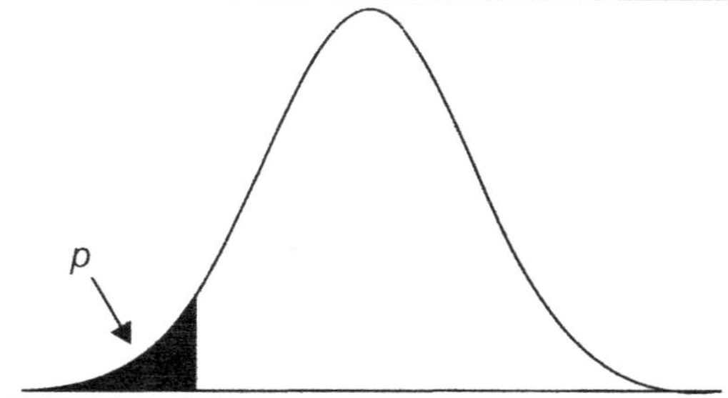

Z	0	0.01	0.02	0.03	0.04	0.05	0.06	0.07	0.08	0.09	Z
0	0.5000	0.4960	0.4920	0.4880	0.4840	0.4801	0.4761	0.4721	0.4681	0.4641	0
–0.1	0.4602	0.4562	0.4522	0.4483	0.4443	0.4404	0.4364	0.4325	0.4286	0.4247	–0.1
–0.2	0.4207	0.4168	0.4129	0.4090	0.4052	0.4013	0.3974	0.3936	0.3897	0.3859	–0.2
–0.3	0.3821	0.3783	0.3745	0.3707	0.3669	0.3632	0.3594	0.3557	0.3520	0.3483	–0.3
–0.4	0.3446	0.3409	0.3372	0.3336	0.3300	0.3264	0.3228	0.3192	0.3156	0.3121	–0.4
–0.5	0.3085	0.3050	0.3015	0.2981	0.2946	0.2912	0.2877	0.2843	0.2810	0.2776	–0.5
–0.6	0.2743	0.2709	0.2676	0.2643	0.2611	0.2578	0.2546	0.2514	0.2483	0.2451	–0.6
–0.7	0.2420	0.2389	0.2358	0.2327	0.2296	0.2266	0.2236	0.2206	0.2177	0.2148	–0.7
–0.8	0.2119	0.2090	0.2061	0.2033	0.2005	0.1977	0.1949	0.1922	0.1894	0.1867	–0.8
–0.9	0.1841	0.1814	0.1788	0.1762	0.1736	0.1711	0.1685	0.1660	0.1635	0.1611	–0.9
–1	0.1587	0.1562	0.1539	0.1515	0.1492	0.1469	0.1446	0.1423	0.1401	0.1379	–1
–1.1	0.1357	0.1335	0.1314	0.1292	0.1271	0.1251	0.1230	0.1210	0.1190	0.1170	–1.1

Z	0	0.01	0.02	0.03	0.04	0.05	0.06	0.07	0.08	0.09	Z
−1.2	0.1151	0.1131	0.1112	0.1093	0.1075	0.1056	0.1038	0.1020	0.1003	0.0985	−1.2
−1.3	0.0968	0.0951	0.0934	0.0918	0.0901	0.0885	0.0869	0.0853	0.0838	0.0823	−1.3
−1.4	0.0808	0.0793	0.0778	0.0764	0.0749	0.0735	0.0721	0.0708	0.0694	0.0681	−1.4
1.5	0.0668	0.0655	0.0643	0.0630	0.0618	0.0606	0.0594	0.0582	0.0571	0.0559	−1.5
−1.5	0.0548	0.0537	0.0526	0.0516	0.0505	0.0495	0.0485	0.0475	0.0465	0.0455	−1.6
−1.7	0.0446	0.0436	0.0427	0.0418	0.0409	0.0401	0.0392	0.0384	0.0375	0.0367	−1.7
−1.8	0.0359	0.0351	0.0344	0.0336	0.0329	0.0322	0.0314	0.0307	0.0301	0.0294	−1.8
−1.9	0.0287	0.0281	0.0274	0.0268	0.0262	0.0256	0.0250	0.0244	0.0239	0.0233	−1.9
−2.0	0.0228	0.0222	0.0217	0.0212	0.0207	0.0202	0.0197	0.0192	0.0188	0.0183	−2
−2.1	0.0179	0.0174	0.0170	0.0166	0.0162	0.0158	0.0154	0.0150	0.0146	0.0143	−2.1
−2.2	0.0139	0.0136	0.0132	0.0129	0.0125	0.0122	0.0119	0.0116	0.0113	0.0110	−2.2
−2.3	0.01072	0.01044	0.01017	0.00990	0.00964	0.00939	0.00914	0.00889	0.00866	0.00842	−2.3
−2.4	0.00820	0.00798	0.00776	0.00755	0.00734	0.00714	0.00695	0.00676	0.00657	0.00639	−2.4
−2.5	0.00621	0.00604	0.00587	0.00570	0.00554	0.00539	0.00523	0.00508	0.00494	0.00480	−2.5
−2.6	0.00466	0.00453	0.00440	0.00427	0.00415	0.00402	0.00391	0.00379	0.00368	0.00357	−2.6
−2.7	0.00347	0.00336	0.00326	0.00317	0.00307	0.00298	0.00289	0.00280	0.00272	0.00264	−2.7
−2.8	0.00256	0.00248	0.00240	0.00233	0.00226	0.00219	0.00212	0.00205	0.00199	0.00193	−2.8
−2.9	0.00187	0.00181	0.00175	0.00169	0.00164	0.00159	0.00154	0.00149	0.00144	0.00139	−2.9

Table C.2 Lower-tail probabilities for positive values of Z.

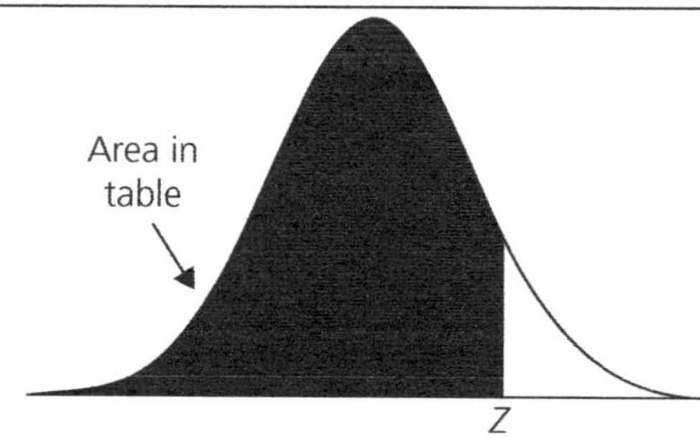

Z	0	0.01	0.02	0.03	0.04	0.05	0.06	0.07	0.08	0.09	Z
0	0.50000	0.50399	0.50798	0.51197	0.51595	0.51994	0.52392	0.52790	0.53188	0.53586	0
0.1	0.53983	0.54380	0.54776	0.55172	0.55567	0.55962	0.56356	0.56749	0.57142	0.57535	0.1
0.2	0.57926	0.58317	0.58706	0.59095	0.59483	0.59871	0.60257	0.60642	0.61026	0.61409	0.2
0.3	0.61791	0.62172	0.62552	0.62930	0.63307	0.63683	0.64058	0.64431	0.64803	0.65173	0.3
0.4	0.65542	0.65910	0.66276	0.66640	0.67003	0.67364	0.67724	0.68082	0.68439	0.68793	0.4
0.5	0.69146	0.69497	0.69847	0.70194	0.70540	0.70884	0.71226	0.71566	0.71904	0.72240	0.5
0.6	0.72575	0.72907	0.73237	0.73565	0.73891	0.74215	0.74537	0.74857	0.75175	0.75490	0.6
0.7	0.75804	0.76115	0.76424	0.76730	0.77035	0.77337	0.77637	0.77935	0.78230	0.78524	0.7
0.8	0.78814	0.79103	0.79389	0.79673	0.79955	0.80234	0.80511	0.80785	0.81057	0.81327	0.8
0.9	0.81594	0.81859	0.82121	0.82381	0.82639	0.82894	0.83147	0.83398	0.83646	0.83891	0.9
1	0.84134	0.84375	0.84614	0.84849	0.85083	0.85314	0.85543	0.85769	0.85993	0.86214	1
1.1	0.86433	0.86650	0.86864	0.87076	0.87286	0.87493	0.87698	0.87900	0.88100	0.88298	1.1

Z	0	0.01	0.02	0.03	0.04	0.05	0.06	0.07	0.08	0.09	Z
1.2	0.88493	0.88686	0.88877	0.89065	0.89251	0.89435	0.89617	0.89796	0.89973	0.90147	1.2
1.3	0.90320	0.90490	0.90658	0.90824	0.90988	0.91149	0.91309	0.91466	0.91621	0.91774	1.3
1.4	0.91924	0.92073	0.92220	0.92364	0.92507	0.92647	0.92785	0.92922	0.93056	0.93189	1.4
1.5	0.93319	0.93448	0.93574	0.93699	0.93822	0.93943	0.94062	0.94179	0.94295	0.94408	1.5
1.6	0.94520	0.94630	0.94738	0.94845	0.94950	0.95053	0.95154	0.95254	0.95352	0.95449	1.6
1.7	0.95543	0.95637	0.95728	0.95818	0.95907	0.95994	0.96080	0.96164	0.96246	0.96327	1.7
1.8	0.96407	0.96485	0.96562	0.96638	0.96712	0.96784	0.96856	0.96926	0.96995	0.97062	1.8
1.9	0.97128	0.97193	0.97257	0.97320	0.97381	0.97441	0.97500	0.97558	0.97615	0.97670	1.9
2	0.97725	0.97778	0.97831	0.97882	0.97932	0.97982	0.98030	0.98077	0.98124	0.98169	2
2.1	0.98214	0.98257	0.98300	0.98341	0.98382	0.98422	0.98461	0.98500	0.98537	0.98574	2.1
2.2	0.98610	0.98645	0.98679	0.98713	0.98745	0.98778	0.98809	0.98840	0.98870	0.98899	2.2
2.3	0.98928	0.98956	0.98983	0.99010	0.99036	0.99061	0.99086	0.99111	0.99134	0.99158	2.3
2.4	0.99180	0.99202	0.99224	0.99245	0.99266	0.99286	0.99305	0.99324	0.99343	0.99361	2.4
2.5	0.99379	0.99396	0.99413	0.99430	0.99446	0.99461	0.99477	0.99492	0.99506	0.99520	2.5
2.6	0.99534	0.99547	0.99560	0.99573	0.99585	0.99598	0.99609	0.99621	0.99632	0.99643	2.6
2.7	0.99653	0.99664	0.99674	0.99683	0.99693	0.99702	0.99711	0.99720	0.99728	0.99736	2.7
2.8	0.99744	0.99752	0.99760	0.99767	0.99774	0.99781	0.99788	0.99795	0.99801	0.99807	2.8
2.9	0.99813	0.99819	0.99825	0.99831	0.99836	0.99841	0.99846	0.99851	0.99856	0.99861	2.9

UPPER-TAIL PROBABILITIES OF THE STANDARD NORMAL DISTRIBUTION

The tables here provide the upper-tail probability of the Z distribution corresponding to values ranging from −3 to 3. For a value of z found in the table, this is equal to $P(Z \geq z)$ or the probability of a Z value greater than or equal to z. The whole number and first decimal place are found in the first column, and rows indicate the second decimal place as shown on top of the tables. For example, to find the probability of Z being greater than or equal to −1.00, the value of −1.0 is found in the left column and the value of .00 is found in the top row. The cell where the −1.0 row intersects the .00 column provides the probability. In this case $P(Z \geq -1.00)$ is found to be 0.8413. Graphically, this probability is represented as shown in Figure C.2.

To find the lower-tail probability, the upper-tail probability is subtracted from 1.

Table C.3 provides the upper-tail probabilities for negative values of z, and Table C.4 provides those of positive values of z.

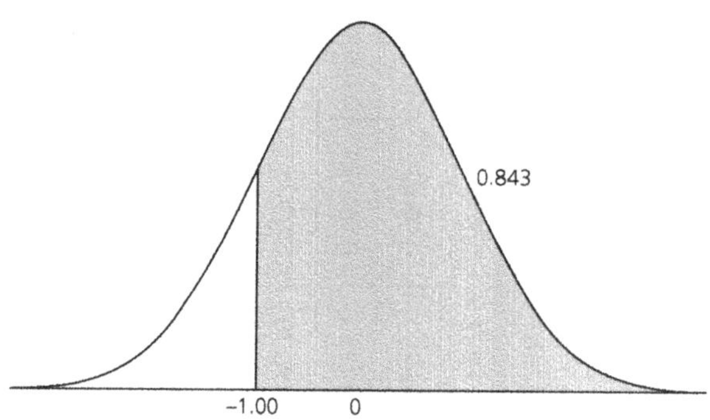

Figure C.2 Graphical illustration of $P(Z \geq -1.00)$.

Table C.3 Upper-tail probabilities for negative values of z.

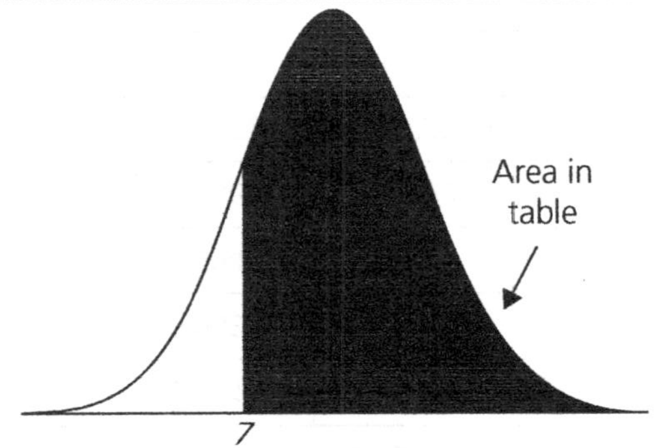

Z	0	0.01	0.02	0.03	0.04	0.05	0.06	0.07	0.08	0.09	Z
0	0.50000	0.50399	0.50798	0.51197	0.51595	0.51994	0.52392	0.52790	0.53188	0.53586	0
–0.1	0.53983	0.54380	0.54776	0.55172	0.55567	0.55962	0.56356	0.56749	0.57142	0.57535	–0.1
–0.2	0.57926	0.58317	0.58706	0.59095	0.59483	0.59871	0.60257	0.60642	0.61026	0.61409	–0.2
–0.3	0.61791	0.62172	0.62552	0.62930	0.63307	0.63683	0.64058	0.64431	0.64803	0.65173	–0.3
–0.4	0.65542	0.65910	0.66276	0.66640	0.67003	0.67364	0.67724	0.68082	0.68439	0.68793	–0.4
–0.5	0.69146	0.69497	0.69847	0.70194	0.70540	0.70884	0.71226	0.71566	0.71904	0.72240	–0.5
–0.6	0.72575	0.72907	0.73237	0.73565	0.73891	0.74215	0.74537	0.74857	0.75175	0.75490	–0.6
–0.7	0.75804	0.76115	0.76424	0.76730	0.77035	0.77337	0.77637	0.77935	0.78230	0.78524	–0.7
–0.8	0.78814	0.79103	0.79389	0.79673	0.79955	0.80234	0.80511	0.80785	0.81057	0.81327	–0.8
–0.9	0.81594	0.81859	0.82121	0.82381	0.82639	0.82894	0.83147	0.83398	0.83646	0.83891	–0.9
–1	0.84134	0.84375	0.84614	0.84849	0.85083	0.85314	0.85543	0.85769	0.85993	0.86214	–1

Table C.3 Upper-tail probabilities for negative values of z. (Continued)

Z	0	0.01	0.02	0.03	0.04	0.05	0.06	0.07	0.08	0.09	Z
–1.1	0.86433	0.86650	0.86864	0.87076	0.87286	0.87493	0.87698	0.87900	0.88100	0.88298	–1.1
–1.2	0.88493	0.88686	0.88877	0.89065	0.89251	0.89435	0.89617	0.89796	0.89973	0.90147	–1.2
–1.3	0.90320	0.90490	0.90658	0.90824	0.90988	0.91149	0.91309	0.91466	0.91621	0.91774	–1.3
–1.4	0.91924	0.92073	0.92220	0.92364	0.92507	0.92647	0.92785	0.92922	0.93056	0.93189	–1.4
–1.5	0.93319	0.93448	0.93574	0.93699	0.93822	0.93943	0.94062	0.94179	0.94295	0.94408	–1.5
–1.6	0.94520	0.94630	0.94738	0.94845	0.94950	0.95053	0.95154	0.95254	0.95352	0.95449	–1.6
–1.7	0.95543	0.95637	0.95728	0.95818	0.95907	0.95994	0.96080	0.96164	0.96246	0.96327	–1.7
–1.8	0.96407	0.96485	0.96562	0.96638	0.96712	0.96784	0.96856	0.96926	0.96995	0.97062	–1.8
–1.9	0.97128	0.97193	0.97257	0.97320	0.97381	0.97441	0.97500	0.97558	0.97615	0.97670	–1.9
-2	0.97725	0.97778	0.97831	0.97882	0.97932	0.97982	0.98030	0.98077	0.98124	0.98169	-2
-2.1	0.98214	0.98257	0.98300	0.98341	0.98382	0.98422	0.98461	0.98500	0.98537	0.98574	-2.1
-2.2	0.98610	0.98645	0.98679	0.98713	0.98745	0.98778	0.98809	0.98840	0.98870	0.98899	-2.2
-2.3	0.98928	0.98956	0.98983	0.99010	0.99036	0.99061	0.99086	0.99111	0.99134	0.99158	-2.3
-2.4	0.99180	0.99202	0.99224	0.99245	0.99266	0.99286	0.99305	0.99324	0.99343	0.99361	-2.4
-2.5	0.99379	0.99396	0.99413	0.99430	0.99446	0.99461	0.99477	0.99492	0.99506	0.99520	-2.5
-2.6	0.99534	0.99547	0.99560	0.99573	0.99585	0.99598	0.99609	0.99621	0.99632	0.99643	-2.6
-2.7	0.99653	0.99664	0.99674	0.99683	0.99693	0.99702	0.99711	0.99720	0.99728	0.99736	-2.7
-2.8	0.99744	0.99752	0.99760	0.99767	0.99774	0.99781	0.99788	0.99795	0.99801	0.99807	-2.8
-2.9	0.99813	0.99819	0.99825	0.99831	0.99836	0.99841	0.99846	0.99851	0.99856	0.99861	-2.9

Table C.4 Upper-tail probabilities for positive values of z.

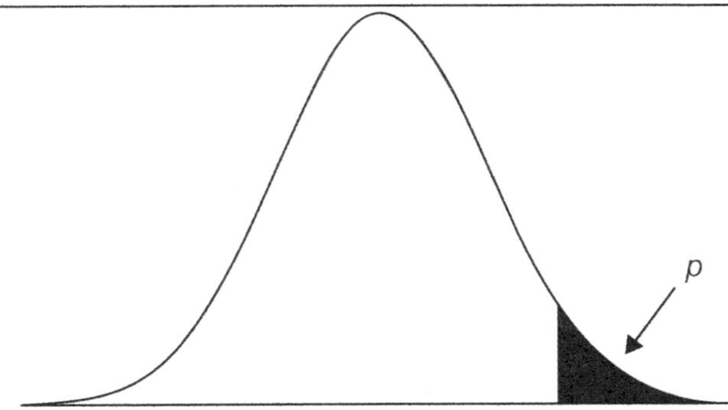

Z	0	0.01	0.02	0.03	0.04	0.05	0.06	0.07	0.08	0.09	Z
0	0.5000	0.4960	0.4920	0.4880	0.4840	0.4801	0.4761	0.4721	0.4681	0.4641	0
0.1	0.4602	0.4562	0.4522	0.4483	0.4443	0.4404	0.4364	0.4325	0.4286	0.4247	0.1
0.2	0.4207	0.4168	0.4129	0.4090	0.4052	0.4013	0.3974	0.3936	0.3897	0.3859	0.2
0.3	0.3821	0.3783	0.3745	0.3707	0.3669	0.3632	0.3594	0.3557	0.3520	0.3483	0.3
0.4	0.3446	0.3409	0.3372	0.3336	0.3300	0.3264	0.3228	0.3192	0.3156	0.3121	0.4
0.5	0.3085	0.3050	0.3015	0.2981	0.2946	0.2912	0.2877	0.2843	0.2810	0.2776	0.5
0.6	0.2743	0.2709	0.2676	0.2643	0.2611	0.2578	0.2546	0.2514	0.2483	0.2451	0.6
0.7	0.2420	0.2389	0.2358	0.2327	0.2296	0.2266	0.2236	0.2206	0.2177	0.2148	0.7
0.8	0.2119	0.2090	0.2061	0.2033	0.2005	0.1977	0.1949	0.1922	0.1894	0.1867	0.8
0.9	0.1841	0.1814	0.1788	0.1762	0.1736	0.1711	0.1685	0.1660	0.1635	0.1611	0.9
1	0.1587	0.1562	0.1539	0.1515	0.1492	0.1469	0.1446	0.1423	0.1401	0.1379	1

Table C.4 Upper-tail probabilities for positive values of z. (Continued)

z	0	0.01	0.02	0.03	0.04	0.05	0.06	0.07	0.08	0.09	*z*
1.1	0.1357	0.1335	0.1314	0.1292	0.1271	0.1251	0.1230	0.1210	0.1190	0.1170	1.1
1.2	0.1151	0.1131	0.1112	0.1093	0.1075	0.1056	0.1038	0.1020	0.1003	0.0985	1.2
1.3	0.0968	0.0951	0.0934	0.0918	0.0901	0.0885	0.0869	0.0853	0.0838	0.0823	1.3
1.4	0.0808	0.0793	0.0778	0.0764	0.0749	0.0735	0.0721	0.0708	0.0694	0.0681	1.4
1.5	0.0668	0.0655	0.0643	0.0630	0.0618	0.0606	0.0594	0.0582	0.0571	0.0559	1.5
1.6	0.0548	0.0537	0.0526	0.0516	0.0505	0.0495	0.0485	0.0475	0.0465	0.0455	1.6
1.7	0.0446	0.0436	0.0427	0.0418	0.0409	0.0401	0.0392	0.0384	0.0375	0.0367	1.7
1.8	0.0359	0.0351	0.0344	0.0336	0.0329	0.0322	0.0314	0.0307	0.0301	0.0294	1.8
1.9	0.0287	0.0281	0.0274	0.0268	0.0262	0.0256	0.0250	0.0244	0.0239	0.0233	1.9
2	0.0228	0.0222	0.0217	0.0212	0.0207	0.0202	0.0197	0.0192	0.0188	0.0183	2
2.1	0.0179	0.0174	0.0170	0.0166	0.0162	0.0158	0.0154	0.0150	0.0146	0.0143	2.1
2.2	0.0139	0.0136	0.0132	0.0129	0.0125	0.0122	0.0119	0.0116	0.0113	0.0110	2.2
2.3	0.01072	0.01044	0.01017	0.00990	0.00964	0.00939	0.00914	0.00889	0.00866	0.00842	2.3
2.4	0.00820	0.00798	0.00776	0.00755	0.00734	0.00714	0.00695	0.00676	0.00657	0.00639	2.4
2.5	0.00621	0.00604	0.00587	0.00570	0.00554	0.00539	0.00523	0.00508	0.00494	0.00480	2.5
2.6	0.00466	0.00453	0.00440	0.00427	0.00415	0.00402	0.00391	0.00379	0.00368	0.00357	2.6
2.7	0.00347	0.00336	0.00326	0.00317	0.00307	0.00298	0.00289	0.00280	0.00272	0.00264	2.7
2.8	0.00256	0.00248	0.00240	0.00233	0.00226	0.00219	0.00212	0.00205	0.00199	0.00193	2.8
2.9	0.00187	0.00181	0.00175	0.00169	0.00164	0.00159	0.00154	0.00149	0.00144	0.00139	2.9

VALUES OF *t* DISTRIBUTION FOR SELECTED LOWER-TAIL PROBABILITIES

Tables C.5 and C.6 provide the values of the *t* distribution for various degrees of freedom corresponding to selected lower-tail probabilities. Each row corresponds to a number of degrees of freedom for the *t* distribution, and each column corresponds to a certain lower-tail probability. The *t* distribution with that number of degrees of freedom has that probability of being less than or equal to the value shown where the row and column intersect. For example, to find the value of a *t* distribution with 10 degrees of freedom with a lower-tail probability of 0.05, the value where the row corresponding to 10 degrees of freedom and the column containing a probability of 0.05 intersect is found to be −1.812. Therefore, $P(T_{10} \leq -1.812)$ is 0.05. Graphically, this probability is represented as shown in Figure C.3.

To find the upper-tail probability, the lower-tail probability is subtracted from 1.

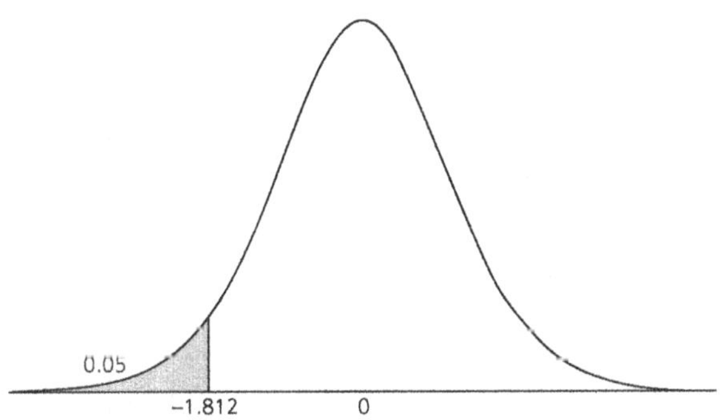

Figure C.3 Graphical illustration of $P(T_{10} \leq -1.812) = 0.05$.

Table C.5 Values of t for lower-tail probabilities greater than 0.5.

df	0.6	0.7	0.8	0.9	0.95	0.975	0.99	0.995	0.999	0.9995	df
1	0.325	0.727	1.376	3.078	6.314	12.706	31.821	63.657	318.309	636.619	1
2	0.289	0.617	1.061	1.886	2.920	4.303	6.965	9.925	22.327	31.599	2
3	0.277	0.584	0.978	1.638	2.353	3.182	4.541	5.841	10.215	12.924	3
4	0.271	0.569	0.941	1.533	2.132	2.776	3.747	4.604	7.173	8.610	4
5	0.267	0.559	0.920	1.476	2.015	2.571	3.365	4.032	5.893	6.869	5
6	0.265	0.553	0.906	1.440	1.943	2.447	3.143	3.707	5.208	5.959	6
7	0.263	0.549	0.896	1.415	1.895	2.365	2.998	3.499	4.785	5.408	7
8	0.262	0.546	0.889	1.397	1.860	2.306	2.896	3.355	4.501	5.041	8
9	0.261	0.543	0.883	1.383	1.833	2.262	2.821	3.250	4.297	4.781	9
10	0.260	0.542	0.879	1.372	1.812	2.228	2.764	3.169	4.144	4.587	10
11	0.260	0.540	0.876	1.363	1.796	2.201	2.718	3.106	4.025	4.437	11
12	0.259	0.539	0.873	1.356	1.782	2.179	2.681	3.055	3.930	4.318	12
13	0.259	0.538	0.870	1.350	1.771	2.160	2.650	3.012	3.852	4.221	13
14	0.258	0.537	0.868	1.345	1.761	2.145	2.624	2.977	3.787	4.140	14
15	0.258	0.536	0.866	1.341	1.753	2.131	2.602	2.947	3.733	4.073	15

df	0.6	0.7	0.8	0.9	0.95	0.975	0.99	0.995	0.999	0.9995	df
16	0.258	0.535	0.865	1.337	1.746	2.120	2.583	2.921	3.686	4.015	16
17	0.257	0.534	0.863	1.333	1.740	2.110	2.567	2.898	3.646	3.965	17
18	0.257	0.534	0.862	1.330	1.734	2.101	2.552	2.878	3.610	3.922	18
19	0.257	0.533	0.861	1.328	1.729	2.093	2.539	2.861	3.579	3.883	19
20	0.257	0.533	0.860	1.325	1.725	2.086	2.528	2.845	3.552	3.850	20
21	0.257	0.532	0.859	1.323	1.721	2.080	2.518	2.831	3.527	3.819	21
22	0.256	0.532	0.858	1.321	1.717	2.074	2.508	2.819	3.505	3.792	22
23	0.256	0.532	0.858	1.319	1.714	2.069	2.500	2.807	3.485	3.768	23
24	0.256	0.531	0.857	1.318	1.711	2.064	2.492	2.797	3.467	3.745	24
25	0.256	0.531	0.856	1.316	1.708	2.060	2.485	2.787	3.450	3.725	25
26	0.256	0.531	0.856	1.315	1.706	2.056	2.479	2.779	3.435	3.707	26
27	0.256	0.531	0.855	1.314	1.703	2.052	2.473	2.771	3.421	3.690	27
28	0.256	0.530	0.855	1.313	1.701	2.048	2.467	2.763	3.408	3.674	28
29	0.256	0.530	0.854	1.311	1.699	2.045	2.462	2.756	3.396	3.659	29
30	0.256	0.530	0.854	1.310	1.697	2.042	2.457	2.750	3.385	3.646	30
40	0.255	0.529	0.851	1.303	1.684	2.021	2.423	2.704	3.307	3.551	40
50	0.255	0.528	0.849	1.299	1.676	2.009	2.403	2.678	3.261	3.496	50
60	0.254	0.527	0.848	1.296	1.671	2.000	2.390	2.660	3.232	3.460	60
80	0.254	0.526	0.846	1.292	1.664	1.990	2.374	2.639	3.195	3.416	80
100	0.254	0.526	0.845	1.290	1.660	1.984	2.364	2.626	3.174	3.390	100
200	0.254	0.525	0.843	1.286	1.653	1.972	2.345	2.601	3.131	3.340	200
500	0.253	0.525	0.842	1.283	1.648	1.965	2.334	2.586	3.107	3.310	500

Table C.6 Values of *t* for lower-tail probabilities less than 0.5.

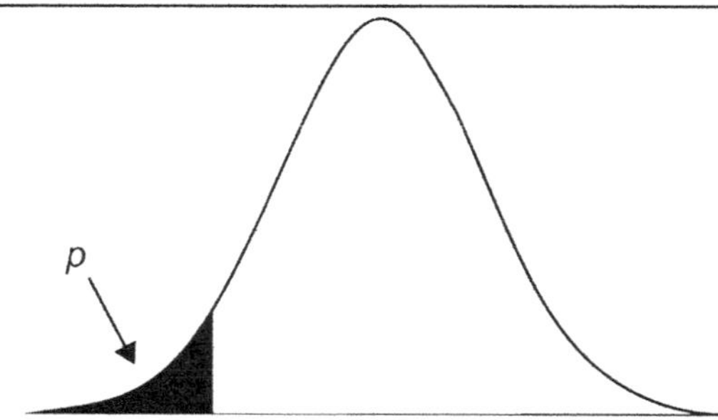

df	0.4	0.3	0.2	0.1	0.05	0.025	0.01	0.005	0.001	0.0005	df
1	−0.325	−0.727	−1.376	−3.078	−6.314	−12.706	−31.821	−63.657	−318.309	−636.619	1
2	−0.289	−0.617	−1.061	−1.886	−2.920	−4.303	−6.965	−9.925	−22.327	−31.599	2
3	−0.277	−0.584	−0.978	−1.638	−2.353	−3.182	−4.541	−5.841	−10.215	−12.924	3
4	−0.271	−0.569	−0.941	−1.533	−2.132	−2.776	−3.747	−4.604	−7.173	−8.610	4
5	−0.267	−0.559	−0.920	−1.476	−2.015	−2.571	−3.365	−4.032	−5.893	−6.869	5
6	−0.265	−0.553	−0.906	−1.440	−1.943	−2.447	−3.143	−3.707	−5.208	−5.959	6
7	0.263	−0.549	−0.896	−1.415	−1.895	−2.365	−2.998	−3.499	−4.785	−5.408	7
8	−0.262	−0.546	−0.889	−1.397	−1.860	−2.306	−2.896	−3.355	−4.501	−5.041	8
9	−0.261	−0.543	−0.883	−1.383	−1.833	−2.262	−2.821	−3.250	−4.297	−4.781	9
10	−0.260	−0.542	−0.879	−1.372	−1.812	−2.228	−2.764	−3.169	−4.144	−4.587	10
11	−0.260	−0.540	−0.876	−1.363	−1.796	−2.201	−2.718	−3.106	−4.025	−4.437	11
12	−0.259	−0.539	−0.873	−1.356	−1.782	−2.179	−2.681	−3.055	−3.930	−4.318	12
13	−0.259	−0.538	−0.870	−1.350	−1.771	−2.160	−2.650	−3.012	−3.852	−4.221	13
14	−0.258	−0.537	−0.868	−1.345	−1.761	−2.145	−2.624	−2.977	−3.787	−4.140	14
15	−0.258	−0.536	−0.866	−1.341	−1.753	−2.131	−2.602	−2.947	−3.733	−4.073	15

df	0.4	0.3	0.2	0.1	0.05	0.025	0.01	0.005	0.001	0.0005	df
16	−0.258	−0.535	−0.865	−1.337	−1.746	−2.120	−2.583	−2.921	−3.686	−4.015	16
17	0.257	−0.534	−0.863	−1.333	−1.740	−2.110	−2.567	−2.898	−3.646	−3.965	17
18	−0.257	−0.534	−0.862	−1.330	−1.734	−2.101	−2.552	−2.878	−3.610	−3.922	18
19	−0.257	−0.533	−0.861	−1.328	−1.729	−2.093	−2.539	−2.861	−3.579	−3.883	19
20	−0.257	−0.533	−0.860	−1.325	−1.725	−2.086	−2.528	−2.845	−3.552	−3.850	20
21	−0.257	−0.532	−0.859	−1.323	−1.721	−2.080	−2.518	−2.831	−3.527	−3.819	21
22	−0.256	−0.532	−0.858	−1.321	−1.717	−2.074	−2.508	−2.819	−3.505	−3.792	22
23	−0.256	−0.532	−0.858	−1.319	−1.714	−2.069	−2.500	−2.807	−3.485	−3.768	23
24	−0.256	−0.531	−0.857	−1.318	−1.711	−2.064	−2.492	−2.797	−3.467	−3.745	24
25	−0.256	−0.531	−0.856	−1.316	−1.708	−2.060	−2.485	−2.787	−3.450	−3.725	25
26	−0.256	−0.531	−0.856	−1.315	−1.706	−2.056	−2.479	−2.779	−3.435	−3.707	26
27	−0.256	−0.531	−0.855	−1.314	−1.703	−2.052	−2.473	−2.771	−3.421	−3.690	27
28	−0.256	−0.530	−0.855	−1.313	−1.701	−2.048	−2.467	−2.763	−3.408	−3.674	28
29	−0.256	−0.530	−0.854	−1.311	−1.699	−2.045	−2.462	−2.756	−3.396	−3.659	29
30	−0.256	−0.530	−0.854	−1.310	−1.697	−2.042	−2.457	−2.750	−3.385	−3.646	30
40	−0.255	−0.529	−0.851	−1.303	−1.684	−2.021	−2.423	−2.704	−3.307	−3.551	40
50	−0.255	−0.528	−0.849	−1.299	−1.676	−2.009	−2.403	−2.678	−3.261	−3.496	50
60	−0.254	−0.527	−0.848	−1.296	−1.671	−2.000	−2.390	−2.660	−3.232	−3.460	60
80	−0.254	−0.526	−0.846	−1.292	−1.664	−1.990	−2.374	−2.639	−3.195	−3.416	80
100	−0.254	−0.526	−0.845	−1.290	−1.660	−1.984	−2.364	−2.626	−3.174	−3.390	100
200	−0.254	−0.525	−0.843	−1.286	−1.653	−1.972	−2.345	−2.601	−3.131	−3.340	200
500	−0.253	−0.525	−0.842	−1.283	−1.648	−1.965	−2.334	−2.586	−3.107	−3.310	500

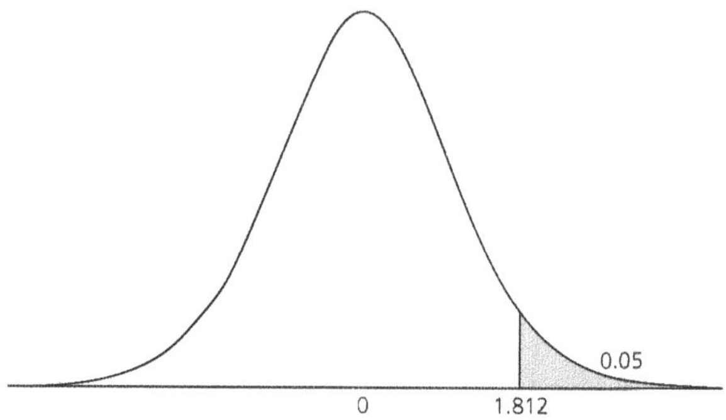

Figure C.4 Graphical illustration of $P(T_{10} \geq 1.812) = 0.05$.

VALUES OF *t* DISTRIBUTION FOR SELECTED UPPER-TAIL PROBABILITIES

The tables here provide the values of the *t* distribution for various degrees of freedom corresponding to selected upper-tail probabilities. Each row corresponds to a number of degrees of freedom for the *t* distribution, and each column corresponds to a certain upper-tail probability. The *t* distribution with that number of degrees of freedom has that probability of being greater than or equal to the value shown where the row and column intersect. For example, to find the value of a *t* distribution with 10 degrees of freedom with an upper-tail probability of 0.05, the value where the row corresponding to 10 degrees of freedom and the column containing a probability of 0.05 intersect is found to be 1.812. Therefore, $P(T_{10} \geq 1.812)$ is 0.05. Graphically, this probability is represented as shown in Figure C.4.

To find the lower-tail probability, the upper-tail probability is subtracted from 1.

VALUES OF CHI-SQUARE DISTRIBUTION FOR SELECTED PROBABILITIES

The tables here provide the values of the Chi-square distribution for various degrees of freedom corresponding to selected upper-tail probabilities. Lower-tail probabilities are simply 1 minus the upper-tail probabilities. Each row corresponds to a number of degrees of freedom for the Chi-square distribution, and each column corresponds to a certain upper-tail probability. The Chi-square distribution with that number of degrees of freedom has that probability of being greater than or equal to the value shown where the row and column intersect. For example, to find the value of a Chi-square distribution with 15 degrees of freedom with an upper-tail probability of 0.1, the value where the row corresponding to 15 degrees of freedom and the column containing a probability of 0.1 intersect is found to be 22.31. Therefore, $P(\chi^2_{15} \geq 22.31)$ is 0.1. Graphically, this probability is represented as shown in Figure C.5.

To find the lower-tail probability, the upper-tail probability is subtracted from 1.

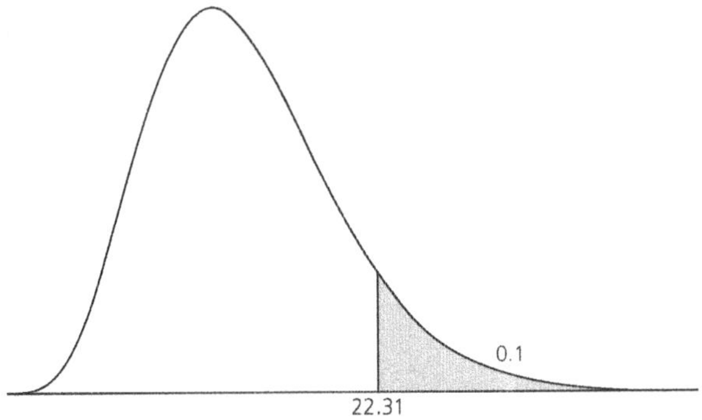

Figure C.5 Graphical illustration of $P(\chi^2_{15} \geq 22.31) = 0.1$.

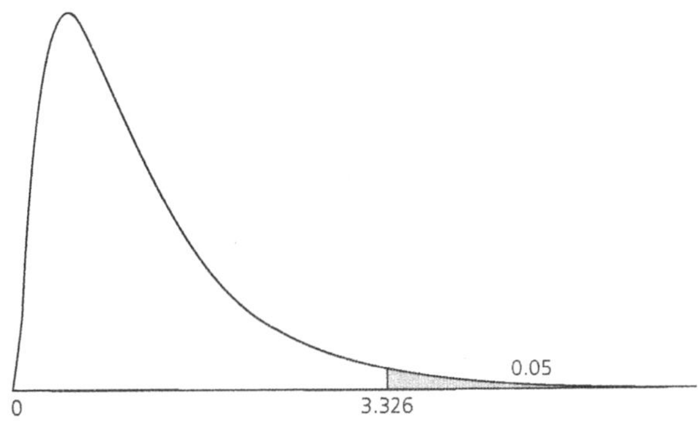

Figure C.6 Graphical illustration of $P(F_{5,10} \geq 3.33) = 0.05$.

VALUES OF F DISTRIBUTION FOR SELECTED PROBABILITIES

Tables C.7–C.19 provide the values of the Chi-square distribution for various degrees of freedom corresponding to selected lower-tail and upper-tail probabilities. Each table is for a specific probability. Within a table, each row corresponds to the denominator degrees of freedom for the F distribution, and each column corresponds to the numerator degrees of freedom. The F distribution with those numerator and denominator degrees of freedom has that table's probability of being less than or equal to the value shown where the row and column intersect when the table corresponds to lower-tail probability. When the table corresponds to upper-tail probability, the F distribution has that probability of being greater than the value shown. For example, to find the value of an F distribution with 5 numerator degrees of freedom, 10 denominator degrees of freedom, and an upper-tail probability of 0.05, first the table is found corresponding to that probability. Then the row corresponding to 10 is found and the column corresponding to 5 is found, and the value where they intersect is 3.33. Therefore, $P(F_{5,10} \geq 3.33)$ is 0.05. Graphically, this probability is represented as shown in Figure C.6.

For a given value, the upper-tail probability is equal to 1 minus the lower-tail probability, and vice versa.

Table C.7 Values of t for upper-tail probabilities greater than 0.5.

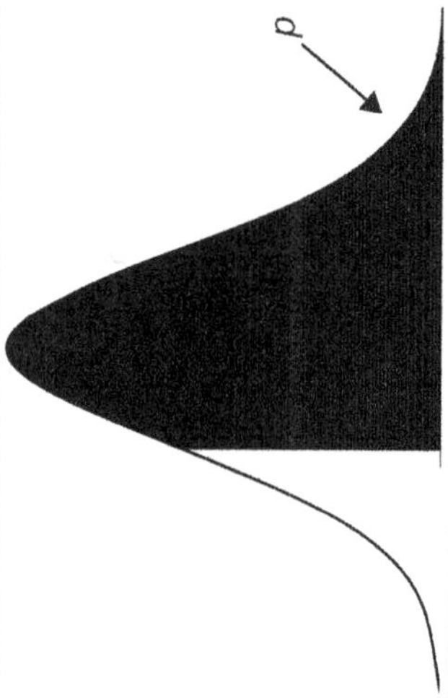

df	0.6	0.7	0.8	0.9	0.95	0.975	0.99	0.995	0.999	0.9995	df
1	−0.325	−0.727	−1.376	−3.078	−6.314	−12.706	−31.821	−63.657	−318.309	−636.619	1
2	−0.289	−0.617	−1.061	−1.886	−2.920	−4.303	−6.965	−9.925	−22.327	−31.599	2
3	−0.277	−0.584	−0.978	−1.638	−2.353	−3.182	−4.541	−5.841	−10.215	−12.924	3
4	−0.271	−0.569	−0.941	−1.533	−2.132	−2.776	−3.747	−4.604	−7.173	−8.610	4
5	−0.267	−0.559	−0.920	−1.476	−2.015	−2.571	−3.365	−4.032	−5.893	−6.869	5
6	−0.265	−0.553	−0.906	−1.440	−1.943	−2.447	−3.143	−3.707	−5.208	−5.959	6
7	0.263	−0.549	−0.896	−1.415	−1.895	−2.365	−2.998	−3.499	−4.785	−5.408	7
8	−0.262	−0.546	−0.889	−1.397	−1.860	−2.306	−2.896	−3.355	−4.501	−5.041	8
9	−0.261	−0.543	−0.883	−1.383	−1.833	−2.262	−2.821	−3.250	−4.297	−4.781	9
10	−0.260	−0.542	−0.879	−1.372	−1.812	−2.228	−2.764	−3.169	−4.144	−4.587	10
11	−0.260	−0.540	−0.876	−1.363	−1.796	−2.201	−2.718	−3.106	−4.025	−4.437	11
12	−0.259	−0.539	−0.873	−1.356	−1.782	−2.179	−2.681	−3.055	−3.930	−4.318	12
13	−0.259	−0.538	−0.870	−1.350	−1.771	−2.160	−2.650	−3.012	−3.852	−4.221	13
14	−0.258	−0.537	−0.868	−1.345	−1.761	−2.145	−2.624	−2.977	−3.787	−4.140	14
15	−0.258	−0.536	−0.866	−1.341	−1.753	−2.131	−2.602	−2.947	−3.733	−4.073	15

df	0.4	0.3	0.2	0.1	0.05	0.025	0.01	0.005	0.001	0.0005	df
16	−0.258	−0.535	−0.865	−1.337	−1.746	−2.120	−2.583	−2.921	−3.686	−4.015	16
17	0.257	−0.534	−0.863	−1.333	−1.740	−2.110	−2.567	−2.898	−3.646	−3.965	17
18	−0.257	−0.534	−0.862	−1.330	−1.734	−2.101	−2.552	−2.878	−3.610	−3.922	18
19	−0.257	−0.533	−0.861	−1.328	−1.729	−2.093	−2.539	−2.861	−3.579	−3.883	19
20	−0.257	−0.533	−0.860	−1.325	−1.725	−2.086	−2.528	−2.845	−3.552	−3.850	20
21	−0.257	−0.532	−0.859	−1.323	−1.721	−2.080	−2.518	−2.831	−3.527	−3.819	21
22	−0.256	−0.532	−0.858	−1.321	−1.717	−2.074	−2.508	−2.819	−3.505	−3.792	22
23	−0.256	−0.532	−0.858	−1.319	−1.714	−2.069	−2.500	−2.807	−3.485	−3.768	23
24	−0.256	−0.531	−0.857	−1.318	−1.711	−2.064	−2.492	−2.797	−3.467	−3.745	24
25	−0.256	−0.531	−0.856	−1.316	−1.708	−2.060	−2.485	−2.787	−3.450	−3.725	25
26	−0.256	−0.531	−0.856	−1.315	−1.706	−2.056	−2.479	−2.779	−3.435	−3.707	26
27	−0.256	−0.531	−0.855	−1.314	−1.703	−2.052	−2.473	−2.771	−3.421	−3.690	27
28	−0.256	−0.530	−0.855	−1.313	−1.701	−2.048	−2.467	−2.763	−3.408	−3.674	28
29	−0.256	−0.530	−0.854	−1.311	−1.699	−2.045	−2.462	−2.756	−3.396	−3.659	29
30	−0.256	−0.530	−0.854	−1.310	−1.697	−2.042	−2.457	−2.750	−3.385	−3.646	30
40	−0.255	−0.529	−0.851	−1.303	−1.684	−2.021	−2.423	−2.704	−3.307	−3.551	40
50	−0.255	−0.528	−0.849	−1.299	−1.676	−2.009	−2.403	−2.678	−3.261	−3.496	50
60	−0.254	−0.527	−0.848	−1.296	−1.671	−2.000	−2.390	−2.660	−3.232	−3.460	60
80	−0.254	−0.526	−0.846	−1.292	−1.664	−1.990	−2.374	−2.639	−3.195	−3.416	80
100	−0.254	−0.526	−0.845	−1.290	−1.660	−1.984	−2.364	−2.626	−3.174	−3.390	100
200	−0.254	−0.525	−0.843	−1.286	−1.653	−1.972	−2.345	−2.601	−3.131	−3.340	200
500	−0.253	−0.525	−0.842	−1.283	−1.648	−1.965	−2.334	−2.586	−3.107	−3.310	500

Table C.8 Values of t for upper-tail probabilities less than 0.5.

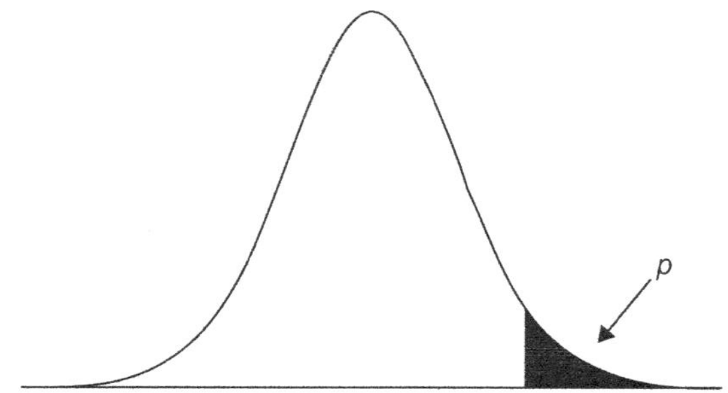

df	0.4	0.3	0.2	0.1	0.05	0.025	0.01	0.005	0.001	0.0005	df
1	0.325	0.727	1.376	3.078	6.314	12.706	31.821	63.657	318.309	636.619	1
2	0.289	0.617	1.061	1.886	2.920	4.303	6.965	9.925	22.327	31.599	2
3	0.277	0.584	0.978	1.638	2.353	3.182	4.541	5.841	10.215	12.924	3
4	0.271	0.569	0.941	1.533	2.132	2.776	3.747	4.604	7.173	8.610	4
5	0.267	0.559	0.920	1.476	2.015	2.571	3.365	4.032	5.893	6.869	5
6	0.265	0.553	0.906	1.440	1.943	2.447	3.143	3.707	5.208	5.959	6
7	0.263	0.549	0.896	1.415	1.895	2.365	2.998	3.499	4.785	5.408	7
8	0.262	0.546	0.889	1.397	1.860	2.306	2.896	3.355	4.501	5.041	8
9	0.261	0.543	0.883	1.383	1.833	2.262	2.821	3.250	4.297	4.781	9
10	0.260	0.542	0.879	1.372	1.812	2.228	2.764	3.169	4.144	4.587	10
11	0.260	0.540	0.876	1.363	1.796	2.201	2.718	3.106	4.025	4.437	11
12	0.259	0.539	0.873	1.356	1.782	2.179	2.681	3.055	3.930	4.318	12
13	0.259	0.538	0.870	1.350	1.771	2.160	2.650	3.012	3.852	4.221	13
14	0.258	0.537	0.868	1.345	1.761	2.145	2.624	2.977	3.787	4.140	14
15	0.258	0.536	0.866	1.341	1.753	2.131	2.602	2.947	3.733	4.073	15

df	0.4	0.3	0.2	0.1	0.05	0.025	0.01	0.005	0.001	0.0005	df
16	0.258	0.535	0.865	1.337	1.746	2.120	2.583	2.921	3.686	4.015	16
17	0.257	0.534	0.863	1.333	1.740	2.110	2.567	2.898	3.646	3.965	17
18	0.257	0.534	0.862	1.330	1.734	2.101	2.552	2.878	3.610	3.922	18
19	0.257	0.533	0.861	1.328	1.729	2.093	2.539	2.861	3.579	3.883	19
20	0.257	0.533	0.860	1.325	1.725	2.086	2.528	2.845	3.552	3.850	20
21	0.257	0.532	0.859	1.323	1.721	2.080	2.518	2.831	3.527	3.819	21
22	0.256	0.532	0.858	1.321	1.717	2.074	2.508	2.819	3.505	3.792	22
23	0.256	0.532	0.858	1.319	1.714	2.069	2.500	2.807	3.485	3.768	23
24	0.256	0.531	0.857	1.318	1.711	2.064	2.492	2.797	3.467	3.745	24
25	0.256	0.531	0.856	1.316	1.708	2.060	2.485	2.787	3.450	3.725	25
26	0.256	0.531	0.856	1.315	1.706	2.056	2.479	2.779	3.435	3.707	26
27	0.256	0.531	0.855	1.314	1.703	2.052	2.473	2.771	3.421	3.690	27
28	0.256	0.530	0.855	1.313	1.701	2.048	2.467	2.763	3.408	3.674	28
29	0.256	0.530	0.854	1.311	1.699	2.045	2.462	2.756	3.396	3.659	29
30	0.256	0.530	0.854	1.310	1.697	2.042	2.457	2.750	3.385	3.646	30
40	0.255	0.529	0.851	1.303	1.684	2.021	2.423	2.704	3.307	3.551	40
50	0.255	0.528	0.849	1.299	1.676	2.009	2.403	2.678	3.261	3.496	50
60	0.254	0.527	0.848	1.296	1.671	2.000	2.390	2.660	3.232	3.460	60
80	0.254	0.526	0.846	1.292	1.664	1.990	2.374	2.639	3.195	3.416	80
100	0.254	0.526	0.845	1.290	1.660	1.984	2.364	2.626	3.174	3.390	100
200	0.254	0.525	0.843	1.286	1.653	1.972	2.345	2.601	3.131	3.340	200
500	0.253	0.525	0.842	1.283	1.648	1.965	2.334	2.586	3.107	3.310	500

Table C.9 Upper-tail probabilities of the Chi-Square distribution.

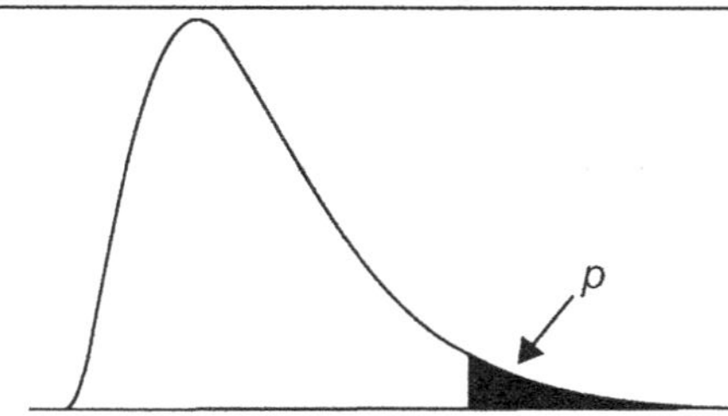

df	0.995	0.990	0.975	0.900	0.750	0.500	0.250	0.100	0.050	0.025	0.010	0.005	df
1	0.00004	0.0002	0.001	0.02	0.10	0.45	1.32	2.71	3.84	5.02	6.63	7.88	1
2	0.01	0.02	0.05	0.21	0.58	1.39	2.77	4.61	5.99	7.38	9.21	10.60	2
3	0.07	0.11	0.22	0.58	1.21	2.37	4.11	6.25	7.81	9.35	11.34	12.84	3
4	0.21	0.30	0.48	1.06	1.92	3.36	5.39	7.78	9.49	11.14	13.28	14.86	4
5	0.41	0.55	0.83	1.61	2.67	4.35	6.63	9.24	11.07	12.83	15.09	16.75	5
6	0.68	0.87	1.24	2.20	3.45	5.35	7.84	10.64	12.59	14.45	16.81	18.55	6
7	0.99	1.24	1.69	2.83	4.25	6.35	9.04	12.02	14.07	16.01	18.48	20.28	7
8	1.34	1.65	2.18	3.49	5.07	7.34	10.22	13.36	15.51	17.53	20.09	21.95	8
9	1.73	2.09	2.70	4.17	5.90	8.34	11.39	14.68	16.92	19.02	21.67	23.59	9
10	2.16	2.56	3.25	4.87	6.74	9.34	12.55	15.99	18.31	20.48	23.21	25.19	10
11	2.60	3.05	3.82	5.58	7.58	10.34	13.70	17.28	19.68	21.92	24.72	26.76	11
12	3.07	3.57	4.40	6.30	8.44	11.34	14.85	18.55	21.03	23.34	26.22	28.30	12
13	3.57	4.11	5.01	7.04	9.30	12.34	15.98	19.81	22.36	24.74	27.69	29.82	13
14	4.07	4.66	5.63	7.79	10.17	13.34	17.12	21.06	23.68	26.12	29.14	31.32	14
15	4.60	5.23	6.26	8.55	11.04	14.34	18.25	22.31	25.00	27.49	30.58	32.80	15
16	5.14	5.81	6.91	9.31	11.91	15.34	19.37	23.54	26.30	28.85	32.00	34.27	16

df	0.995	0.990	0.975	0.900	0.750	0.500	0.250	0.100	0.050	0.025	0.010	0.005	df
17	5.70	6.41	7.56	10.09	12.79	16.34	20.49	24.77	27.59	30.19	33.41	35.72	17
18	6.26	7.01	8.23	10.86	13.68	17.34	21.60	25.99	28.87	31.53	34.81	37.16	18
19	6.84	7.63	8.91	11.65	14.56	18.34	22.72	27.20	30.14	32.85	36.19	38.58	19
20	7.43	8.26	9.59	12.44	15.45	19.34	23.83	28.41	31.41	34.17	37.57	40.00	20
21	8.03	8.90	10.28	13.24	16.34	20.34	24.93	29.62	32.67	35.48	38.93	41.40	21
22	8.64	9.54	10.98	14.04	17.24	21.34	26.04	30.81	33.92	36.78	40.29	42.80	22
23	9.26	10.20	11.69	14.85	18.14	22.34	27.14	32.01	35.17	38.08	41.64	44.18	23
24	9.89	10.86	12.40	15.66	19.04	23.34	28.24	33.20	36.42	39.36	42.98	45.56	24
25	10.52	11.52	13.12	16.47	19.94	24.34	29.34	34.38	37.65	40.65	44.31	46.93	25
26	11.16	12.20	13.84	17.29	20.84	25.34	30.43	35.56	38.89	41.92	45.64	48.29	26
27	11.81	12.88	14.57	18.11	21.75	26.34	31.53	36.74	40.11	43.19	46.96	49.64	27
28	12.46	13.56	15.31	18.94	22.66	27.34	32.62	37.92	41.34	44.46	48.28	50.99	28
29	13.12	14.26	16.05	19.77	23.57	28.34	33.71	39.09	42.56	45.72	49.59	52.34	29
30	13.79	14.95	16.79	20.60	24.48	29.34	34.80	40.26	43.77	46.98	50.89	53.67	30
40	20.71	22.16	24.43	29.05	33.66	39.34	45.62	51.81	55.76	59.34	63.69	66.77	40
50	27.99	29.71	32.36	37.69	42.94	49.33	56.33	63.17	67.50	71.42	76.15	79.49	50
60	35.53	37.48	40.48	46.46	52.29	59.33	66.98	74.40	79.08	83.30	88.38	91.95	60
70	43.28	45.44	48.76	55.33	61.70	69.33	77.58	85.53	90.53	95.02	100.43	104.21	70
80	51.17	53.54	57.15	64.28	71.14	79.33	88.13	96.58	101.88	106.63	112.33	116.32	80
90	59.20	61.75	65.65	73.29	80.62	89.33	98.65	107.57	113.15	118.14	124.12	128.30	90
100	67.33	70.06	74.22	82.36	90.13	99.33	109.14	118.50	124.34	129.56	135.81	140.17	100

Table C.10 F distribution values with upper-tail probability of 0.01.

Denominator degrees of freedom	Numerator degrees of freedom: 1	2	3	4	5	6	7	8	9
1	4052.2	4999.5	5403.4	5624.6	5763.6	5859.0	5928.4	5981.1	6022.5
2	98.50	99.00	99.17	99.25	99.30	99.33	99.36	99.37	99.39
3	34.12	30.82	29.46	28.71	28.24	27.91	27.67	27.49	27.35
4	21.20	18.00	16.69	15.98	15.52	15.21	14.98	14.80	14.66
5	16.26	13.27	12.06	11.39	10.97	10.67	10.46	10.29	10.16
6	13.75	10.92	9.78	9.15	8.75	8.47	8.26	8.10	7.98
7	12.25	9.55	8.45	7.85	7.46	7.19	6.99	6.84	6.72
8	11.26	8.65	7.59	7.01	6.63	6.37	6.18	6.03	5.91
9	10.56	8.02	6.99	6.42	6.06	5.80	5.61	5.47	5.35
10	10.04	7.56	6.55	5.99	5.64	5.39	5.20	5.06	4.94
12	9.33	6.93	5.95	5.41	5.06	4.82	4.64	4.50	4.39
15	8.68	6.36	5.42	4.89	4.56	4.32	4.14	4.00	3.89
20	8.10	5.85	4.94	4.43	4.10	3.87	3.70	3.56	3.46
24	7.82	5.61	4.72	4.22	3.90	3.67	3.50	3.36	3.26
30	7.56	5.39	4.51	4.02	3.70	3.47	3.30	3.17	3.07
40	7.31	5.18	4.31	3.83	3.51	3.29	3.12	2.99	2.89
60	7.08	4.98	4.13	3.65	3.34	3.12	2.95	2.82	2.72
120	6.85	4.79	3.95	3.48	3.17	2.96	2.79	2.66	2.56
	1	2	3	4	5	6	7	8	9
	Numerator degrees of freedom								

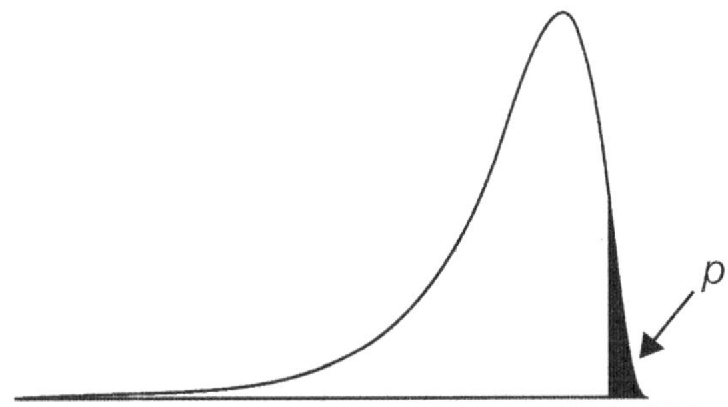

Denominator degrees of freedom	10	12	15	20	24	30	40	60	120
1	6055.8	6106.3	6157.3	6208.7	6234.6	6260.6	6286.8	6313.0	6339.4
2	99.40	99.42	99.43	99.45	99.46	99.47	99.47	99.48	99.49
3	27.23	27.05	26.87	26.69	26.60	26.50	26.41	26.32	26.22
4	14.55	14.37	14.20	14.02	13.93	13.84	13.75	13.65	13.56
5	10.05	9.89	9.72	9.55	9.47	9.38	9.29	9.20	9.11
6	7.87	7.72	7.56	7.40	7.31	7.23	7.14	7.06	6.97
7	6.62	6.47	6.31	6.16	6.07	5.99	5.91	5.82	5.74
8	5.81	5.67	5.52	5.36	5.28	5.20	5.12	5.03	4.95
9	5.26	5.11	4.96	4.81	4.73	4.65	4.57	4.48	4.40
10	4.85	4.71	4.56	4.41	4.33	4.25	4.17	4.08	4.00
12	4.30	4.16	4.01	3.86	3.78	3.70	3.62	3.54	3.45
15	3.80	3.67	3.52	3.37	3.29	3.21	3.13	3.05	2.96
20	3.37	3.23	3.09	2.94	2.86	2.78	2.69	2.61	2.52
24	3.17	3.03	2.89	2.74	2.66	2.58	2.49	2.40	2.31
30	2.98	2.84	2.70	2.55	2.47	2.39	2.30	2.21	2.11
40	2.80	2.66	2.52	2.37	2.29	2.20	2.11	2.02	1.92
60	2.63	2.50	2.35	2.20	2.12	2.03	1.94	1.84	1.73
120	2.47	2.34	2.19	2.03	1.95	1.86	1.76	1.66	1.53

Numerator degrees of freedom

Table C.11 F distribution values with upper-tail probability of 0.025.

	Numerator degrees of freedom								
Denominator degrees of freedom	1	2	3	4	5	6	7	8	9
1	647.79	799.5	864.16	899.58	921.85	937.11	948.22	956.66	963.28
2	38.506	39	39.165	39.248	39.298	39.331	39.355	39.373	39.387
3	17.443	16.044	15.439	15.101	14.885	14.735	14.624	14.54	14.473
4	12.218	10.649	9.9792	9.6045	9.3645	9.1973	9.0741	8.9796	8.9047
5	10.007	8.4336	7.7636	7.3879	7.1464	6.9777	6.8531	6.7572	6.6811
6	8.8131	7.2599	6.5988	6.2272	5.9876	5.8198	5.6955	5.5996	5.5234
7	8.0727	6.5415	5.8898	5.5226	5.2852	5.1186	4.9949	4.8993	4.8232
8	7.5709	6.0595	5.416	5.0526	4.8173	4.6517	4.5286	4.4333	4.3572
9	7.2093	5.7147	5.0781	4.7181	4.4844	4.3197	4.197	4.102	4.026
10	6.9367	5.4564	4.8256	4.4683	4.2361	4.0721	3.9498	3.8549	3.779
12	6.5538	5.0959	4.4742	4.1212	3.8911	3.7283	3.6065	3.5118	3.4358
15	6.1995	4.765	4.1528	3.8043	3.5764	3.4147	3.2934	3.1987	3.1227
20	5.8715	4.4613	3.8587	3.5147	3.2891	3.1283	3.0074	2.9128	2.8365
24	5.7166	4.3187	3.7211	3.3794	3.1548	2.9946	2.8738	2.7791	2.7027
30	5.5675	4.1821	3.5894	3.2499	3.0265	2.8667	2.746	2.6513	2.5746
40	5.4239	4.051	3.4633	3.1261	2.9037	2.7444	2.6238	2.5289	2.4519
60	5.2856	3.9253	3.3425	3.0077	2.7863	2.6274	2.5068	2.4117	2.3344
120	5.1523	3.8046	3.2269	2.8943	2.674	2.5154	2.3948	2.2994	2.2217
	1	2	3	4	5	6	7	8	9
	Numerator degrees of freedom								

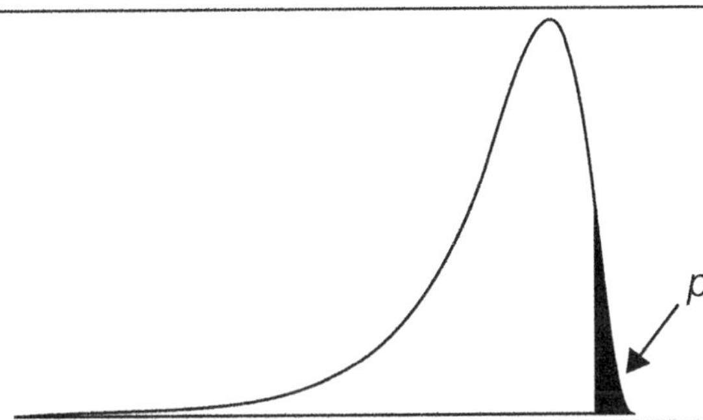

Numerator degrees of freedom									
10	12	15	20	24	30	40	60	120	Denominator degrees of freedom
968.63	976.71	984.87	993.1	997.25	1001.4	1005.6	1009.8	1014	1
39.398	39.415	39.431	39.448	39.456	39.465	39.473	39.481	39.49	2
14.419	14.337	14.253	14.167	14.124	14.081	14.037	13.992	13.947	3
8.8439	8.7512	8.6565	8.5599	8.5109	8.4613	8.4111	8.3604	8.3092	4
6.6192	6.5245	6.4277	6.3286	6.278	6.2269	6.175	6.1225	6.0693	5
5.4613	5.3662	5.2687	5.1684	5.1172	5.0652	5.0125	4.9589	4.9044	6
4.7611	4.6658	4.5678	4.4667	4.415	4.3624	4.3089	4.2544	4.1989	7
4.2951	4.1997	4.1012	3.9995	3.9472	3.894	3.8398	3.7844	3.7279	8
3.9639	3.8682	3.7694	3.6669	3.6142	3.5604	3.5055	3.4493	3.3918	9
3.7168	3.62609	3.5217	3.4185	3.3654	3.311	3.2554	3.1984	3.1399	10
3.3736	3.2773	3.1772	3.0728	3.0187	2.9633	2.9063	2.8478	2.7874	12
3.0602	2.9633	2.8621	2.7559	2.7006	2.6437	2.585	2.5242	2.4611	15
2.7737	2.6758	2.5731	2.4645	2.4076	2.3486	2.2873	2.2234	2.1562	20
2.6396	2.5411	2.4374	2.3273	2.2693	2.209	2.146	2.0799	2.0099	24
2.5112	2.412	2.3072	2.1952	2.1359	2.0739	2.0089	1.94	1.8664	30
2.3882	2.2882	2.1819	2.0677	2.0069	1.9429	1.8752	1.8028	1.7242	40
2.2702	2.1692	2.0613	1.9445	1.8817	1.8152	1.744	1.6668	1.581	60
2.157	2.0548	1.945	1.8249	1.7597	1.6899	1.6141	1.5299	1.4327	120
10	12	15	20	24	30	40	60	120	
Numerator degrees of freedom									

Table C.12 F distribution values with upper-tail probability of 0.05.

Denominator degrees of freedom	Numerator degrees of freedom								
	1	2	3	4	5	6	7	8	9
1	161.4	199.5	215.7	224.6	230.2	234.0	236.8	238.9	240.5
2	18.51	19.00	19.16	19.25	19.30	19.33	19.35	19.37	19.38
3	10.13	9.55	9.28	9.12	9.01	8.94	8.89	8.85	8.81
4	7.71	6.94	6.59	6.39	6.26	6.16	6.09	6.04	6.00
5	6.61	5.79	5.41	5.19	5.05	4.95	4.88	4.82	4.77
6	5.99	5.14	4.76	4.53	4.39	4.28	4.21	4.15	4.10
7	5.59	4.74	4.35	4.12	3.97	3.87	3.79	3.73	3.68
8	5.32	4.46	4.07	3.84	3.69	3.58	3.50	3.44	3.39
9	5.12	4.26	3.86	3.63	3.48	3.37	3.29	3.23	3.18
10	4.96	4.10	3.71	3.48	3.33	3.22	3.14	3.07	3.02
12	4.75	3.89	3.49	3.26	3.11	3.00	2.91	2.85	2.80
15	4.54	3.68	3.29	3.06	2.90	2.79	2.71	2.64	2.59
20	4.35	3.49	3.10	2.87	2.71	2.60	2.51	2.45	2.39
24	4.26	3.40	3.01	2.78	2.62	2.51	2.42	2.36	2.30
30	4.17	3.32	2.92	2.69	2.53	2.42	2.33	2.27	2.21
40	4.08	3.23	2.84	2.61	2.45	2.34	2.25	2.18	2.12
60	4.00	3.15	2.76	2.53	2.37	2.25	2.17	2.10	2.04
120	3.92	3.07	2.68	2.45	2.29	2.18	2.09	2.02	1.96
	1	2	3	4	5	6	7	8	9
	Numerator degrees of freedom								

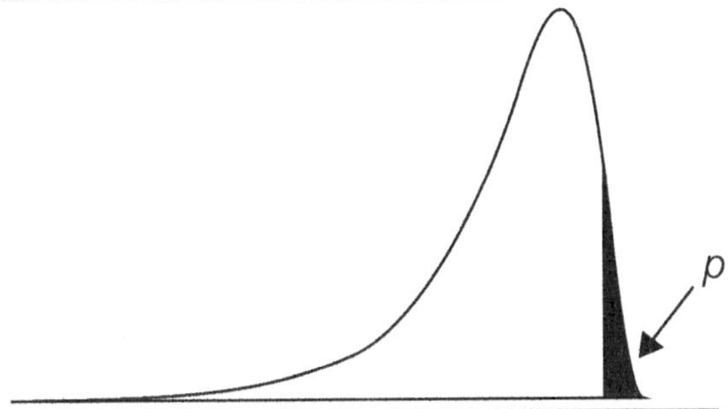

Numerator degrees of freedom									
10	12	15	20	24	30	40	60	120	Denominator degrees of freedom
241.9	243.9	245.9	248.0	249.1	250.1	251.1	252.2	253.3	1
19.40	19.41	19.43	19.45	19.45	19.46	19.47	19.48	19.49	2
8.79	8.74	8.70	8.66	8.64	8.62	8.59	8.57	8.55	3
5.96	5.91	5.86	5.80	5.77	5.75	5.72	5.69	5.66	4
4.74	4.68	4.62	4.56	4.53	4.50	4.46	4.43	4.40	5
4.06	4.00	3.94	3.87	3.84	3.81	3.77	3.74	3.70	6
3.64	3.57	3.51	3.44	3.41	3.38	3.34	3.30	3.27	7
3.35	3.28	3.22	3.15	3.12	3.08	3.04	3.01	2.97	8
3.14	3.07	3.01	2.94	2.90	2.86	2.83	2.79	2.75	9
2.98	2.91	2.85	2.77	2.74	2.70	2.66	2.62	2.58	10
2.75	2.69	2.62	2.54	2.51	2.47	2.43	2.38	2.34	12
2.54	2.48	2.40	2.33	2.29	2.25	2.20	2.16	2.11	15
2.35	2.28	2.20	2.12	2.08	2.04	1.99	1.95	1.90	20
2.25	2.18	2.11	2.03	1.98	1.94	1.89	1.84	1.79	24
2.16	2.09	2.01	1.93	1.89	1.84	1.79	1.74	1.68	30
2.08	2.00	1.92	1.84	1.79	1.74	1.69	1.64	1.58	40
1.99	1.92	1.84	1.75	1.70	1.65	1.59	1.53	1.47	60
1.91	1.83	1.75	1.66	1.61	1.55	1.50	1.43	1.35	120
10	12	15	20	24	30	40	60	120	
Numerator degrees of freedom									

Table C.13 F distribution values with upper-tail probability of 0.1.

Denominator degrees of freedom	Numerator degrees of freedom: 1	2	3	4	5	6	7	8	9
1	39.9	49.5	53.6	55.8	57.2	58.2	58.9	59.4	59.9
2	8.53	9.00	9.16	9.24	9.29	9.33	9.35	9.37	9.38
3	5.54	5.46	5.39	5.34	5.31	5.28	5.27	5.25	5.24
4	4.54	4.32	4.19	4.11	4.05	4.01	3.98	3.95	3.94
5	4.06	3.78	3.62	3.52	3.45	3.40	3.37	3.34	3.32
6	3.78	3.46	3.29	3.18	3.11	3.05	3.01	2.98	2.96
7	3.59	3.26	3.07	2.96	2.88	2.83	2.78	2.75	2.72
8	3.46	3.11	2.92	2.81	2.73	2.67	2.62	2.59	2.56
9	3.36	3.01	2.81	2.69	2.61	2.55	2.51	2.47	2.44
10	3.29	2.92	2.73	2.61	2.52	2.46	2.41	2.38	2.35
12	3.18	2.81	2.61	2.48	2.39	2.33	2.28	2.24	2.21
15	3.07	2.70	2.49	2.36	2.27	2.21	2.16	2.12	2.09
20	2.97	2.59	2.38	2.25	2.16	2.09	2.04	2.00	1.96
24	2.93	2.54	2.33	2.19	2.10	2.04	1.98	1.94	1.91
30	2.88	2.49	2.28	2.14	2.05	1.98	1.93	1.88	1.85
40	2.84	2.44	2.23	2.09	2.00	1.93	1.87	1.83	1.79
60	2.79	2.39	2.18	2.04	1.95	1.87	1.82	1.77	1.74
120	2.75	2.35	2.13	1.99	1.90	1.82	1.77	1.72	1.68
	1	2	3	4	5	6	7	8	9
	Numerator degrees of freedom								

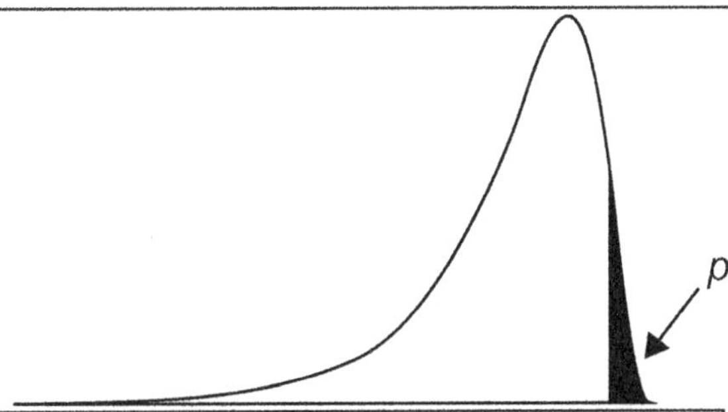

Numerator degrees of freedom									
10	**12**	**15**	**20**	**24**	**30**	**40**	**60**	**120**	**Denominator degrees of freedom**
60.2	60.7	61.2	61.7	62.0	62.3	62.5	62.8	63.1	**1**
9.39	9.41	9.42	9.44	9.45	9.46	9.47	9.47	9.48	**2**
5.23	5.22	5.20	5.18	5.18	5.17	5.16	5.15	5.14	**3**
3.92	3.90	3.87	3.84	3.83	3.82	3.80	3.79	3.78	**4**
3.30	3.27	3.24	3.21	3.19	3.17	3.16	3.14	3.12	**5**
2.94	2.90	2.87	2.84	2.82	2.80	2.78	2.76	2.74	**6**
2.70	2.67	2.63	2.59	2.58	2.56	2.54	2.51	2.49	**7**
2.54	2.50	2.46	2.42	2.40	2.38	2.36	2.34	2.32	**8**
2.42	2.38	2.34	2.30	2.28	2.25	2.23	2.21	2.18	**9**
2.32	2.28	2.24	2.20	2.18	2.16	2.13	2.11	2.08	**10**
2.19	2.15	2.10	2.06	2.04	2.01	1.99	1.96	1.93	**12**
2.06	2.02	1.97	1.92	1.90	1.87	1.85	1.82	1.79	**15**
1.94	1.89	1.84	1.79	1.77	1.74	1.71	1.68	1.64	**20**
1.88	1.83	1.78	1.73	1.70	1.67	1.64	1.61	1.57	**24**
1.82	1.77	1.72	1.67	1.64	1.61	1.57	1.54	1.50	**30**
1.76	1.71	1.66	1.61	1.57	1.54	1.51	1.47	1.42	**40**
1.71	1.66	1.60	1.54	1.51	1.48	1.44	1.40	1.35	**60**
1.65	1.60	1.55	1.48	1.45	1.41	1.37	1.32	1.26	**120**
10	**12**	**15**	**20**	**24**	**30**	**40**	**60**	**120**	

Numerator degrees of freedom

Table C.14 F distribution values with upper-tail probability of 0.2.

Denominator degrees of freedom	Numerator degrees of freedom 1	2	3	4	5	6	7	8	9
1	9.4721	12	13.064	13.644	14.008	14.258	14.439	14.577	14.685
2	3.5556	4	4.1563	4.2361	4.2844	4.3168	4.3401	4.3576	4.3712
3	2.6822	2.886	2.9359	2.9555	2.9652	2.9707	2.9741	2.9763	2.9779
4	2.3507	2.4721	2.4847	2.4826	2.478	2.4733	2.4691	2.4654	2.4623
5	2.1782	2.2591	2.253	2.2397	2.2275	2.2174	2.209	2.2021	2.1963
6	2.0729	2.1299	2.1126	2.0924	2.0755	2.0619	2.0508	2.0417	2.0342
7	2.002	2.0434	2.0186	1.9937	1.9736	1.9575	1.9445	1.9339	1.9251
8	1.9511	1.9814	1.9513	1.923	1.9005	1.8826	1.8682	1.8564	1.8466
9	1.9128	1.9349	1.9007	1.87	1.8455	1.8262	1.8107	1.7979	1.7874
10	1.8829	1.8986	1.8614	1.8286	1.8027	1.7823	1.7658	1.7523	1.7411
12	1.8393	1.846	1.8042	1.7684	1.7403	1.7182	1.7003	1.6856	1.6734
15	1.7972	1.7952	1.749	1.7103	1.6801	1.6561	1.6368	1.6209	1.6076
20	1.7565	1.7462	1.6958	1.6543	1.6218	1.596	1.5752	1.5581	1.5436
24	1.7367	1.7224	1.6699	1.6269	1.5933	1.5667	1.5451	1.5272	1.5122
30	1.7172	1.699	1.6445	1.6001	1.5654	1.5378	1.5154	1.4968	1.4812
40	1.698	1.676	1.6195	1.5737	1.5379	1.5093	1.4861	1.4668	1.4505
60	1.6792	1.6534	1.595	1.5478	1.5108	1.4813	1.4572	1.4371	1.4201
120	1.6606	1.6312	1.5709	1.5222	1.4841	1.4536	1.4287	1.4078	1.3901
	1	2	3	4	5	6	7	8	9
	Numerator degrees of freedom								

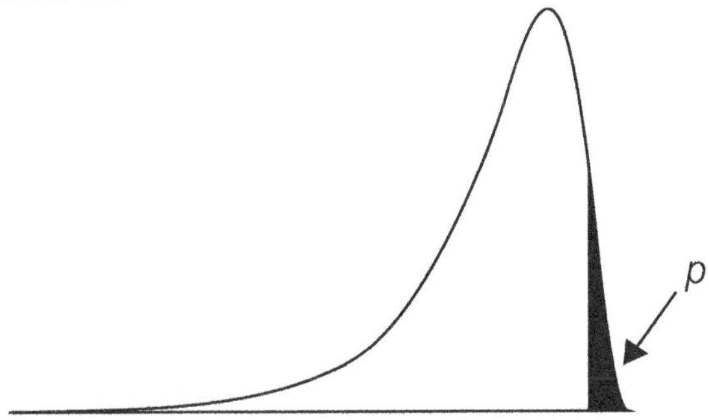

Numerator degrees of freedom										
10	12	15	20	24	30	40	60	120		Denominator degrees of freedom
14.772	14.904	15.037	15.171	15.238	15.306	15.374	15.442	15.511	1	
4.3822	4.3986	4.4151	4.4316	4.4399	4.4482	4.4565	4.4648	4.4731	2	
2.9791	2.9806	2.9819	2.983	2.9834	2.9838	2.9841	2.9843	2.9845	3	
2.4596	2.4552	2.4503	2.445	2.4421	2.4392	2.4361	2.4329	2.4295	4	
2.1914	2.1835	2.1751	2.166	2.1612	2.1562	2.1511	2.1457	2.1402	5	
2.0278	2.0177	2.0068	1.9951	1.9889	1.9825	1.9758	1.9689	1.9618	6	
1.9176	1.9058	1.893	1.8793	1.8721	1.8646	1.8567	1.8486	1.8402	7	
1.8383	1.8251	1.8109	1.7956	1.7875	1.7791	1.7703	1.7611	1.7515	8	
1.7784	1.7642	1.7488	1.7321	1.7233	1.7141	1.7044	1.6944	1.6839	9	
1.7316	1.7164	1.7	1.6823	1.6728	1.6629	1.6525	1.6417	1.6303	10	
1.663	1.6464	1.6284	1.6088	1.5983	1.5873	1.5757	1.5635	1.5506	12	
1.5964	1.5782	1.5584	1.5367	1.525	1.5127	1.4996	1.4858	1.4711	15	
1.5313	1.5115	1.4897	1.4656	1.4525	1.4385	1.4237	1.4078	1.3906	20	
1.4994	1.4787	1.4558	1.4302	1.4163	1.4014	1.3855	1.3682	1.3495	24	
1.4678	1.4461	1.422	1.3949	1.3801	1.3641	1.3469	1.328	1.3073	30	
1.4365	1.4137	1.3883	1.3596	1.3437	1.3264	1.3076	1.2869	1.2635	40	
1.4055	1.3816	1.3547	1.3241	1.3069	1.2881	1.2674	1.2441	1.2172	60	
1.3748	1.3496	1.3211	1.2282	1.2696	1.2489	1.2257	1.1989	1.1666	120	
10	12	15	20	24	30	40	60	120		
Numerator degrees of freedom										

Table C.15 F distribution values with lower-tail probability of 0.01.

Denominator degrees of freedom	Numerator degrees of freedom								
	1	2	3	4	5	6	7	8	9
1	0.0002	0.01	0.029	0.0472	0.0615	0.0728	0.0817	0.0888	0.0947
2	0.0002	0.01	0.032	0.0556	0.0753	0.0915	0.1047	0.1156	0.1247
3	0.0002	0.01	0.034	0.0599	0.0829	0.1023	0.1183	0.1317	0.143
4	0.0002	0.01	0.035	0.0626	0.0878	0.1093	0.1274	0.1427	0.1557
5	0.0002	0.01	0.035	0.0644	0.0912	0.1143	0.134	0.1508	0.1651
6	0.0002	0.01	0.036	0.0658	0.0937	0.1181	0.1391	0.157	0.1724
7	0.0002	0.01	0.036	0.0668	0.0956	0.1211	0.143	0.1619	0.1782
8	0.0002	0.01	0.036	0.0676	0.0972	0.1234	0.1462	0.1659	0.1829
9	0.0002	0.01	0.037	0.0682	0.0984	0.1254	0.1488	0.1692	0.1869
10	0.0002	0.01	0.037	0.0687	0.0995	0.127	0.1511	0.172	0.1902
12	0.0002	0.01	0.037	0.0696	0.1011	0.1296	0.1546	0.1765	0.1956
15	0.0002	0.01	0.037	0.0704	0.1029	0.1323	0.1584	0.1813	0.2015
20	0.0002	0.01	0.037	0.0713	0.1047	0.1352	0.1625	0.1866	0.208
24	0.0002	0.01	0.038	0.0718	0.1056	0.1367	0.1646	0.1894	0.2115
30	0.0002	0.01	0.038	0.0723	0.1066	0.1383	0.1669	0.1924	0.2151
40	0.0002	0.01	0.038	0.0728	0.1076	0.14	0.1692	0.1955	0.219
60	0.0002	0.01	0.038	0.0732	0.1087	0.1417	0.1717	0.1987	0.2231
120	0.0002	0.01	0.038	0.0738	0.1097	0.1435	0.1743	0.2022	0.2274
	1	2	3	4	5	6	7	8	9
	Numerator degrees of freedom								

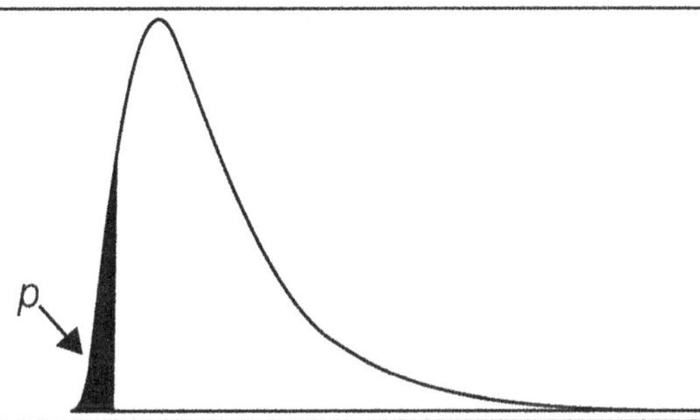

Numerator degrees of freedom									
10	12	15	20	24	30	40	60	120	Denominator degrees of freedom
0.0996	0.1072	0.1152	0.1235	0.1278	0.1322	0.1367	0.1413	0.146	1
0.1323	0.1444	0.1573	0.171	0.1781	0.1855	0.1931	0.2009	0.2089	2
0.1526	0.168	0.1846	0.2025	0.212	0.2217	0.2319	0.2424	0.2532	3
0.1668	1.1848	0.2044	0.2257	0.2371	0.2489	0.2612	0.274	0.2874	4
0.1774	0.1975	0.2195	0.2437	0.2567	0.2703	0.2846	0.2995	0.3151	5
0.1857	0.2074	0.2316	0.2853	0.2727	0.2879	0.3039	0.3206	0.3383	6
0.1923	0.2155	0.2415	0.2704	0.286	0.3026	0.3201	0.3386	0.3582	7
0.1978	0.2223	0.2497	0.2806	0.2974	0.3152	0.3341	0.3542	0.3755	8
0.2023	0.2279	0.2568	0.2893	0.3071	0.3261	0.3463	0.3679	0.3908	9
0.2062	0.2328	0.2628	0.2969	0.3156	0.3357	0.3571	0.38	0.4045	10
0.2125	0.2407	0.2728	0.3095	0.3299	0.3517	0.3753	0.4006	0.428	12
0.2194	0.2494	0.2839	0.3238	0.3462	0.3703	0.3966	0.4251	0.4563	15
0.227	0.2592	0.2966	0.3404	0.3652	0.3924	0.4221	0.455	0.4915	20
0.2311	0.2645	0.3036	0.3497	0.3761	0.405	0.4371	0.4727	0.5128	24
0.2355	0.2702	0.3111	0.3599	0.388	0.4191	0.4538	0.493	0.5376	30
0.2401	0.2763	0.3193	0.3711	0.4012	0.4349	0.473	0.5165	0.5673	40
0.245	0.2828	0.3282	0.3835	0.4161	0.4529	0.4952	0.5446	0.604	60
0.2502	0.2899	0.3379	0.3973	0.4329	0.4738	0.5216	0.5793	0.6523	120
10	12	15	20	24	30	40	60	120	
Numerator degrees of freedom									

Table C.16 F distribution values with lower-tail probability of 0.025.

Denominator degrees of freedom	Numerator degrees of freedom 1	2	3	4	5	6	7	8	9
1	0.0015	0.026	0.0573	0.0818	0.0999	0.1135	0.1239	0.1321	0.1387
2	0.0013	0.0256	0.0623	0.0939	0.1186	0.1377	0.1529	0.165	0.175
3	0.0012	0.0255	0.0648	0.1002	0.1288	0.1515	0.1698	0.1846	0.1969
4	0.0011	0.0255	0.0662	0.1041	0.1354	0.1606	0.1811	0.1979	0.212
5	0.0011	0.0254	0.0672	0.1068	0.1399	0.167	0.1892	0.2076	0.223
6	0.0011	0.0254	0.0679	0.1087	0.1433	0.1718	0.1954	0.215	0.2315
7	0.0011	0.0254	0.0684	0.1102	0.1459	0.1756	0.2002	0.2208	0.2383
8	0.001	0.0254	0.0688	0.114	0.148	0.1786	0.2041	0.2256	0.2438
9	0.001	0.0254	0.0691	0.1123	0.1497	0.181	0.2073	0.2295	0.2484
10	0.001	0.0254	0.0694	0.1131	0.1511	0.1831	0.21	0.2328	0.2523
12	0.001	0.0254	0.0698	0.1143	0.1533	0.1864	0.2143	0.2381	0.2585
15	0.001	0.0254	0.0702	0.1155	0.1556	0.1898	0.2189	0.2438	0.2653
20	0.001	0.0253	0.0706	0.1168	0.158	0.1935	0.2239	0.25	0.2727
24	0.001	0.0253	0.0708	0.1175	0.1593	0.1954	0.2265	0.2533	0.2767
30	0.001	0.0253	0.071	0.1182	0.1606	0.1974	0.2292	0.2568	0.2809
40	0.001	0.0253	0.0712	0.1189	0.1619	0.1995	0.2321	0.2604	0.2853
60	0.001	0.0253	0.0715	0.1196	0.1633	0.2017	0.2351	0.2642	0.2899
120	0.001	0.0253	0.0717	0.1203	0.1648	0.2039	0.2382	0.2682	0.2948
	1	2	3	4	5	6	7	8	9
	Numerator degrees of freedom								

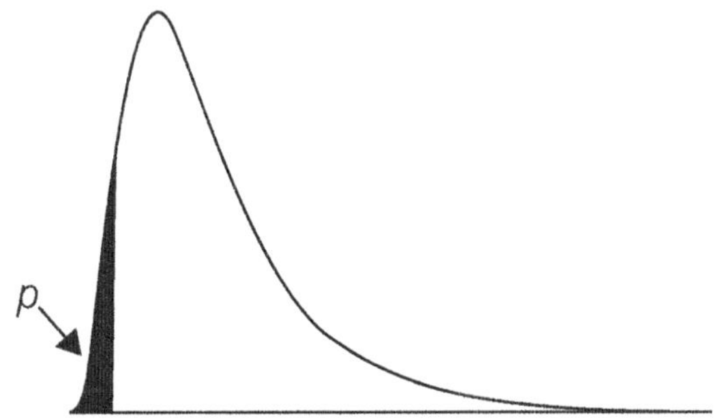

Numerator degrees of freedom									
10	12	15	20	24	30	40	60	120	Denominator degrees of freedom
0.1442	0.1526	0.1613	0.1703	0.1749	0.1796	0.1844	0.1892	0.1941	1
0.1833	0.1962	0.2099	0.2242	0.2315	0.2391	0.2469	0.2548	0.2628	2
0.2072	0.2235	0.2408	0.2592	0.2687	0.2786	0.2887	0.2992	0.3099	3
0.2238	0.2426	0.2629	0.2845	0.2959	0.3077	0.3199	0.3325	0.3455	4
0.2361	0.257	0.2796	0.304	0.317	0.3304	0.3444	0.3589	0.374	5
0.2456	0.2682	0.2929	0.3197	0.3339	0.3488	0.3644	0.3806	0.3976	6
0.2532	0.2773	0.3036	0.3325	0.348	0.3642	0.3811	0.3989	0.4176	7
0.2594	0.2848	0.3126	0.3433	0.3598	0.3772	0.3954	0.4147	0.4349	8
0.2646	0.291	0.3202	0.3525	0.37	0.3884	0.4078	0.4284	0.4501	9
0.269	0.2964	0.3268	0.3605	0.3788	0.3982	0.4187	0.4405	0.4636	10
0.2762	0.3051	0.3375	0.3737	0.3935	0.4146	0.437	0.461	0.4867	12
0.284	0.3147	0.3494	0.3886	0.4103	0.4334	0.4583	0.4851	0.5141	15
0.2925	0.3254	0.3629	0.4058	0.4297	0.4555	0.4836	0.5143	0.548	20
0.2971	0.3313	0.3703	0.4154	0.4407	0.4682	0.4983	0.5314	0.5683	24
0.302	0.3375	0.3783	0.4258	0.4527	0.4822	0.5147	0.5509	0.5917	30
0.3072	0.3441	0.3868	0.4372	0.466	0.4978	0.5333	0.5734	0.6195	40
0.3127	0.3512	0.3962	0.4498	0.4808	0.5155	0.5547	0.6	0.6536	60
0.3185	0.3588	0.4063	0.4638	0.4975	0.5358	0.58	0.6325	0.698	120
10	12	15	20	24	30	40	60	120	
Numerator degrees of freedom									

Table C.17 F distribution values with lower-tail probability of 0.05.

Denominator degrees of freedom	Numerator degrees of freedom 1	2	3	4	5	6	7	8	9
1	0.0062	0.054	0.0987	0.1297	0.1513	0.167	0.1788	0.1881	0.1954
2	0.005	0.0526	0.1047	0.144	0.1728	0.1944	0.2111	0.2243	0.2349
3	0.0046	0.0522	0.1078	0.1517	0.1849	0.2102	0.2301	0.2459	0.2589
4	0.0045	0.052	0.1097	0.1565	0.1926	0.2206	0.2427	0.2606	0.2752
5	0.0043	0.0518	0.1109	0.1598	0.198	0.2279	0.2518	0.2712	0.2872
6	0.0043	0.0517	0.1118	0.1623	0.202	0.2334	0.2587	0.2793	0.2964
7	0.0042	0.0517	0.1125	0.1641	0.2051	0.2377	0.2641	0.2857	0.3037
8	0.0042	0.0516	0.1131	0.1655	0.2075	0.2411	0.2684	0.2909	0.3096
9	0.0042	0.0516	0.1135	0.1667	0.2095	0.244	0.272	0.2951	0.3146
10	0.0041	0.0516	0.1138	0.1677	0.2112	0.2463	0.275	0.2988	0.3187
12	0.0041	0.0515	0.1144	0.1692	0.2138	0.25	0.2797	0.3045	0.3254
15	0.0041	0.0515	0.1149	0.1707	0.2165	0.2539	0.2848	0.3107	0.3327
20	0.004	0.0514	0.1155	0.1723	0.2194	0.2581	0.2903	0.3174	0.3405
24	0.004	0.0514	0.1158	0.1732	0.2209	0.2603	0.2932	0.321	0.3448
30	0.004	0.0514	0.1161	0.174	0.2224	0.2626	0.2962	0.3247	0.3492
40	0.004	0.0514	0.1164	0.1749	0.224	0.265	0.2994	0.3286	0.3539
60	0.004	0.0513	0.1167	0.1758	0.2257	0.2674	0.3026	0.3327	0.3588
120	0.0039	0.0513	0117	0.1767	0.2274	0.2699	0.306	0.337	0.364
	1	2	3	4	5	6	7	8	9
	Numerator degrees of freedom								

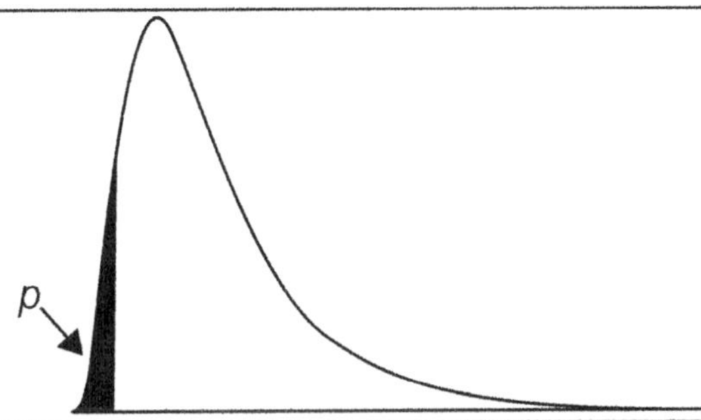

Numerator degrees of freedom									
10	12	15	20	24	30	40	60	120	Denominator degrees of freedom
0.2014	0.2106	0.2201	0.2298	0.2348	0.2398	0.2448	0.2499	0.2551	1
0.2437	0.2574	0.2716	0.2863	0.2939	0.3016	0.3094	0.3174	0.3255	2
0.2697	0.2865	0.3042	0.3227	0.3324	0.3422	0.3523	0.3626	0.3731	3
0.2875	0.3068	0.3273	0.3489	0.3602	0.3718	0.3837	0.396	0.4086	4
0.3007	0.322	0.3447	0.3689	0.3816	0.3947	0.4083	0.4222	0.4367	5
0.3108	0.3338	0.3584	0.3848	0.3987	0.4131	0.4281	0.4436	0.4598	6
0.3189	0.3432	0.3695	0.3978	0.4128	0.4284	0.4446	0.4616	0.4792	7
0.3256	0.3511	0.3787	0.4087	0.4246	0.4413	0.4587	0.4769	0.4959	8
0.3311	0.3576	0.3865	0.4179	0.4347	0.4523	0.4708	0.4902	0.5105	9
0.3358	0.3632	0.3931	0.4259	0.4435	0.462	0.4814	0.5019	0.5234	10
0.3433	0.3722	0.404	0.4391	0.458	0.478	0.4991	0.5215	0.5453	12
0.3515	0.3821	0.4161	0.4539	0.4745	0.4963	0.5196	0.5445	0.5713	15
0.3605	0.3931	0.4296	0.4708	0.4934	0.5177	0.5438	0.5721	0.6029	20
0.3653	0.3991	0.4371	0.4802	0.5041	0.5298	0.5577	0.5882	0.6217	24
0.3704	0.4055	0.4451	0.4904	0.5157	0.5432	0.5733	0.6064	0.6434	30
0.3758	0.4122	0.4537	0.5016	0.5286	0.5581	0.5907	0.6272	0.6688	40
0.3815	0.4194	0.4629	0.5138	0.5428	0.5749	0.6108	0.6518	0.6998	60
0.3876	0.4272	0.473	0.5273	0.558	0.594	0.6343	0.6815	0.7397	120
10	12	15	20	24	30	40	60	120	
Numerator degrees of freedom									

Table C.18 F distribution values with lower-tail probability of 0.1.

Denominator degrees of freedom	Numerator degrees of freedom								
	1	2	3	4	5	6	7	8	9
1	0.0251	0.1173	0.1806	0.22	0.2463	0.2648	0.2786	0.2892	0.2976
2	0.0202	0.1111	0.1831	0.2312	0.2646	0.2887	0.307	0.3212	0.3326
3	0.0187	0.1091	0.1855	0.2386	0.2763	0.3041	0.3253	0.342	0.3555
4	0.0179	0.1082	0.1872	0.2435	0.2841	0.3144	0.3378	0.3563	0.3714
5	0.0175	0.1076	0.1884	0.2469	0.2896	0.3218	0.3468	0.3668	0.3831
6	0.0172	0.1072	0.1892	0.2494	0.2937	0.3274	0.3537	0.3748	0.392
7	0.017	0.107	0.1899	0.2513	0.2969	0.3317	0.3591	0.3811	0.3992
8	0.0168	0.1068	0.1904	0.2528	0.2995	0.3352	0.3634	0.3862	0.405
9	0.0167	0.1066	0.1908	0.2541	0.3015	0.3381	0.367	0.3904	0.4098
10	0.0166	0.1065	0.1912	0.2551	0.3033	0.3405	0.37	0.394	0.4139
12	0.0165	0.1063	0.1917	0.2567	0.306	0.3443	0.3748	0.3997	0.4204
15	0.0163	0.1061	0.1923	0.2584	0.3088	0.3483	0.3799	0.4058	0.4274
20	0.0162	0.1059	0.1929	0.2601	0.3119	0.3526	0.3854	0.4124	0.4351
24	0.0161	0.1058	0.1932	0.261	0.3134	0.3548	0.3883	0.416	0.4392
30	0.0161	0.1057	0.1935	0.262	0.3151	0.3571	0.3913	0.4196	0.4435
40	0.016	0.1056	0.1938	0.2629	0.3167	0.3596	0.3945	0.4235	0.448
60	0.0159	0.1055	0.1941	0.2639	0.3184	0.3621	0.3977	0.4275	0.4528
120	0.0159	0.1055	0.1945	0.2649	0.3202	0.3647	0.4012	0.4317	0.4578
	1	2	3	4	5	6	7	8	9
	Numerator degrees of freedom								

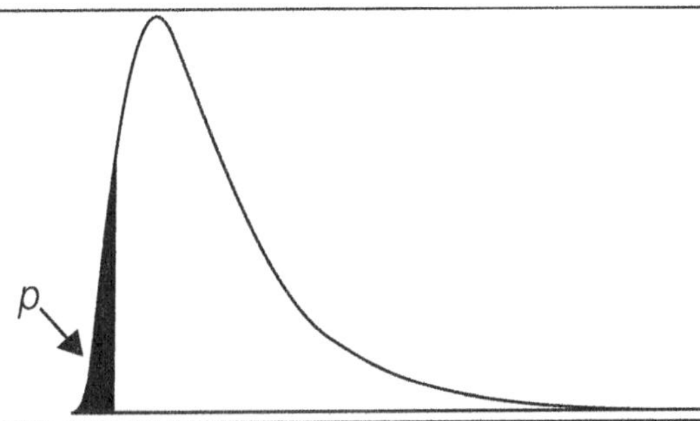

Numerator degrees of freedom									
10	12	15	20	24	30	40	60	120	Denominator degrees of freedom
0.3044	0.3148	0.3254	0.3362	0.3416	0.3471	0.3527	0.3583	0.3639	1
0.3419	0.3563	0.371	0.3862	0.394	0.4018	0.4098	0.4178	0.426	2
0.3666	0.3838	0.4016	0.4202	0.4297	0.4394	0.4492	0.4593	0.4695	3
0.3838	0.4032	0.4235	0.4447	0.4556	0.4668	0.4783	0.49	0.5019	4
0.3966	0.4177	0.4399	0.4633	0.4755	0.488	0.5008	0.514	0.5275	5
0.4064	0.429	0.4529	0.4782	0.4917	0.505	0.519	0.5334	0.5483	6
0.4143	0.4381	0.4634	0.4903	0.5044	0.519	0.534	05496	0.5658	7
0.4207	0.4455	0.472	0.5004	0.5153	0.5308	0.5468	0.5634	0.5807	8
0.426	0.4518	0.4793	0.5089	0.5246	0.5408	0.5578	0.5754	0.5937	9
0.4306	0.4571	0.4856	0.5163	0.5326	0.5496	0.5673	0.5858	0.6052	10
0.4378	0.4657	0.4958	0.5284	0.5459	0.5641	0.5832	0.6033	0.6245	12
0.4457	0.4751	0.507	0.542	0.5608	0.5806	0.6015	0.6237	0.6472	15
0.4544	0.4855	0.5197	0.5575	0.578	0.5998	0.623	0.6479	0.6747	20
0.459	0.4912	0.5266	0.566	0.5876	0.6106	0.6353	0.6619	0.691	24
0.4639	0.4971	0.534	0.5753	0.598	0.6225	0.6489	0.6777	0.7095	30
0.4691	0.5035	0.5419	0.5854	0.6095	0.6356	0.6642	0.6957	0.7312	40
0.4746	0.5103	0.5504	0.5964	0.6222	0.6504	0.6816	0.7167	0.7574	60
0.4804	0.5175	0.5597	0.6085	0.6364	0.6672	0.7019	0.7421	0.7908	120
10	12	15	20	24	30	40	60	120	
Numerator degrees of freedom									

Table C.19 F distribution values with lower-tail probability of 0.2.

Denominator degrees of freedom	Numerator degrees of freedom 1	2	3	4	5	6	7	8	9
1	0.1056	0.2813	0.3728	0.4254	0.4591	0.4824	0.4995	0.5125	0.5228
2	0.0833	0.25	0.3465	0.4045	0.4426	0.4695	0.4894	0.5047	0.5168
3	0.0765	0.2406	0.3406	0.4025	0.4438	0.4734	0.4954	0.5125	0.5261
4	0.0733	0.2361	0.3384	0.4028	0.4465	0.4779	0.5016	0.52	0.5348
5	0.0714	0.2334	0.3372	0.4036	0.4489	0.4818	0.5067	0.5262	0.5418
6	0.0701	0.2317	0.3366	0.4043	0.451	0.485	0.5109	0.5312	0.5476
7	0.0693	0.2304	0.3362	0.405	0.4527	0.4876	0.5143	0.5353	0.5523
8	0.0686	0.2295	0.336	0.4056	0.4541	0.4898	0.5171	0.5387	0.5562
9	0.0681	0.2288	0.3358	0.4061	0.4553	0.4916	0.5195	0.5415	0.5595
10	0.0677	0.2282	0.3357	0.4066	0.4563	0.4931	0.5215	0.544	0.5623
12	0.0671	0.2273	0.3355	0.4073	0.458	0.4956	0.5247	0.5479	0.5668
15	0.0665	0.2265	0.3354	0.4081	0.4598	0.4983	0.5283	0.5522	0.5718
20	0.0659	0.2257	0.3352	0.409	0.4617	0.5012	0.5321	0.5569	0.5773
24	0.0656	0.2252	0.3352	0.4095	0.4627	0.5028	0.5342	0.5594	0.5803
30	0.0653	0.2248	0.3351	0.41	0.4638	0.5044	0.5363	0.5621	0.5834
40	0.065	0.2244	0.3351	0.4105	0.4649	0.5061	0.5386	0.5649	0.5867
60	0.0648	0.224	0.3351	0.411	0.466	0.5079	0.5409	0.5678	0.5902
120	0.0645	0.2236	0.3351	0.4116	0.4672	0.5097	0.5434	0.5709	0.5939
	1	2	3	4	5	6	7	8	9
	Numerator degrees of freedom								

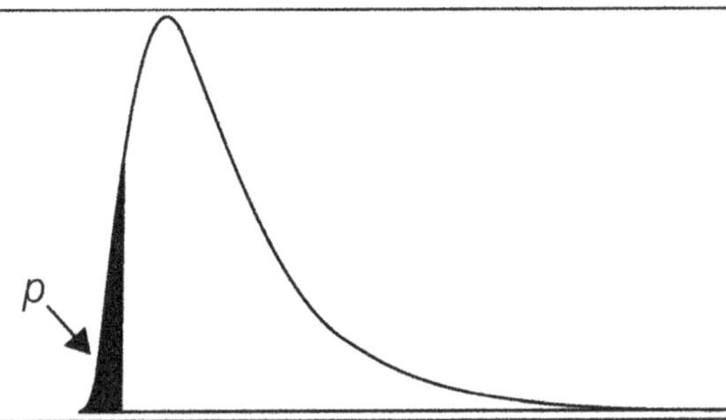

Numerator degrees of freedom									
10	12	15	20	24	30	40	60	120	Denominator degrees of freedom
0.5311	0.5437	0.5564	0.5693	0.5758	0.5823	0.5889	0.5955	0.6022	1
0.5267	0.5417	0.5571	0.5727	0.5806	0.5886	0.5967	0.6048	0.613	2
0.5372	0.5543	0.5717	0.5897	0.5988	0.6081	0.6175	0.627	0.6366	3
0.5469	0.5655	0.5847	0.6045	0.6146	0.625	0.6354	0.6461	0.6569	4
0.5547	0.5746	0.5952	0.6166	0.6276	0.6388	0.6503	0.6619	0.6738	5
0.5611	0.582	0.6038	0.6265	0.6383	0.6503	0.6625	0.6751	0.6879	6
0.5663	0.5881	0.6109	0.6348	0.6472	0.6599	0.6729	0.6862	0.7	7
0.5707	0.5932	0.6169	0.6418	0.6548	0.6681	0.6817	0.6958	0.7103	8
0.5744	0.5976	0.622	0.6478	0.6613	0.6751	0.6894	0.7042	0.7194	9
0.5775	0.6013	0.6264	0.653	0.6669	0.6813	0.6961	0.7115	0.7274	10
0.5826	0.6074	0.6336	0.6616	0.6763	0.6915	0.7073	0.7238	0.741	12
0.5882	0.6141	0.6417	0.6713	0.6869	0.7032	0.7203	0.7382	0.7569	15
0.5944	0.6216	0.6507	0.6823	0.6992	0.7169	0.7355	0.7552	0.7763	20
0.5978	0.6257	0.6557	0.6885	0.7061	0.7246	0.7442	0.7652	0.7877	24
0.6014	0.63	0.6611	0.6951	0.7136	0.7331	0.7539	07763	0.8007	30
0.6051	0.6346	0.6668	0.7024	0.7218	0.7425	0.7647	0.789	0.8158	40
0.6091	0.6396	0.673	0.7103	0.7309	0.753	0.7771	0.8038	0.8341	60
0.6134	0.6449	0.6798	0.7191	0.741	0.7649	0.7914	0.8215	0.8572	120
10	12	15	20	24	30	40	60	120	
Numerator degrees of freedom									

Appendix D
Statistical Process Control Constants

Table D.1 Control chart biasing constants (created using Minitab statistical software).

n	d_2	D_3	D_4	c_4	B_3	B_4	n
2	1.128	0	3.26729	0.7979	0	2.60632	2
3	1.693	0	2.57425	0.8862	0	2.27598	3
4	2.059	0	2.28188	0.9213	0	2.08775	4
5	2.326	0	2.11449	0.94	0	1.96363	5
6	2.534	0	2.00395	0.9515	0.0289	1.87417	6
7	2.704	0.0755917	1.92441	0.9594	0.1129	1.80583	7
8	2.847	0.1361433	1.86386	0.965	0.1786	1.75144	8
9	2.97	0.1840404	1.81596	0.9693	0.2318	1.70683	9
10	3.078	0.2230994	1.7769	0.9727	0.2759	1.66937	10
11	3.173	0.2556256	1.74437	0.9754	0.3134	1.63734	11
12	3.258	0.2831492	1.71685	0.9776	0.3456	1.60954	12
13	3.336	0.3071942	1.69281	0.9794	0.3737	1.58511	13
14	3.407	0.3281479	1.67185	0.981	0.3985	1.56343	14
15	3.472	0.3466014	1.6534	0.9823	0.4206	1.54401	15
16	3.532	0.3630521	1.63695	0.9835	0.4405	1.52648	16
17	3.588	0.3778428	1.62216	0.9845	0.4585	1.51055	17
18	3.64	0.3912637	1.60874	0.9854	0.4748	1.496	18
19	3.689	0.4034969	1.5965	0.9862	0.4898	1.48264	19
20	3.735	0.4146988	1.5853	0.9869	0.5036	1.4703	20
21	3.778	0.4249338	1.57507	0.9876	0.5163	1.45888	21
22	3.819	0.4344855	1.56551	0.9882	0.5281	1.44825	22
23	3.858	0.4433126	1.55669	0.9887	0.5391	1.43834	23
24	3.895	0.4515276	1.54847	0.9892	0.5493	1.42906	24
25	3.931	0.4593742	1.54063	0.9896	0.5589	1.42035	25
26	3.964	0.466873	1.53313	0.9901	0.568	1.41215	26
27	3.997	0.4734404	1.52656	0.9904	0.5765	1.40442	27
28	4.027	0.4794597	1.52054	0.9908	0.5845	1.39711	28
29	4.057	0.4853382	1.51466	0.9911	0.592	1.39018	29
30	4.086	0.4909552	1.50904	0.9914	0.5992	1.38361	30
n	d_2	D_3	D_4	c_4	B_3	B_4	n

Appendix E
A Tour of Minitab

Figure E.1 shows a Minitab screen with the Minitab terminology. Figure E.2 shows an opened project manager, and Figure E.3 shows a recommended configuration for working in Minitab.

Selecting a statistical method in the menu options will often open a gallery (see Figure E.4) or a dialog box (see Figure E.5). The results will either be in the session window or shown as a graph (see Figure E.6).

Minitab offers many keyboard shortcuts. It is also possible to create new keyboard shortcuts by going to **Tools > Customize** and selecting **Keyboard**. The default shortcuts include:

- Ctrl + M: Move to session window
- Ctrl + D: Move between worksheets
- Ctrl + I: Open project manager
- Ctrl + N: Create a new worksheet
- Ctrl + Home: Go to first column and row in a worksheet
- Ctrl + C: Copy selected text

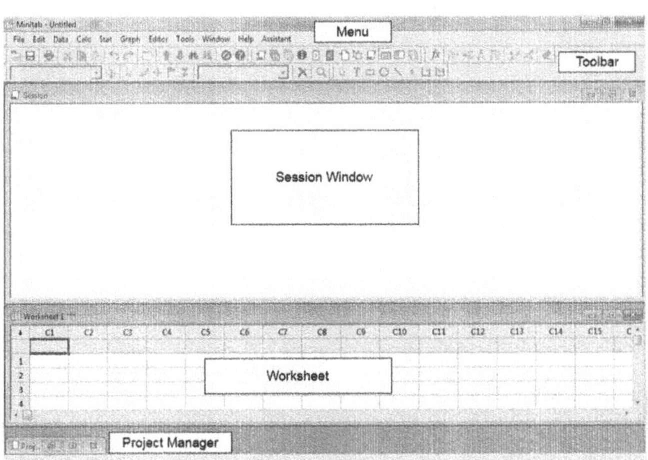

Figure E.1 Minitab screen.

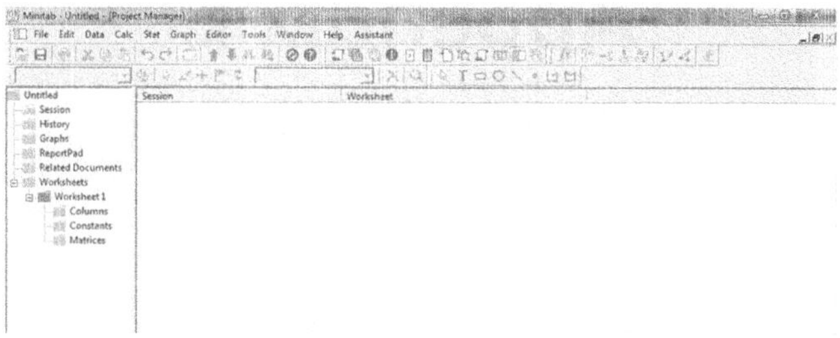

Figure E.2 Project manager.

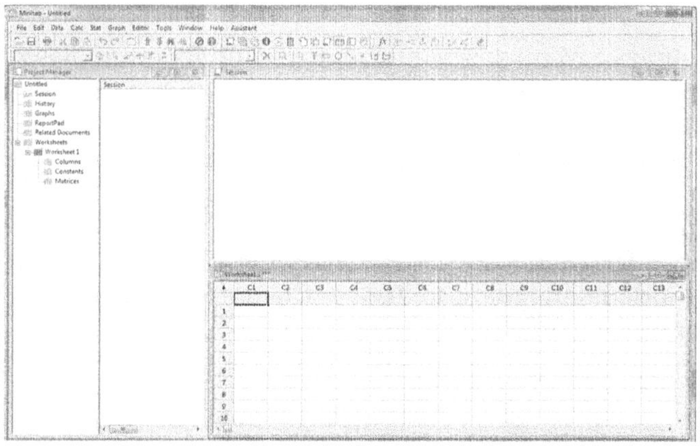

Figure E.3 Example of Minitab set up for use.

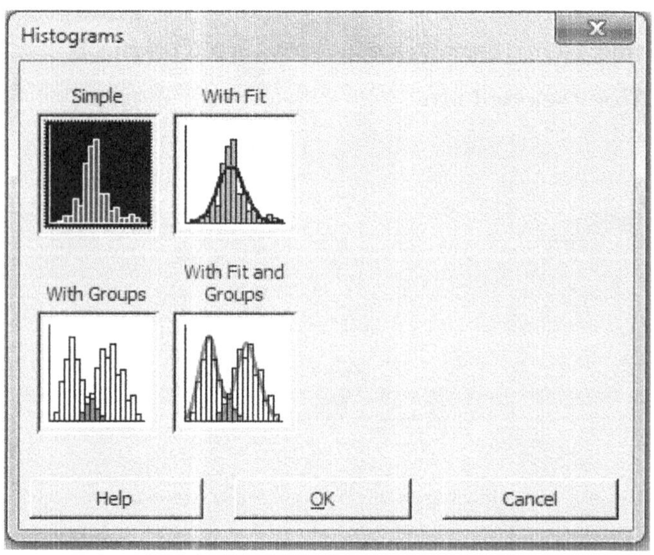

Figure E.4 Gallery for Graph > Histogram.

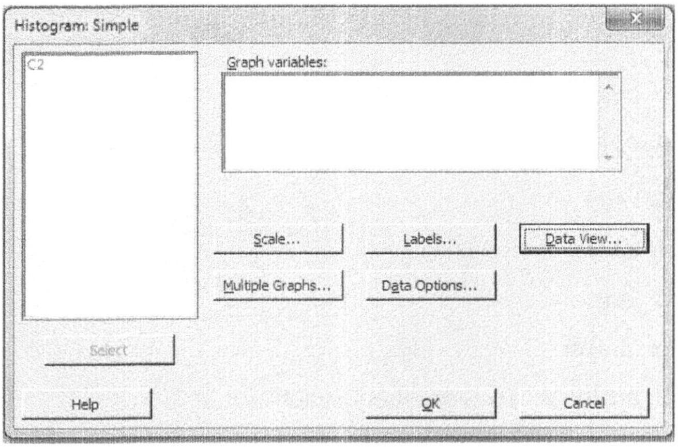

Figure E.5 Dialog box for Histogram.

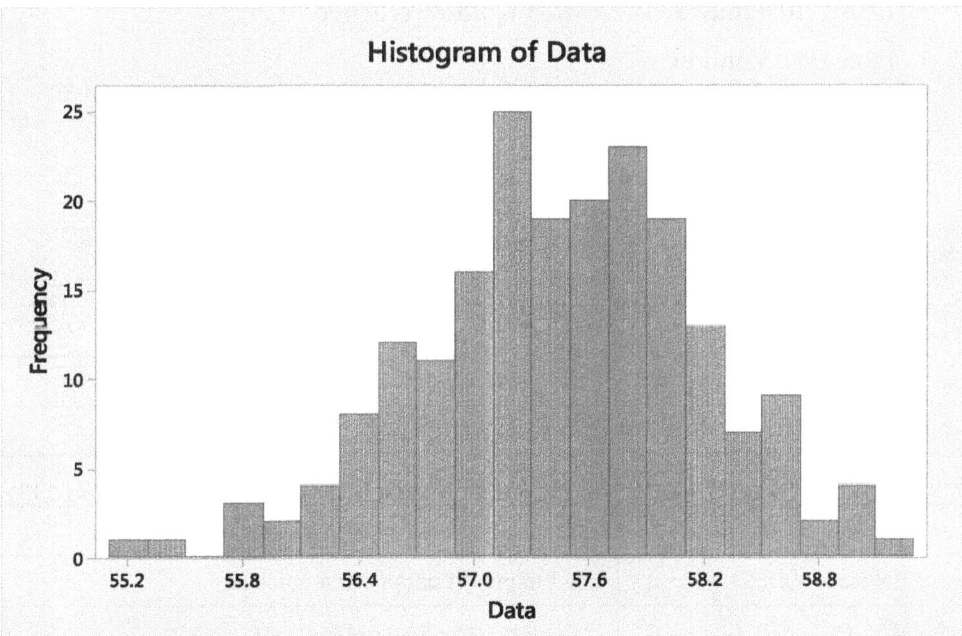

Figure E.6 Example of a graph.

- Ctrl+X: Cut selected text
- Ctrl+V: Paste copied text
- Ctrl+F: Find
- Ctrl+H: Find and replace
- F3: Reset dialog box to default
- Ctrl+E: Open last dialog box

The many options provided by the Minitab menus are presented in the tables below:

- Table E.1: File
- Table E.2: Edit
- Table E.3: Data
- Table E.4: Calc
- Table E.5: Stat
- Table E.6: Graph
- Table E.7: Editor when worksheet is active
- Table E.8: Tools
- Table E.9: Editor when graph is active
- Table E.10: Editor when session window is active
- Table E.11: Window
- Table E.12: Help
- Table E.13: Assistant

Table E.1 File.

New	Opens a new project or worksheet
Open	Opens an existing file
Save Project	To save a project
Save Project As	To save a project using a specific name
Save (Worksheet, Graph, etc.) AS	Saves the active window in a variety of formats
Query Database (ODBC)	Used to import data from a database
Print (Window, Graph, etc,)	Prints the active item in the project
Description	Project related details such a name, date, and comments can be added
Exit	Ends the Minitab session
Recent Files	Opens recently accessed Minitab files

Table E.2 Edit.

Undo	Used to undo the previous action
Redo	Used to redo actions that have been undone
Clear	Removes values from highlighted cells
Delete	Removes highlighted selection
Copy	Copies to clipboard
Cut	Removes the selected material and stores it on the clipboard
Paste	Pastes material from the clipboard
Paste Links	Pastes copied links
Worksheet Links	Used to manage Dynamic Data Exchange (DDE) links
Select All	Selects all cells in the active window
Edit Last Dialog	Reopens the last dialog
Command Line Editor	Allows the user to submit command language

Table E.3 Data.

Subset Worksheet	Creates a new worksheet using a condition, row numbers, or brushed rows
Split Worksheet	Splits the current worksheet into new worksheets based on a variable
Merge Worksheets	Merges separate worksheets into one worksheet
Copy	Provides many copying options, including from column to column and worksheet to worksheet, and allows the user to subset the data while doing so
Unstack Columns	Unstacks data from one or more columns into multiple columns based on a variable
Stack	Stacks columns into one column
Transpose Columns	Converts columns to rows and rows to columns
Sort	Provides multiple options to sort data form lowest to highest value or highest to lowest value
Rank	Ranks the values in a column from smallest to largest
Delete Rows	Removes selected rows
Erase Variables	Removes selected data
Conditional Formatting	Allows cells to be formatted based on a variety of conditions
Recode	Provides many options for coding data such as converting individual values and ranges of values
Change Data Type	Offers many options to change the type of data in a column
Date/Time	Used to round dates as well as extract select parts of dates
Concatenate	Combines multiple columns of text into one column
Display data	Used to display data such as from a column in the session window

Table E.4 Calc.

Calculator	Used to perform calculations with data in columns
Column Statistics	Provides selected statistics such as the mean and standard deviation for a column
Row Statistics	Provides selected statistics such as the mean and standard deviation for a row
Standardize	Standardizes data—for example, by subtracting the mean and dividing by the standard deviation
Make Patterned Data	Creates patterned data such as strings of numbers or lists of text
Make Mesh Data	Used to create a grid of x,y coordinates
Make Indicator Variables	Creates columns of 1s and 0s using categorical variables
Set Base	Used to set the starting point for Minitab's random number generator
Random Data	Generates random data either from a specified distribution or by sampling from columns
Probability Distributions	Used to perform probability calculations from the chosen distribution
Matrices	Used to create and modify matrices and perform matrix algebra

Table E.5 Stat.

Basic Statistics	Descriptive statistics as well as various types of statistical hypothesis tests
Regression	Multiple options for performing a regression analysis
ANOVA	Contains ANOVA and variations on ANOVA
DOE	To design and analyze the results of experiments such as factorial, response surface, mixture, and Taguchi
Control Charts	Statistical process control options for various types of data and uses, and an option for data transformations
Quality Tools	Contains basic tools such as Pareto chart, run chart, multi-vari chart, and cause and effect diagrams, as well as individual distribution identification, capability analyses, and measurement system analysis
Reliability/Survival	Offers various reliability and survival study methods
Multivariate	Multiple options for multivariate analysis
Time Series	Contains many tools for time series analysis, including autoregressive integrated moving average and autocorrelation
Tables	Many tabulation tools including the Chi-square test
Nonparametrics	Multiple options for performing nonparametric hypothesis tests
Equivalence Tests	Hypothesis tests to determine whether means are within a specified range from one another
Power and Sample Size	Used for identifying required sample sizes as well as determining the power of a statistical test

Table E.6 Graph.

Scatterplot	Displays relationship between continuous X and Y variables
Matrix Plot	Displays relationship between multiple continuous X and Y variables
Bubble Plot	Displays relationship between continuous X and Y variables with a third continuous variable displayed as the bubble size
Marginal Plot	Displays relationship between continuous X and Y variables along with the distributions of the values
Histogram	Multiple options for displaying a histogram of a data set
Dotplot	Used to compare distributions with values on a number line
Stem-and-Leaf	Displays the shape and spread of data arranged from lowest to highest
Probability Plot	Probability plots for assessing the distribution of data
Empirical CDF	Compares the cumulative distribution of the data to that expected from a specified distribution
Probability Distribution Plot	Used to generate and compare various types of probability distributions
Boxplot	Generates boxplots of continuous data
Interval Plot	Plots the mean and confidence interval of one or more continuous variables
Individual Value Plot	Plots the individual data points in a data set
Line Plot	Compares means of a continuous variable at values of one or more categorical variables to identify interactions
Bar Chart	Creates a bar chart of data, including clustered and stacked bar charts
Pie Chart	Creates a pie chart of the data
Time Series Plot	Provides multiple options to plot data in the order they were collected
Area Graph	Plots the contribution of multiple time series to their sum
Contour Plot	Displays the relationship between a response variable and two predictor variables using contours
3D Scatter Plot	Displays the relationship between a response variable and two predictor variables by plotting the data on three axes
3D Surface Plot	Displays the relationship between a response variable and two predictor variables using a three dimensional surface

Table E.7 Editor when worksheet is active.

Find and Replace	Used to find or find and replace the contents of cells or the session window; can also be used for formatting cells
Sort Columns	Sorts columns
Cell Properties	Cells can be formatted and comments can be added to the cells
Format Column	Column format can be changed
Column Properties	Properties of columns can be changed
Conditional Formatting	Many options for formatting cells based on their values
Formulas	Formulas can be assigned to cells
Subset Worksheet	Data can be included or excluded from worksheets using individual values, ranges, or formatting
Worksheet Properties	Changes the direction the active cell moves when Enter is pressed and allows a description to be entered
Insert Cells	Adds cells to the worksheet
Insert Rows	Adds rows to the worksheet
Insert Columns	Adds columns to the worksheet
Move Columns	Moves selected columns to a different location in the work sheet

Table E.8 Tools.

Quality Trainer	Opens a link to the Quality Trainer
Microsoft Calculator	Opens the Microsoft calculator
Notepad	Opens the notepad
Windows Explorer	Opens a Windows Explorer
Run an Exec	Runs an executable file with Minitab commands
Toolbars	Select which toolbars are displayed
Status Bar	Adds or removes the status bar at the bottom of the screen
Customize	Used to add and remove menus and toolbars as well as specific items from a menu, submenu, or tool bar; provides additional tools such as EDA and various graph options
Options	Sets default behavior for many graphs and tools, changes languages (when languages are installed), and changes default settings for file and macro locations
Manage Profiles	Sets up and manages different profiles with different toolbar and option setups
File Security	Provides options to set passwords for opening or modifying project files; there is also an option to initiate a notice recommending that the file be "read only"

Table E.9 Editor when graph is active.

Select	Used to select parts of a graph with the cursor
Brush	Used to find the row corresponding to a point or points in a graph
Crosshairs	Creates a set of crosshairs in a graph
Edit	Used to edit the currently selected item
Select Item	Used to select part of a graph
Add	Provides many graph modification options
Panel	Used to edit graph panels
Graph Options	Used to change special options available for specific graphs like 3D plots
Layout Tool	Used to combine multiple graph into one panel
Copy Test	Copies the text displayed on a tooltip in the graph
Copy Command Language	Copies the last command executed
Duplicate Graph	Creates a duplicate of the selected graph
Make Similar Graph	Creates a new version of the selected graph, but with new variables
Annotation	Activates or deactivates graph annotation tools in the toolbar
Update	Updates graph when the data in the worksheet changes
Zoom	Used to zoom in or zoom out in the selected graph

Table E.10 Editor when session window is active.

Next Command	Moves the cursor to the next set of session window output
Previous Command	Moves the cursor to the previous set of session window output
Expand All	Expands all sections of output to display
Collapse All	Collapses all results in the session window
Show Command Line	Activates the command line for typing in commands
Add Note	Adds a comments section to the current location in the session window
Delete Note	Deletes a previously added comments section
Decimal Places	Changes the number of decimal places in what is selected in the session window
Zoom	Used to zoom in or zoom out in the session window

Table E.11 Window.

Cascade	Layers open windows
Tile	Arranges open windows side by side
Minimize All	Minimizes all open windows
Restore Icons	Opens all minimized windows
Arrange Icons	Moves all minimized windows to the bottom of the screen
Refresh	Refreshes the open window
Close All Graphs	Closes all open graphs; graphs must be re-created once closed
Update All Graphs Now	Updates all graphs with current data from the worksheet
1 Session	Opens the session window on top of any other open items
2 Project Manager	Opens the project manager
3 Worksheet 1	Use to select the desired worksheet or graph; additional options will appear as more worksheets or graphs are opened

Table E.12 Help.

Help	Provides a searchable list of guidance for the use of Minitab
Sample Data	Opens a link to sample Minitab data sets used in help files
Getting Started	Opens a link to details on getting started using Minitab
Keyboard Map	Explains Minitab commands and provides hotkeys where applicable
Check for Updates	Determines if updates are available and assists in installing updates
Minitab on the Web	Links to Minitab's homepage
Contact Us	Links to contact information at the Minitab website
About Minitab	Lists installed components and provides system information

Table E.13 Assistant.

Measurement System Analysis (MSA)	Provides guidance and a decision tree to assist in performing an MSA
Capability Analysis	Presents guidance and a decision tree to assist in performing a capability analysis
Graphical Analysis	Helps in the selection of various types of graphs based on intended use
Hypothesis Tests	Assists in identifying the correct type of statistical test to perform and presents guidance on performing the test
Regression	Used to choose between types of regression analysis deepening on the number of variables and the objective of the analysis and guidance on results
DOE	Provides both guidance and assistance in planning and analyzing design of experiments (DOEs)
Before/After Capability Analysis	Assists in performing capability studies to compare before and after results
Before/After Control Charts	Provides assistance for creating control charts to compare before and after results
Control Charts	Provides a decision tree to assist in selecting and using control charts based on the type of data and subgroup size

References

Barrentine, Larry B. 1999. *An Introduction to Design of Experiments: A Simplified Approach*. Milwaukee, WI: ASQ Quality Press.

Barsalou, Matthew A. 2015a. *Root Cause Analysis: A Step-by-Step Guide to Using the Right Tool and the Right Time*. Boca Raton, FL: CRC Press.

Barsalou, Matthew A. 2015b. *Statistics for Six Sigma Black Belts*. Milwaukee, WI: ASQ Quality Press.

Benbow, Donald W., and T. M. Kubiak. 2009. *The Certified Six Sigma Black Belt Handbook*. Milwaukee, WI: ASQ Quality Press.

Borror, Connie, M., ed. 2009. *The Certified Quality Engineer Handbook*, 3rd ed. Milwaukee, WI: ASQ Quality Press.

Boslaugh, Sarah, and Paul Andrew Watters. 2008. *Statistics in a Nutshell: A Desktop Quick Reference*. Sebastopol, CA: O'Reilly Media.

Box, George, Soren Bisgaard, and Conrad Fung. 1988. "An Exploration and Critique of Taguchi's Contributions to Quality Engineering." *Quality and Reliability Engineering International* 4: 123–131.

Box, George E. P., J. Stuart Hunter, and William G. Hunter. 2005. *Statistics for Experimenters: Design, Innovation, and Discovery*, 2nd ed. Hoboken, NJ: John Wiley & Sons.

Breyfogle, Forrest, III. 2003. *Implementing Six Sigma: Smarter Solutions® Using Statistical Methods*, 2nd ed. Hoboken, NJ: John Wiley & Sons.

Breyfogle, Forrest, III. 2014. "30,000-Foot-Level Performance Reporting." *Six Sigma Forum Magazine* 13 (2): 18–32.

Chrysler, Ford, and General Motors. 2010. *Measurements System Analysis*, 4th ed. Southfield, MI: Automotive Industry Action Group.

Clarke, G. M., and D. A. Cooke. 1992. *Basic Course in Statistics*, 3rd ed. Kent, England: Edward Arnold.

Committee E-11 on Quality and Statistics. 2002. *Manual on Presentation of Data and Control Chart Analysis*, 7th ed. West Conshohocken, PA: ASTM International.

Conover, W. J. 1999. *Practical Nonparametric Statistics*, 3rd ed. Hoboken, NJ: John Wiley & Sons.

D'Agostino, Ralph B., and Michael A. Stephens, eds. 1986. *Goodness-of-Fit Techniques*. New York: Marcel Dekker.

Daniel, Wayne. 1990. *Applied Nonparametric Statistics*, 2nd ed. Boston: PWS-Kent.

de Mast, Jeroen, and Albert Trip. 2007. "Quality-Improvement Projects," *Journal of Quality Technology* 39 (4): 301–311.

Durivage, Mark Allen. 2015. *Practical Engineering, Process, and Reliability Statistics*. Milwaukee, WI: ASQ Quality Press.

Efron, Bradley. 1969. "Student's t-Test under Symmetry Conditions." *Journal of the American Statistical Association* 64 (328): 1278–1302.

Griffith, Gary K. 2003. *The Quality Technician's Handbook*, 5th ed. Upper Saddle River, NJ: Prentice Hall.

Hoerl, Roger, and Ron Snee. 2009. *Statistical Thinking, Strategies and Tools*. Mason, OH: Cengage Learning.

Hoerl, Roger, and Ron Snee. 2012. *Statistical Thinking: Improving Business Performance*, 2nd ed. Hoboken, NJ: John Wiley & Sons.

Ishikawa, Kaoru. 1991. *Guide to Quality Control*, 2nd ed. Translated by Asian Productivity Organization. Tokyo: Asian Productivity Organization.

Juran, J. M. 2005. "The Non-Pareto Principle—Mea Culpa." In *Juran, Quality, and a Century of Improvement*, edited by K. S. Stephens. Milwaukee, WI: ASQ Quality Press.

Kahn, M. Rehman. 2013. *Problem Solving and Data Analysis Using Minitab: A Clear and Easy Guide to Six Sigma Methodology*. Chichester, UK: John Wiley & Sons.

Kenney, James M. 1988. "Hypothesis Testing: Guilty or Innocent." *Quality Progress* 21 (1): 55–58.

Lawson, John, and John Erjavec. 2001. *Modern Statistics for Engineering and Quality Improvement*. Pacific Grove, CA: Wadsworth Group.

McClave, James T., and Frank H. Dietrich, II. 1991. *Statistics*, 5th ed. New York: Dellen Publishing.

McClave, James T., and Terry Sincich. 2009. *Statistics*, 11th ed. Upper Saddle River, NJ: Pearson Education.

Minitab. "Attribute Control Charts." Minitab Assistant White Paper. Accessed February 15, 2016. http://support.minitab.com/en-us/minitab/17/Assistant_Attribute_Control_Charts.pdf.

Minitab. "Multiple Regression." Minitab Assistant White Paper. Accessed February 15, 2016. http://support.minitab.com/en-us/minitab/17/Assistant_Multiple_Regression.pdf.

Minitab. "1-Sample Standard Deviation Test." Minitab Assistant White Paper. Accessed February 15, 2016. http://support.minitab.com/en-us/minitab/17/Assistant_One_Sample_Variance.pdf.

Minitab. "1-Sample t-Test." Minitab Assistant White Paper. Accessed February 15, 2016. http://support.minitab.com/en-us/minitab/17/Assistant_One_Sample_t.pdf.

Minitab. "Tests for Standard Deviations (Two or More Samples)." Minitab Assistant White Paper. Accessed February 15, 2016. http://support.minitab.com/en-us/minitab/17/Assistant_Test_for_Standard_Deviations.pdf.

Minitab. "2-Sample t-Test." Minitab Assistant White Paper. Accessed February 15, 2016. http://support.minitab.com/en-us/minitab/17/Assistant_Two_Sample_t.pdf.

Minitab. "Understanding Test for Equal Variances." Minitab Assistant White Paper. Accessed June 2, 2016. http://support.minitab.com/en-us/minitab/17/topic-library/modeling-statistics/anova/basics/understanding-test-for-equal-variances/.

Minitab. "Variables Control Charts." Minitab Assistant White Paper. Accessed February 15, 2016. http://support.minitab.com/en-us/minitab/17/Assistant_Variables_Control_Charts.pdf.

Montgomery, Douglas C. 1997. *Design and Analysis of Experiments*, 4th ed. New York: John Wiley & Sons.

Montgomery, Douglas C., and George C. Runger. 1999. *Applied Statistics and Probability for Engineers*, 2nd ed. New York: John Wiley & Sons.

Montgomery, Douglas C., George C. Runger, and Norma F. Hubele. 2001. *Engineering Statistics*, 2nd ed. New York: John Wiley & Sons.

Moore, David. S. 1991. *Statistics: Concepts and Controversies*, 3rd ed. New York: W. H. Freeman.

Nelson, Lloyd S. 1984. "Technical Aids," *Journal of Quality Technology* 16 (4): 238–239.

Nelson, Peter R., Marie Coffin, and Karen A. F. Copeland. 2003. *Introductory Statistics for Engineering Experimentation*. San Diego, CA: Elsevier.

NIST/SEMATECH. 2012. *e-Handbook of Statistical Methods*. http://www.itl.nist.gov/div898/handbook/.

O'Connor, Patrick D. T., and Andre Kleyner. 2012. *Practical Reliability Engineering*, 5th ed. West Sussex, UK: John Wiley & Sons.

Owen, Frank, and Ron Jones. 1990. *Statistics*, 3rd ed. London: Pitman Publishing.

Preis, Kim H. 2009. *Six Sigma for the New Millennium*, 2nd ed. Milwaukee, WI: ASQ Quality Press.

ReVelle, Jack B. 2004. *Quality Essentials: A Reference Guide from A to Z*. Milwaukee, WI: ASQ Quality Press.

Ryan, Thomas A., Jr., and Brian L. Joiner. 1976. "Normal Probability Plots and Tests for Normality." Technical Report. Statistics Department, Pennsylvania State University.

Shewhart, Walter A. 1980. *Economic Control of Quality of Manufactured Product*, 50th Anniversary Commemorative Reissue. Milwaukee, WI: ASQ Quality Press.

Silver, Mick. 1997. *Business Statistics*, 2nd ed. Berkshire, England: McGraw-Hill.

Simmons, J. P., L. D. Nelson, and U. Simonsohn. 2011. "False-Positive Psychology: Undisclosed Flexibility in Data Collection and Analysis Allows Presenting Anything as Significant." *Psychological Science* 22 (11): 1359–1366.

Stephens, Larry J. 2004. *Advanced Statistics Demystified: A Self-Teaching Guide*. New York: McGraw-Hill.

Tague, Nancy R. 2005. *The Quality Toolbox*, 2nd ed. Milwaukee, WI: ASQ Quality Press.

Tukey, John W. 1977. *Exploratory Data Analysis*. Reading, MA: Addison-Wesley.

Vining, Geoffrey, and Scott M. Kowalski. 2006. *Statistical Methods for Engineers*, 2nd ed. Belmont, CA: Thompson Higher Education.

Witte, Robert S. 1993. *Statistics*, 4th ed. New York: Harcourt Brace College Publishers.

Yap, B. W., and C. H. Sim. 2011. "Comparison of Various Types of Normality Tests." *Journal of Statistical Computation and Simulation* 81 (12): 2141–2155.

Yeo, In-Kwaon and Richard A. Johnson. 2000. "A New Family of Power Transformations to Improve Normality or Symmetry." *Biometrika* 87 (4): 954–959.

Index

Note: Page numbers followed by *f* or *t* refer to figures or tables, respectively.